NEUROMETHODS

AF891838

Series Editor
Wolfgang Walz
University of Saskatchewan
Saskatoon, SK, Canada

For further volumes:
http://www.springer.com/series/7657

Neuromethods publishes cutting-edge methods and protocols in all areas of neuroscience as well as translational neurological and mental research. Each volume in the series offers tested laboratory protocols, step-by-step methods for reproducible lab experiments and addresses methodological controversies and pitfalls in order to aid neuroscientists in experimentation. *Neuromethods* focuses on traditional and emerging topics with wide-ranging implications to brain function, such as electrophysiology, neuroimaging, behavioral analysis, genomics, neurodegeneration, translational research and clinical trials. *Neuromethods* provides investigators and trainees with highly useful compendiums of key strategies and approaches for successful research in animal and human brain function including translational "bench to bedside" approaches to mental and neurological diseases.

Experimental and Clinical Methods in Hepatic Encephalopathy Research

Edited by

Tiziano Balzano

HM CINAC (Centro Integral de Neurociencias Abarca Campal), Hospital Universitario HM Puerta del Sur, HM Hospitales, Madrid, Spain

Humana Press

Editor
Tiziano Balzano
HM CINAC (Centro Integral de
Neurociencias Abarca Campal)
Hospital Universitario HM Puerta del Sur, HM Hospitales
Madrid, Spain

ISSN 0893-2336 ISSN 1940-6045 (electronic)
Neuromethods
ISBN 978-1-0716-4356-3 ISBN 978-1-0716-4354-9 (eBook)
https://doi.org/10.1007/978-1-0716-4354-9

This Humana imprint is published by the registered company Springer Science+Business Media, LLC, part of Springer Nature.
The registered company address is: 1 New York Plaza, New York, NY 10004, U.S.A.

Preface to the Series

Experimental life sciences have two basic foundations: concepts and tools. The Neuromethods series focuses on the tools and techniques unique to the investigation of the nervous system and excitable cells. It will not, however, shortchange the concept side of things as care has been taken to integrate these tools within the context of the concepts and questions under investigation. In this way, the series is unique in that it not only collects protocols but also includes theoretical background information and critiques which led to the methods and their development. Thus it gives the reader a better understanding of the origin of the techniques and their potential future development. The *Neuromethods* publishing program strikes a balance between recent and exciting developments like those concerning new animal models of disease, imaging, in vivo methods, and more established techniques, including, for example, immunocytochemistry and electrophysiological technologies. New trainees in neurosciences still need a sound footing in these older methods in order to apply a critical approach to their results.

Under the guidance of its founders, Alan Boulton and Glen Baker, the Neuromethods series has been a success since its first volume published through Humana Press in 1985. The series continues to flourish through many changes over the years. It is now published under the umbrella of Springer Protocols. While methods involving brain research have changed a lot since the series started, the publishing environment and technology have changed even more radically. Neuromethods has the distinct layout and style of the Springer Protocols program, designed specifically for readability and ease of reference in a laboratory setting.

The careful application of methods is potentially the most important step in the process of scientific inquiry. In the past, new methodologies led the way in developing new disciplines in the biological and medical sciences. For example, Physiology emerged out of Anatomy in the nineteenth century by harnessing new methods based on the newly discovered phenomenon of electricity. Nowadays, the relationships between disciplines and methods are more complex. Methods are now widely shared between disciplines and research areas. New developments in electronic publishing make it possible for scientists that encounter new methods to quickly find sources of information electronically. The design of individual volumes and chapters in this series takes this new access technology into account. Springer Protocols makes it possible to download single protocols separately. In addition, Springer makes its print-on-demand technology available globally. A print copy can therefore be acquired quickly and for a competitive price anywhere in the world.

Saskatoon, SK, Canada ***Wolfgang Walz***

Preface

Hepatic encephalopathy (HE) is a complex neuropsychiatric syndrome that occurs in patients with acute or chronic liver disease and ranks as the third most serious complication of decompensated liver cirrhosis. Affecting at least 30% of cirrhotic patients, HE presents a broad spectrum of neuropsychiatric abnormalities, ranging from subtle cognitive impairments to severe conditions such as marked disorientation, confusion, coma, and, in extreme cases, death.

Despite its significance, the pathophysiology of HE remains only partially understood. Although ammonia has traditionally been considered the primary factor in the development of HE, emerging research suggests that the condition is likely multifactorial. Factors such as peripheral inflammation, disruptions in intestinal and microbiome functions, neuroinflammation, oxidative stress, mitochondrial dysfunction, disturbances in cellular bioenergetics, and impaired neurotransmission are now recognized as playing significant roles. This complexity underscores the need for advanced research methodologies to better understand and manage HE.

This volume aims to address these complexities by exploring the latest experimental and clinical protocols used to investigate HE. The book is divided into three main parts, each focusing on different aspects of HE research.

The first part includes five protocols for diagnosing and monitoring HE in clinical practice. In Chapter 1, Gallego and collaborators discuss psychometric tools and tests essential for diagnosing and tracking minimal HE (MHE), shedding light on how these tools can identify subtle cognitive impairments and monitor their progression. In Chapter 2, Higuera-de-la-Tijera and colleagues provide insights into the use of auditory P300 event-related potentials, a neurophysiological marker that has shown promise in detecting MHE. Their chapter offers valuable information on the diagnostic accuracy and practical applications of this technique. Chapter 3, by Martí-Aguado and Martí-Bonmatí, explores how advanced MRI techniques are employed to quantify liver steatosis, iron overload, and fibrosis. This noninvasive approach provides valuable information on the extent of liver damage and its progression. In Chapter 4, López-Aguirre and El-Khatib-Núñez detail acquisition protocols and basic preprocessing techniques essential for obtaining and analyzing neuroimaging data in HE research and clinical practice. These imaging techniques are crucial for visualizing and understanding the structural and functional brain changes associated with HE. In Chapter 5, Carvalho and collaborators describe the use of immunoenzymatic assays to detect and measure inflammatory markers in both preclinical and clinical studies of HE. By targeting central and peripheral inflammation, these assays contribute to a deeper understanding of the inflammatory processes involved in HE and aid in evaluating potential therapeutic targets.

The second part of the book shifts focus to diverse experimental approaches used to study HE. This section includes four protocols and explores various animal models and experimental techniques designed to investigate the pathogenesis of HE and evaluate potential treatments. In Chapter 6, Martínez-García and colleagues provide an in-depth examination of different models used to replicate various aspects of HE. The chapter discusses the advantages and limitations of each model, offering insights into their relevance for studying HE's pathophysiology and testing interventions.

Complementing the discussion on animal models, Chapter 7, by El-Mansoury and collaborators, outlines various tests used to assess cognitive and motor functions in animal models of HE. These tests are essential for evaluating the behavioral and neurological impacts of HE, providing critical information on how the syndrome affects brain function and how potential therapies might mitigate these effects. Chapter 8 introduces an innovative approach where Becker-Kojić and collaborators describe a novel technique involving human liver spheroids derived from peripheral blood. These spheroids offer a unique platform for studying HE, potentially providing more relevant insights into human liver pathology and facilitating the development of new treatment strategies. In Chapter 9, Balzano details a method for accurately estimating Purkinje cell numbers in the cerebellum of mouse models. This stereological technique is valuable for studying neurodegeneration associated with HE and helps in understanding the extent of neuronal loss and damage.

The third part of the book focuses on uncovering the underlying mechanisms and pathophysiological processes involved in HE. This section includes four protocols and presents advanced techniques and strategies for investigating the complex interactions that contribute to HE. Chapter 10, by Romaní-Pérez and colleagues, discusses commonly used techniques, methods, and recommendations for developing tailored microbiome-based strategies that could aid in the prevention or treatment of HE. Complementing the discussion on the role of gut microbiota in HE pathogenesis, Sepehrinezhad and Shahbazi provide a comprehensive overview of techniques used to evaluate damage across the gut-liver-brain axis. This section highlights the systemic nature of HE and the importance of assessing multiorgan interactions to gain a complete understanding of the disease process. In Chapter 12, Heidari and collaborators review various techniques for measuring mitochondrial dysfunction and oxidative stress in the brain, both of which are critical factors in the pathophysiology of HE. These methods are essential for understanding the biochemical disturbances associated with HE and for evaluating potential therapeutic interventions. Finally, Chapter 13 includes a detailed examination of ammonia detection in formalin-fixed paraffin-embedded mammalian tissue. Delgado and colleagues describe the Nessler staining method used for detecting ammonia in specific regions of the mammalian brain, liver, and other organs. This technique is critical for studying the distribution and impact of ammonia in HE-related tissue changes, offering valuable insights into the biochemical alterations that occur in HE.

Together, these parts of the book provide a comprehensive exploration of both experimental and clinical methods and the mechanisms underlying hepatic encephalopathy. They offer valuable resources for researchers and clinicians seeking to understand and address this complex condition.

Madrid, Spain *Tiziano Balzano*

Contents

Contributors

Ana Agustí • *Microbiome, Nutrition & Health Research Unit, Institute of Agrochemistry and Food Technology, Spanish National Research Council (IATA-CSIC), Valencia, Spain*

Yaiza M. Arenas • *Laboratory of Neurobiology, Principe Felipe Research Centre, Valencia, Spain*

Tiziano Balzano • *HM CINAC (Centro Integral de Neurociencias Abarca Campal), Hospital Universitario HM Puerta del Sur, HM Hospitales, Madrid, Spain*

Zorica A. Becker-Kojić • *ACA CELL Biotech GmbH, Heidelberg, Germany*

Brener Cunha Carvalho • *Laboratório de Toxicologia, Departamento de Análises Clínicas e Toxicológicas, Faculdade de Farmácia, Universidade Federal de Minas Gerais, Belo Horizonte, Brazil*

Franc Casanova-Ferrer • *Laboratory of Neurological Impairment. Fundación Investigación Hospital Clínico, INCLIVA Biomedical Research Institute, Valencia, Spain*

Heliana de Barros Fernandes • *Laboratório de Neurobiologia "Conceição Machado", Departamento de Morfologia, Instituto de Ciências Biológicas, Universidade Federal de Minas Gerais, Belo Horizonte, Brazil*

Eliana Cristina de Brito Toscano • *Laboratório Integrado de Pesquisas em Patologia, Departamento de Patologia, Faculdade de Medicina, Universidade Federal de Juiz de Fora, Juiz de Fora, Brazil*

Aline Silva de Miranda • *Laboratório de Neurobiologia "Conceição Machado", Departamento de Morfologia, Instituto de Ciências Biológicas, Universidade Federal de Minas Gerais, Belo Horizonte, Brazil*

María del Carmen Cenit • *Microbiome, Nutrition & Health Research Unit, Institute of Agrochemistry and Food Technology, Spanish National Research Council (IATA-CSIC), Valencia, Spain*

Javier de las Heras • *Biobizkaia Health Research Institute, Barakaldo, Bizkaia, Spain; Division of Paediatric Metabolism, CIBERER, MetabERN, Cruces University Hospital, Barakaldo, Bizkaia, Spain; Department of Paediatrics, University of the Basque Country (UPV/EHU), Leioa, Bizkaia, Spain*

Teresa C. Delgado • *Biobizkaia Health Research Institute, Barakaldo, Bizkaia, Spain; Liver Disease Laboratory, Center for Cooperative Research in Biosciences (CIC bioGUNE), Basque Research and Technology Alliance (BRTA), Derio, Bizkaia, Spain; IKERBASQUE, Basque Foundation for Science, Bilbao, Spain*

Omar El Hiba • *Nutritional Physiopathologies, Neuroscience and Toxicology Team, Laboratory of Anthropogenic, Biotechnology and Health, Faculty of Sciences, Chouaib Doukkali University, El Jadida, Morocco*

Abdelaati El Khiat • *Nutritional Physiopathologies, Neuroscience and Toxicology Team, Laboratory of Anthropogenic, Biotechnology and Health, Faculty of Sciences, Chouaib Doukkali University, El Jadida, Morocco; Higher Institute of Nursing Professions and Health Techniques, Ministry of Health, Ouarzazate, Morocco*

Jorge El-Khatib-Núñez • *Servicio de Radiodiagnóstico. Hospital Universitario Puerta de Hierro, Majadahonda, Spain*

Bilal El-Mansoury • *Nutritional Physiopathologies, Neuroscience and Toxicology Team, Laboratory of Anthropogenic, Biotechnology and Health, Faculty of Sciences, Chouaib Doukkali University, El Jadida, Morocco*

Alessandra Fiorillo • *Laboratory of Neurological Impairment. Fundación Investigación Hospital Clínico, INCLIVA Biomedical Research Institute, Valencia, Spain*

Juan-José Gallego • *Laboratory of Neurological Impairment. Fundación Investigación Hospital Clínico, INCLIVA Biomedical Research Institute, Valencia, Spain*

Reza Heidari • *Pharmaceutical Sciences Research Center, Shiraz University of Medical Sciences, Shiraz, Iran*

Vicente Hernández-Rabaza • *Departamento Ciencias Biomédicas, Universidad Cardenal Herrera-CEU, Valencia, Spain*

Fátima Higuera-de-la-Tijera • *Head of Gastroenterology and Hepatology Department, Mexico's General Hospital "Dr. Eduardo Liceaga", Mexico City, Mexico*

Paula Izquierdo-Altarejos • *Laboratory of Neurobiology, Principe Felipe Research Centre, Valencia, Spain*

Arumugam Jayakumar • *Miller School of Medicine, University of Miami, Miami, FL, USA*

Marta Llansola • *Laboratory of Neurobiology, Principe Felipe Research Centre, Valencia, Spain*

Miguel López-Aguirre • *HM CINAC MADRID (Centro Integral de Neurociencias Abarca Campal). Hospital Universitario HM Puerta del Sur, HM Hospitales, Madrid, Spain; Instituto de Investigación Sanitaria HM Hospitales, Madrid, Spain; Centro de Investigación Biomédica en Red de Enfermedades Neurodegenerativas (CIBERNED), Instituto de Salud Carlos III, Madrid, Spain*

Adrià López-Gramaje • *Laboratory of Neurological Impairment. Fundación Investigación Hospital Clínico, INCLIVA Biomedical Research Institute, Valencia, Spain*

David Martí-Aguado • *Digestive Disease Department, Clinic University Hospital, INCLIVA Health Research Institute, Valencia, Spain*

Luis Martí-Bonmatí • *Radiology Department, La Fe University and Polytechnic Hospital, Valencia, Spain. Biomedical Imaging Research Group (GIBI230), La Fe Health Research Institute, and Imaging La Fe node at Distributed Network for Biomedical Imaging (ReDIB) Unique Scientific and Technical Infrastructures (ICTS), Valencia, Spain*

María Luz Martínez-Chantar • *Liver Disease Laboratory, Center for Cooperative Research in Biosciences (CIC bioGUNE), Basque Research and Technology Alliance (BRTA), Derio, Bizkaia, Spain; Centro de Investigación Biomédica en Red de Enfermedades Hepáticas y Digestivas (CIBERehd), Madrid, Spain*

Mar Martínez-García • *Laboratory of Neurobiology, Principe Felipe Research Centre, Valencia, Spain*

Gergana Mincheva • *Laboratory of Neurobiology, Principe Felipe Research Centre, Valencia, Spain*

Carmina Montoliu • *Laboratory of Neurological Impairment. Fundación Investigación Hospital Clínico, INCLIVA Biomedical Research Institute, Valencia, Spain; Departamento de Patología, Universidad Valencia, Valencia, Spain*

Hossein Niknahad • *Pharmaceutical Sciences Research Center, Shiraz University of Medical Sciences, Shiraz, Iran; Department of Pharmacology and Toxicology, Faculty of Pharmacy, Shiraz University of Medical Sciences, Shiraz, Iran*

Mohammad Mehdi Ommati • *Henan Key Laboratory of Environmental and Animal Product Safety, College of Animal Science and Technology, Henan University of Science and Technology, Luoyang, Henan, China*

JOSÉ LUIS PÉREZ-HERNÁNDEZ • *Liver Clinic from Gastroenterology and Hepatology Department, Mexico's General Hospital "Dr. Eduardo Liceaga", Mexico City, Mexico*
LUCIE PICARD • *Departamento Ciencias Biomédicas, Universidad Cardenal Herrera-CEU, Valencia, Spain*
MARINA ROMANÍ-PÉREZ • *Microbiome, Nutrition & Health Research Unit, Institute of Agrochemistry and Food Technology, Spanish National Research Council (IATA-CSIC), Valencia, Spain*
DANIEL SANTANA-VARGAS • *Sleep Disorders Clinic, Faculty of Medicine, National Autonomous University of Mexico, Mexico City, Mexico*
ANNE-KATHRIN SCHOTT • *ACA CELL Biotech GmbH, Heidelberg, Germany*
ALI SEPEHRINEZHAD • *Neuroscience Research Center, Mashhad University of Medical Sciences, Mashhad, Iran; Department of Neuroscience, Faculty of Medicine, Mashhad University of Medical Sciences, Mashhad, Iran*
ALI SHAHBAZI • *Cellular and Molecular Research Center, Iran University of Medical Sciences, Tehran, Iran; Department of Neuroscience, Faculty of Advanced Technologies in Medicine, Iran University of Medical Sciences, Tehran, Iran*
ANTONIO LUCIO TEIXEIRA • *The Glenn Biggs Institute for Alzheimer's & Neurodegenerative Diseases, Lozano Long School of Medicine, University of Texas Health Science Center at San Antonio, San Antonio, TX, USA*
AMPARO URIOS • *Laboratory of Neurological Impairment. Fundación Investigación Hospital Clínico, INCLIVA Biomedical Research Institute, Valencia, Spain*
IVAN ZIPANČIĆ • *Departamento Ciencias Biomédicas, Universidad Cardenal Herrera-CEU, Valencia, Spain*

Part I

Diagnostic and Monitoring Protocols

Chapter 1

Psychometric Methods for Diagnosing and Monitoring Minimal Hepatic Encephalopathy

Juan-José Gallego, Alessandra Fiorillo, Franc Casanova-Ferrer, Adrià López-Gramaje, Amparo Urios, and Carmina Montoliu

Abstract

Hepatic encephalopathy (HE) is a complex neuropsychiatric syndrome that exhibits a wide spectrum of nonspecific neurological impairments, ranging from subclinical alterations to coma and death. Minimal HE (MHE) is the initial stage of HE and is considered the greatest risk factor for its development. Patients with MHE show mild cognitive impairment, attention deficits, psychomotor slowing, and impaired motor coordination. These cognitive and motor alterations can lead to an increase in the risk of accidents and hospitalizations, which in turn reduce life quality and span. Therefore, MHE is a major health, social, and economic problem, whose diagnosis is essential for treatment and to prevent progression to HE. Currently, the MHE diagnosis is carried out mainly by psychometric tests, such as the Psychometric Hepatic Encephalopathy Score (PHES) and critical flicker frequency (CFF). Since the 11th World Congress of Gastroenterology (1998), PHES has been proposed as a gold standard for diagnosis of MHE. The spectrum of cognitive and motor impairments presented by patients with MHE is broad and cannot be assessed by PHES or CFF only. Other psychometric tests should be used to evaluate the full spectrum of cognitive and motor impairments, such as the Stroop test, the d2 test, the oral Symbol Digit Modalities Test, the digit span, letter-number sequencing test, bimanual coordination test, and visuo-motor coordination test. This chapter will detail the methodology necessary for performing psychometric tests used for MHE diagnosis, as well as other more sensitive tests for assessing associated cognitive and motor impairments.

Key words Minimal hepatic encephalopathy, Cognitive impairment, Motor impairment, Diagnosis, Psychometric tests, Psychometric Hepatic Encephalopathy Score, Critical flicker frequency, Stroop test, d2 test, Motor coordination test

1 Introduction

Patients with liver diseases may show hepatic encephalopathy (HE), a complex neuropsychiatric syndrome that leads to alterations in personality, cognitive function, motor activity and coordination, consciousness, and may lead to coma and death [1, 2]. Around 40% of cirrhotic patients without clinical symptoms of HE show minimal HE (MHE), which can be detected using psychometric

Tiziano Balzano (ed.), *Experimental and Clinical Methods in Hepatic Encephalopathy Research*, Neuromethods, vol. 219, https://doi.org/10.1007/978-1-0716-4354-9_1,

tests [2–5]. Patients with MHE show mild cognitive impairment, characterized by attention deficits, psychomotor slowing, impaired visuo-motor skills, and bimanual coordination [1–3]. MHE reduces the quality of life and lifespan of patients and increases the risk of accidents, falls, fractures, hospitalizations, and associated costs, thereby imposing a substantial economic burden on health systems [6–8]. MHE affects more than two million people in the USA and a similar number in the EU, representing a significant health, social, and economic problem. MHE places considerable financial and socioeconomic burden on patients and their caregivers, adversely affecting the work capacity of patients [9].

We have previously shown that the reduced mental processing speed, attention deficits, and alterations in visual-spatial and motor coordination contribute to impaired driving ability in MHE [10], and that attention and coordination deficits affect balance in these patients, contributing to an increased risk of falls [7].

Detection and diagnosis of MHE is essential to prevent progression to overt HE. MHE diagnosis is based on assessment the cognitive and motor functions using psychometric tests. The tests used for the diagnosis of MHE are not specific to this pathology, which should therefore be performed in patients with liver disease after ruling out signs of HE or other neurological diseases, or medication, drugs or alcohol use that could alter results [11]. Tests ruling out the presence of MHE should nonetheless be repeated after six months to monitor patient progress [12].

Currently, the two most widely used tests for the diagnosis of MHE are the Psychometric Hepatic Encephalopathy Score (PHES) and the critical flicker frequency (CFF) [13, 14].

The PHES is the gold standard test for MHE diagnosis and has been recommended in multicenter studies [15]. The PHES battery is composed of five subtests that assess different cognitive and motor functions, such as attention, visuo-spatial perception, visuo-spatial construction, psychomotor speed, and motor accuracy. A test specifically assessing good attention focus is not included, as this function is required to perform the tests [13, 15].

Since the 11th World Congress of Gastroenterology in 1998, PHES has been considered by consensus the common gold standard for diagnosing MHE [5, 13]. Hannover Medical School holds the copyright for the test and provides the test material via Dr. Karin Weissenborn.

Generally, a subject is diagnosed with MHE when the global score obtained in the PHES is equal to or lower than −4 [13], although different cut-offs may be applied in other populations. This battery is paper-pencil based, which is an inexpensive method to assess the spectrum of cognitive and motor impairments that affect HE without requiring specialized equipment. The main disadvantages are the longer time required to obtain diagnosis and the potential for age and education level-related bias, although this drawback can be overcome using data adjusted for age and level of education for different populations.

Critical flicker frequency (CFF) analysis has been proposed for easy, sensitive, and objective quantification of MHE [14].

The CFF is the mean frequency at which patients can perceive a flicker of a light, as distinguished from continuous light. A CFF of less than 39 Hz is considered indicative of brain dysfunction [14], although other studies propose that a cut-off point of 38 Hz achieves 77.2% specificity and 72.4% sensitivity in MHE diagnosis [16], improving on the threshold of 39 Hz. The HEPAtonorm analyzer is the most frequently used tool for determining CFF.

The main advantages of assessing CFF are that it is barely affected by the training subject effect, level of education, age, or the time of day of assessment. The disadvantages are that this test cannot be performed on patients without stereoscopic vision or who are color blind (as they will not perceive the light stimulus optimally), that it does not take into account motor impairments and that it is less cost-effective than PHES in terms of equipment needed.

Another test less frequently used for MHE diagnosis is the Stroop test, which assesses the processing speed and inhibition systems of an automated response [17, 18]. This test was used as a reference to develop a smartphone application called EncephalApp, which several studies suggest is a useful tool for screening MHE [19, 20].

It has been demonstrated that the PHES is not sensitive enough to detect early neurological alterations, since patients who are not diagnosed with MHE already present cognitive and motor impairments [21]. For this reason, more sensitive tests are required to monitor and characterize the cognitive and motor impairment associated with these patients. These tests can be used to evaluate various executive functions impaired in patients with MHE: processing speed and automated response inhibition systems with Stroop test [3, 17, 18, 21]; processing speed, sustained and selective attention, and concentration with d2 test [21–23]; selective attention and mental processing speed with Oral Symbol Digit Modalities Test (SDMT) [21, 24, 25]; working memory with digit span test and the letter-number sequencing test [21, 26, 27]; and motor coordination with the bimanual coordination test [3, 21, 28] and visuo-motor coordination test [3, 21, 29].

We previously showed that combining PHES with d2, oral SDMT, and visuo-motor and bimanual coordination tests significantly increased the number of patients with MHE detected. We also concluded that patients classified as without MHE by PHES but with alterations in these additional tests have an increased risk of clinical complications and must be diagnosed and clinically followed [21].

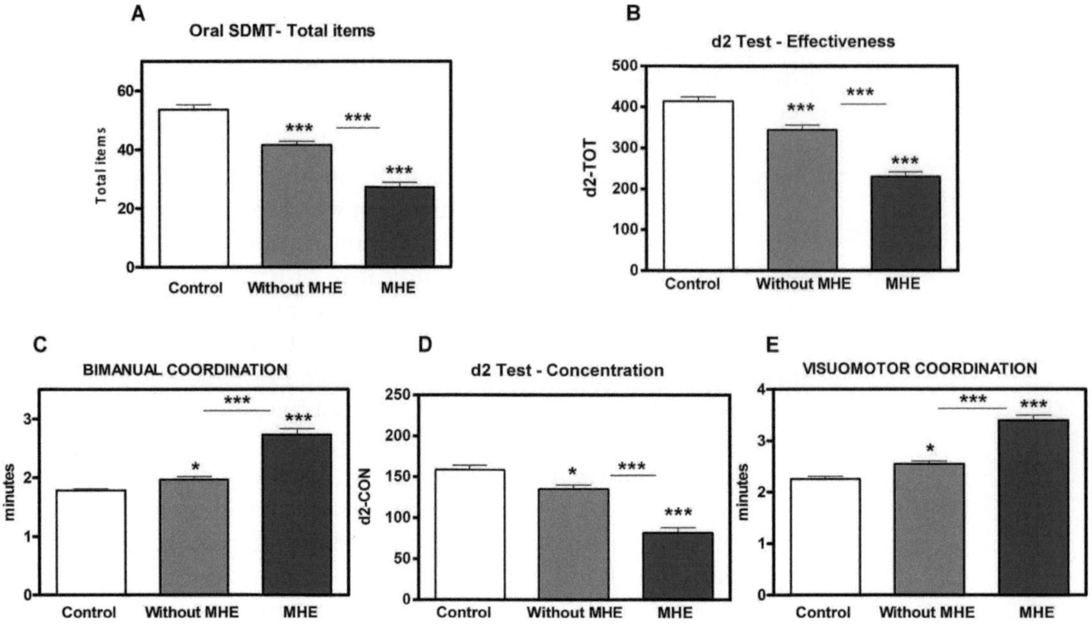

Fig. 1 Example of results obtained in psychometric test in healthy subjects and cirrhotic patients with or without MHE. Modified from Giménez-Garzó et al. *PLoS One*, 12 (2):e0171211 (2017) (22). (**a**) Total items in Oral SDMT, (**b**) effectiveness index of d2 test, (**c**) time to complete bimanual coordination test, (**d**) concentration index of d2 test and (**e**) time to complete visuo-motor coordination test. Values are the mean ± SEM. Values significantly different are indicated by asterisks (*) (*$p < 0.05$; ***$p < 0.001$)

The initial neurological changes are different depending on the patients. Some patients present coordination alterations before mild cognitive impairment as detected by PHES [21, 30], while in other patients, concentration and attention deficits appear before coordination impairment [3, 21]. Figures 1 and 2 show examples of results obtained from evaluation of different tests in cirrhotic patients.

2 Materials

1. PHES, a booklet with tests: DST, NCTA (pre-test); NCTA (test), NCTB, SD, LTT (pre-test) and LTT (test). Test material provided by Dr. Karin Weissenborn.
2. HEPAtonorm analyzer (D-72127, Accelab GmbH; Kusterdingen, Germany).
3. Stroop test (30250, Stoelting Co.; Wood Dale, Illinois, United States).
4. d2 test (19509, Hogrefe Publishing; Göttingen, Germany).
5. Oral SDMT (W-129, WPS; Torrance, California, United States).

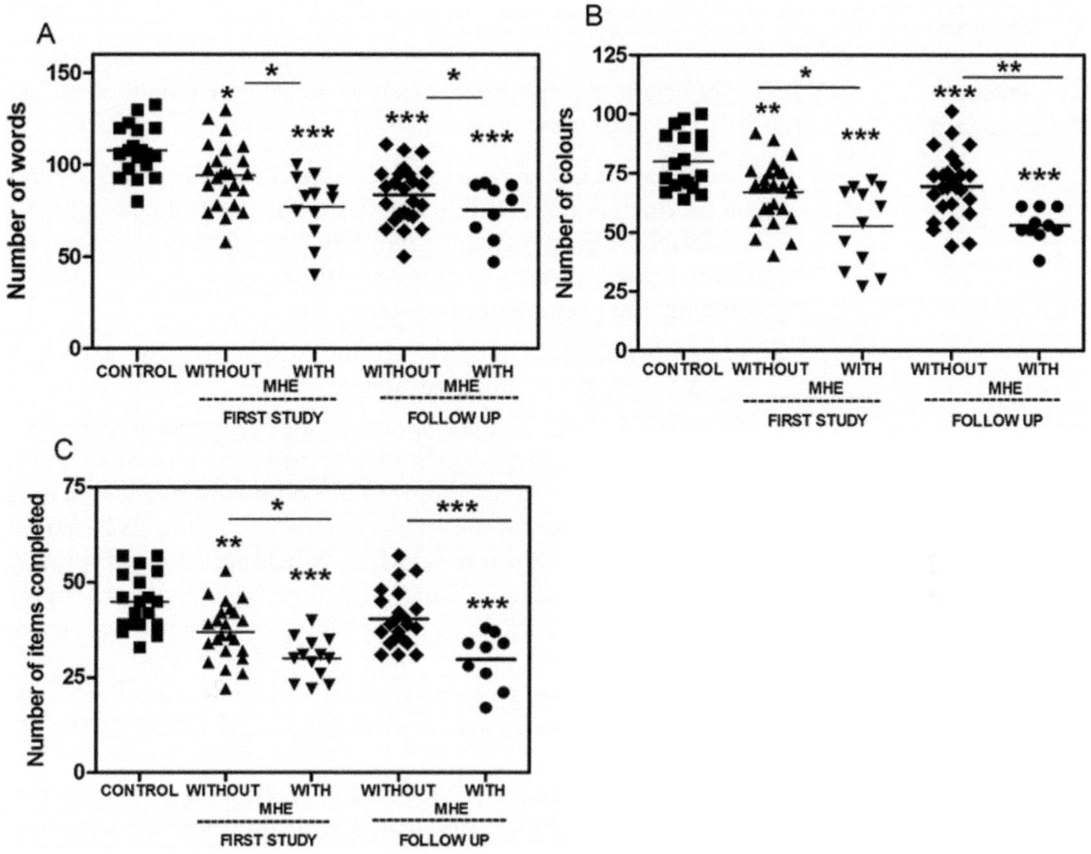

Fig. 2 Example of results obtained in Stroop test in healthy subjects and cirrhotic patients with or without MHE and at follow-up after one year. Modified from Felipo et al. *Hepatology*, 55 (2):530–539 (2012) (3). Correct answers in (**a**) congruent task, (**b**) neutral task and (**c**) incongruent task. Values are the mean ± SEM. Values significantly different are indicated by asterisks (*) (*$p < 0.05$; **$p < 0.01$; ***$p < 0.001$)

6. Digit span test (0158980832, Pearson PLC; London, United Kingdom).
7. Letter-number sequencing test (0158980832, Pearson PLC; London, United Kingdom).
8. Bimanual coordination test (Yela, TEA Ediciones, Madrid, Spain).
9. Visuo-motor coordination test (Yela, TEA Ediciones, Madrid, Spain).
10. Black and red fiber tip pens with a fine nib (1 mm).
11. Results record sheet and pen.
12. Timer.
13. Normative tables for each psychometric test.

3 Methods

3.1 Previous Considerations

It is important to minimize external factors that may alter the results of psychometric tests:

1. *Control of space and timing of test administration*: Tests should be conducted in a well-lit room with only the evaluator and the evaluated subject present, avoiding any distractions such as photographs, posters, windows with an external view, people coming and going from the room, etc.
2. *Visual difficulties*: Subjects should use corrective glasses if needed during test administration.
3. *Utilization of normalization tables*: After each test is corrected, it is recommended to obtain scaled scores from normalization tables specific to an age-appropriate population. These scores enable comparison of the results of evaluated subjects based on the normal population to which they belong, adjusted for age. Normalization tables should therefore be utilized whenever possible to derive scaled scores for the psychometric tests used.

3.2 PHES

Composed of five subtests that assess different executive functions, each with specific instructions and correction procedures.

3.2.1 Digit Symbol Test (DST)

This test evaluates processing speed and working memory. In this subtest, a reference table is provided with numbers from 1 to 9, each associated with a symbol. In the test cells, subjects must draw the symbol corresponding to the number indicated in each cell based on the reference table. First, a training phase is conducted to familiarize the patient with the methodology, followed by a 90-s test phase:

Training Phase: Consists of a row with 20 cells, where the subject must draw the symbol corresponding to each number. The training ends when the subject completes the row.

Test Phase: Consists of four rows with 20 cells each, where the subject must draw the symbol corresponding to each number. The subject has 90 s to complete as many cells as possible. The score obtained in this test derives from the number of correct cells in the test phase.

3.2.2 Number Connection Test-a (NCT-A)

This test evaluates processing speed and attention. This subtest consists of a series of randomly arranged circles numbered from 1 to 25 on the sheet. The task is to connect the numbers with lines in ascending order. Any mistakes made during the test must be corrected before the subject can continue to the next item. A training phase is first conducted to familiarize the patient with the methodology, followed by the test phase on a separate sheet:

Training Phase: Subjects must connect the numbers with lines in ascending order (1-2-3…23-24-25) using either curved or straight lines according to personal preference of the evaluated subject.

Test Phase: Subjects must connect numbers from 1 to 25 with lines in ascending order, arranged differently in this phase than in the training phase. The test score is based on the time taken to complete the test phase.

3.2.3 Number Connection Test-B (NCT-B)

This test evaluates processing speed and attention. This subtest consists of a series of circles numbered from 1 to 13 and circles with letters from A to L arranged randomly on the sheet. The task is to connect the circles alternating between numbers (in ascending order) and letters (in alphabetical order), starting with 1 and ending with L (1-A-2-B…K-12-L-13). Any mistakes made during the test must be corrected before the subject can continue to the next item. Unlike NCT-A, there is no training phase in this subtest. The score obtained is based on the time taken to complete the test.

3.2.4 Serial Dotting Test (SDT)

This test evaluates visuo-spatial coordination. This subtest consists of a series of circles. The subjects are required to place a dot as centrally as possible within each circle. A training phase is conducted prior to the test phase to acquaint the subject with the methodology.

Training Phase: Consists of two rows with ten circles each and the subject must place a dot as centrally as possible within each circle from left to right and from top to bottom (*see* **Note 1**).

Test Phase: Consists of ten rows with ten circles each, and subjects must place a dot as centrally as possible within each circle from left to right and from top to bottom. The score obtained is based on the time taken to complete the test phase.

3.2.5 Line Tracing Test (LTT)

This test evaluates visuospatial coordination. This subtest involves tracing a line marked by parallel lines with straight and curved segments, as well as abrupt or curved turns. The subject must draw a continuous line from the beginning to the end of the parallel lines, trying not to draw on or outside these lines. Two rules apply to the test: (1) once the test begins, the fiber tip pen cannot be lifted from the paper, and (2) the paper cannot be moved to facilitate the exercise. The training and test phases are conducted on separate sheets.

Training Phase: Involves a short course where the subject must trace a continuous line from beginning to end (*see* **Note 2**).

Test Phase: Involves a longer course where the subject must trace a continuous line from beginning to end. The time taken to complete the test phase is recorded.

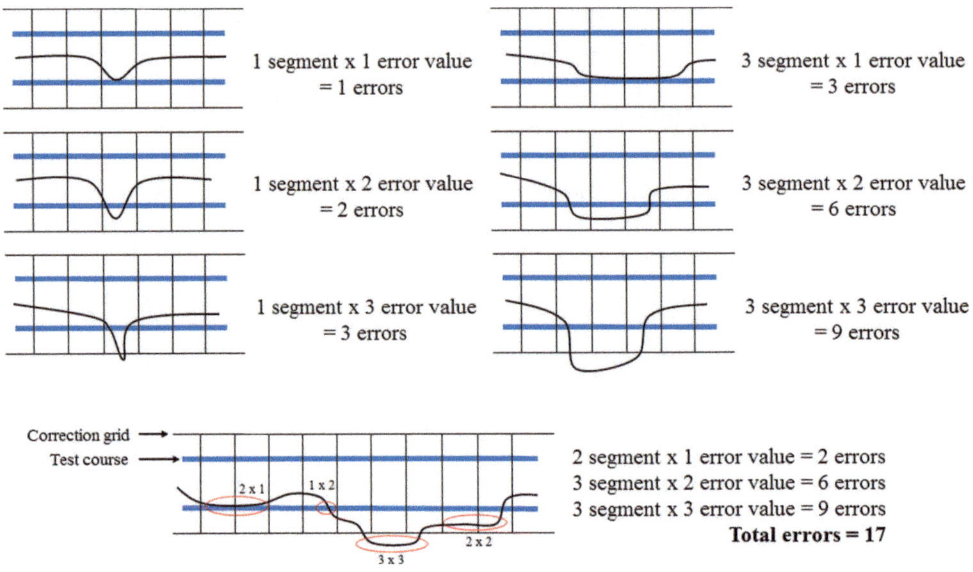

Fig. 3 Example of LTT correction

LTT correction: The test score is the sum of the time in seconds taken to complete the test plus number of errors made. This test is corrected using a template consisting of a grid of equal length but wider and divided into segments. Errors are calculated based on the number of crossed out segments in the grid: x1 if touching the course wall, x2 if outside the course wall, and x3 if outside the walls of the grid (Fig. 3).

3.2.6 PHES Score Calculation

The PHES score is calculated using normalization tables adapted to the population studied. These tables include the standard deviation (SD) and formulas for calculating expected result in each subtest according to age and years of education. The subject score for each subtest is calculated using the following formula: (*true result—expected result)/SD*. The score is obtained as an integer, either positive or negative, depending on the subject's performance in the subtest. This score represents the number of SDs by which the actual result deviates from the expected result. The final PHES score is obtained by summing the scores of the subtests. Normative tables for different populations should also be used to calculate the PHES score: Spanish [31], Thai [32], Chinese [33], Romanian [34], Korean [35], etc.

Table 1 shows an example of PHES score calculation using values for the Spanish population [31], and an application for such calculation was developed (http://www.redeh.org/TEST_phes.htm).

Table 1
Formulas and SD values of reference for Spanish population to calculate PHES Score

Test	*SD*	*Formulas for expected result*			
DST	10.45	56.800 – (0.614 × age (years)) + (1.317 × education (years in the education system))			
NCT-A	21.30	26.772 + (0.596 × age (years)) – (1.519 × education (years in the education system))			
NCT-B	44.74	31.638 + (1.856 × age (years)) – (2.809 × education (years in the education system))			
SDT	23.30	61.346 + (0.236 × age (years)) – (1.279 × education (years in the education system))			
LTT	25.62	51.303 + (1.043 × age (years)) – (0.990 × education (years in the education system))			
		Example for Spanish patients aged **50** and with **10** years of education			
Test	*SD*	*Formulas for expected result*	*Expected result*	*True result*	*Score*
DST	10.45	56.800 – (0.614 × **50**) + (1.317 × **10**)	39.27	9	(39.27–9)/10.45 = –2
NCT-A	21.30	26.772 + (0.596 × **50**) – (1.519 × **10**)	41.38	122	(41.38–122)/21.30 = –3
NCT-B	44.74	31.638 + (1.856 × **50**) – (2.809 × **10**)	96.35	160	(96.35–160)/44.74 = –1
SDT	23.30	61.346 + (0.236 × **50**) – (1.279 × **10**)	60.36	65	(60.36–65)/23.30 = 0
LTT	25.62	51.303 + (1.043 × **50**) – (0.990 × **10**)	93.55	115	(93.55–115)/25.62 = 0
PHES					–6

Adapted from Romero et al. *Med Clin (Barc.)*, 127, 246–249 (2006) [31]. *PHES* Psychometric Hepatic Encephalopathy Score; *DST* Digit Symbol Test; *NCT-A* Number Connection Test-A; *NCT-B* Number Connection Test-B; *SDT* Serial Dotting Test; *LTT* Line Tracking Test; *SD* Standard Deviation

3.3 HEPAtonorm Analyzer

The HEPAtonorm analyzer consists of a control for the evaluator, a set of head-mounted goggles, and a stop button for the patient (Fig. 4a). The device generates a red-light stimulus through an LED on a white background at a virtual distance of 12 meters. When the evaluator activates the device, the light begins to flicker at a frequency of 60 Hz, gradually decreasing to 25 Hz. The subject should press the stop button as soon as they notice the light starting to flicker. The light stimulus must be clearly observable: if the subject requires glasses, they should use them during the test to focus on the stimulus. The test should be conducted in low ambient light, without distracting noises, with clean apparatus and lenses, and the device should be operating at optimal voltage so that when low battery is indicated, it should be replaced with a new one. Procedure:

1. Placing the head-mounted goggles on the subject, check that they are correctly attached to the head (*see* **Note 3**) (Fig. 4b), and allowing a few seconds for the vision of subject to adjust to the lack of light.

A B

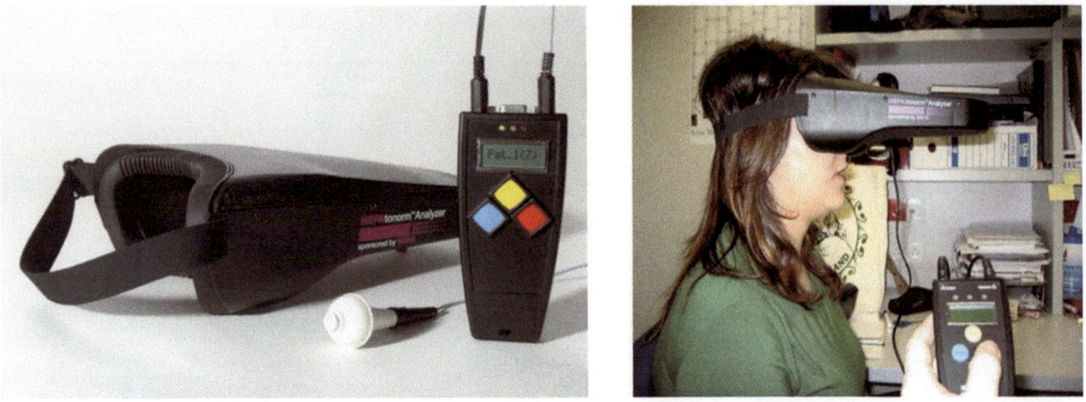

Fig. 4 HEPAtonorm Analyzer for CFF measurement. (**a**) Control for the evaluator, a set of head-mounted goggles, and a stop button for the patient. (**b**) Example of the correct position of the head-mounted goggles to perform the test

3.3.1 Training Phase

1. In the initial phase, the head-mounted goggles are activated and the subject observes the stimulus until it stops automatically (at a flicker frequency of 25 Hz) so that they can clearly identify the flickering.
2. The subject is asked to confirm that they can clearly see the light stimulus and flickering.
3. The subject is instructed that the stimulus will be activated, and they should press the stop button at the moment they begin to perceive the flickering of the light. The Hz at which the subject stops the instrument is recorded.
4. The previous step is repeated a further three times, making a total of four tests.

3.3.2 Test Phase

1. The subject is clearly instructed that the stimulus will be activated, and they should press the stop button just when they begin to notice the flickering of the light. The frequency at which the subject stops the instrument is recorded.
2. The previous step is repeated eight more times, making a total of nine test phases (*see* **Note 4**).
3. The CFF of patient is obtained by calculating the average of the nine test phases.

3.4 Stroop Test

The Stroop test comprises three tasks (congruent, neutral, and incongruent). Each task consists of an A4-sized sheet containing 100 items divided into five columns of 20 items each. The items should be named following the order of top to bottom and left to right (*see* **Note 5**). The items are different in each task.

3.4.1 Congruent Task

The presented items are the names of colors (blue, green and red) written in black ink. The subject is requested to read as many words as possible in 45 s, following the order on the sheet. This task evaluates processing speed.

3.4.2 Neutral Task

The items presented consist of four Xs (XXXX), with each item printed in different colors (blue, green, and red). The subject must name as many colors as possible in 45 s. This task evaluates processing speed.

3.4.3 Incongruent Task

The presented items are words that are colors but printed in a different color than the one they represent (e.g., the word "red" printed in blue ink) (*see* **Note 6**). The subject has 45 s to indicate the color of the ink for each word. Any errors made (reading the word instead of naming the color of the ink), must be corrected before the subject can proceed. This task evaluates cognitive flexibility and inhibition of automated response, i.e., the ability to inhibit the normal response (reading the word) in favor of the less habitual response (naming the color of the ink).

3.4.4 Correction and Scoring of the Stroop Test

From this test, three direct scores are obtained: number of words (P), number of colors (C), and number of colored words named (PC), corresponding to each task of the Stroop test. Direct scores are corrected in relation to age and the interference parameter is calculated from these three scores. To determine the interference score, the difference between the PC and the estimated PC (PC') is calculated with the following formula: *Interference* $= PC - PC'$, *where* $PC' = (C \times P) / (C + P)$. Scaled scores are obtained from normality tables for each score obtained. The number of words in congruent task and colors in neutral task assess processing speed, while the number of colored words in incongruent task and interference parameter evaluate cognitive flexibility and the ability to inhibit automatic responses, and are used as estimators of inhibitory and attentional control ability.

3.5 d2 Test

This test evaluates several executive functions, including processing speed, selective attention, sustained selective attention, and concentration. The test consists of an A4-sized sheet with 14 rows, each containing 47 items. These items are the letters "d" or "p" accompanied by 1–4 dashes above and/or below the item. In this test, the subject must cross out all the target items, "d" with two dashes (Fig. 5), while ignoring the remaining items.

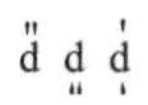

Fig. 5 Target items in d2 test. Subject must mark the target items: "d" with two dashes above and/or below the item

3.5.1 Training Phase

The subject is instructed to cross out the "d" items with two dashes (Fig. 5), as shown in the explanation sheet, and is given a practice row of 22 items, which is corrected with the subject, and errors corrected to ensure instructions are clear.

3.5.2 Testing Phase

The subject is given 20 s per row to cross out the maximum number of target items possible from left to right. Rows should be processed consecutively without pauses. The evaluator controls the time on each row and informs the subject when to move to the next line.

3.5.3 Correction of the d2 Test Correction

The d2 test is corrected using a template indicating the position of all items in each row and highlighting the target items within squares. The following parameters are obtained after test correction:

1. *Total responses (TR):* The position of the last crossed out item is determined for each row, and these numbers are summed to obtain the TR variable. This variable evaluates processing speed and the quantity of work in the test.
2. *Total correct responses (TA):* The number of correctly crossed out target items is determined in each row (elements marked within the template squares). Summing the correctly crossed out items from each row results in the TA variable, which measures attention and accuracy in processing.
3. *Omission errors (O):* The number of target items not crossed out (items within the template squares not crossed out) is determined for each row up to the last crossed out element. The sum of these errors is the O variable.
4. *Commission errors (C):* The number of incorrectly crossed out items (items crossed out outside the template squares) is determined for each row, up to the last crossed out element. The sum of these errors is the C variable.
5. *Effectiveness index (TOT):* $TOT = TR - (O + C)$. This variable is used to study the relationship between speed and accuracy in the test, thus evaluating sustained selective attention.
6. *Concentration index (CON):* $CON = TA - C$. This variable indicates the balance between speed and accuracy in the test, thus evaluating concentration.

3.6 Oral SDMT

This test evaluates selective attention, with partial assessment of processing speed and working memory. The test consists of an A4-sized sheet with a reference table containing nine symbols, each associated with a number from 1 to 9, and eight rows, each with 15 symbols. During the test, the subject must orally state the number corresponding to each symbol according to the reference table, following a specific order along the sheet: from left to right and top to bottom.

3.6.1 Training Phase

The subject orally states the numbers corresponding to the first ten symbols in the first row.

3.6.2 Test Phase

For 90 s, the subject must orally state the number corresponding to each symbol, continuing from where the training phase ended.

3.6.3 Correction of Oral SDMT

The test provides various parameters, including total number of responses, number of correct responses, and number of errors. The number of correct responses is the parameter that evaluates selective attention.

3.7 Digit Span Test

Extracted from the Wechsler Adult Intelligence Scale (WAIS), this test assesses immediate and working memory. The test involves listening to number sequences and repeating them in the same or reverse order, depending on the test phase.

3.7.1 Digit Forward Phase

This phase includes eight series, each comprising two randomly generated number sequences. The first block starts with sequence of two numbers, and one more number is added to each subsequent sequence, reaching a sequence with nine numbers in the last block. The subject listens to the sequences and must repeat them back in the same order in which numbers were recited; for example, sequence: 5-8-2, correct response: 5-8-2. The test concludes when the subject fails to correctly repeat both sequences in a single block. Correctly repeated sequences are recorded.

3.7.2 Digit Backward Phase

This phase follows the same format as the digit forward phase but consists of seven series. The subject listens to the number sequences and must repeat them in reverse order; for example, sequence: 6-2-9, correct response: 9-2-6. The test concludes when the subject fails to correctly repeat both sequences in a single block. Correctly repeated sequences are recorded.

3.8 Letter-Number Sequencing Test

Extracted from the WAIS battery, this test assesses working memory more precisely. It consists of seven series, each comprising three sequences of randomly generated numbers and letters. The first block starts with series of two items, and with each subsequent block one more number is added to the series, reaching a total of series with eight items in the last block. The subject listens to the sequence, memorizes the letters and numbers, and must repeat them in a specific order. The repetition order is first numbers in ascending order and then letters in alphabetical order; for example, series: T-9-A-3, correct response: 3-9-A-T. The test concludes when the subject fails to correctly recite all three sequences in a single block. Correctly repeated sequences are recorded.

3.9 Bimanual Coordination Test

This test assesses motor coordination and consists of two elements: a pegboard and 40 pegs to be inserted into the holes. The pegboard measures 26 × 26 cm and has two sets of holes separated by a 2.7 cm space. Each set of holes comprises four rows with ten holes each spaced 6 mm apart. The metal pegs weigh 15.8 g and are 6.5 cm long with a 6.5 mm diameter.

3.9.1 First Phase

The pegboard is placed in front of the subject, about 6 cm from the edge of the table. The test begins with the pegs inserted into the set of holes closest to the subject. Using both hands and moving from the extremes toward the center of the row, the subject must transfer the pegs from one side of the board to the corresponding position on the other side, starting with the furthest row. Figure 6 illustrates the order and movement of the pegs. The time taken by the subject to complete the task is recorded.

3.9.2 Second Phase

Without moving the board, the subject must return the pegs to their original position, starting with the last row moved, reversing the action performed in the first phase. The pegs should be moved in the same way as in the first phase: using both hands and moving from the extremes toward the center of the row. The time taken by the subject to complete this task is recorded.

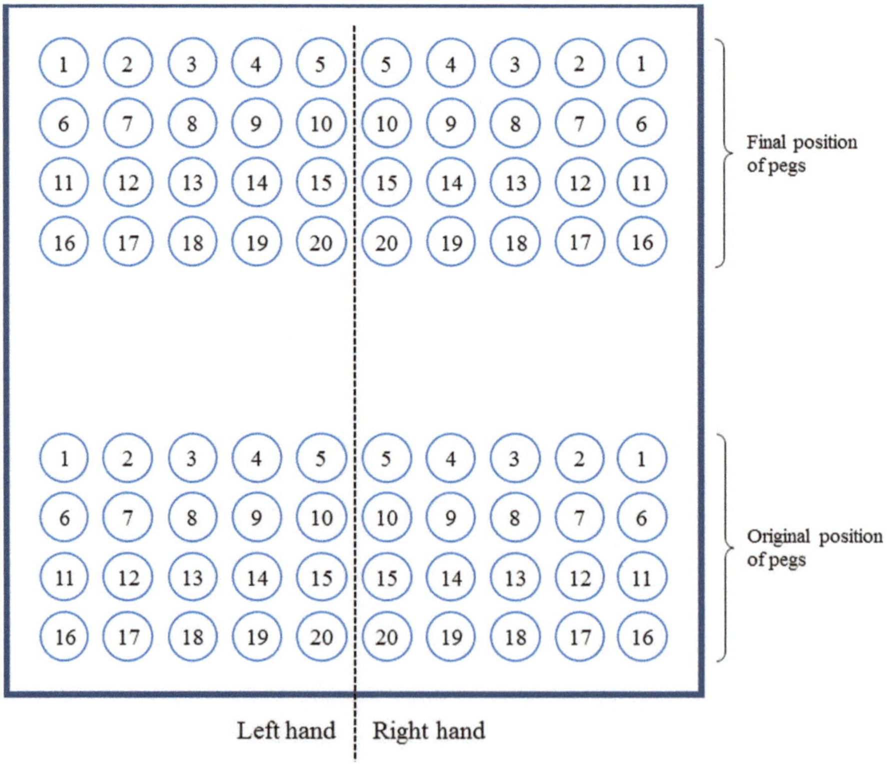

Fig. 6 Bimanual Coordination Test. The square pegboard of the bimanual coordination test is shown, indicating the initial position of the pegs and the position they should be moved to, with which hand, and the order (marked by numbers) in which the pegs should be moved

3.9.3 Third and Fourth Phase

The procedure from the first and second phases is repeated. The time taken by the subject to complete all phases is summed to obtain the final bimanual coordination parameter.

3.10 Visuo-Motor Coordination Test

This test evaluates visuo-motor coordination. This test consists of two elements, a box with a board with holes in one part, and 36 prisms that are inserted into the holes. The board is square, measuring 17 × 17 cm, with holes of 18.2 × 16.3 mm with different orientations, arranged in six rows of six holes each. The prisms are metallic, weighing 72.5 g, and measuring 30 × 18.2 × 16.3 mm.

3.10.1 First Phase

The box is placed open, centered in front of the subject about 6 cm from the edge of the table, with the board with holes on the side opposite to the dominant hand and the prisms on the side closer to the dominant hand. Using only the dominant hand, the subject must insert the prisms into the holes of the board, completing this task from left to right and from top to bottom. The time it takes the subject to complete this task is recorded.

3.10.2 Second Phase

The evaluator returns the prisms to the side of the box opposite the board with holes, and the same methodology followed in the first phase is repeated. The time it takes the subject to complete both phases of the test is summed to obtain the final parameter of visuo-motor coordination.

4 Notes

1. Use of fiber tip pens is recommended for this subtest, as touching the paper draws the dot and the subject does not lose seconds drawing the dot.
2. Using a fiber tip pen for this subtest allows a continuous line to be drawn more smoothly than with a ballpoint pen.
3. Avoid abrupt movements or tremors during the test that may hinder perceiving the flicker. If necessary, the test can be conducted lying down or with additional support to prevent head tremors.
4. If there are variations among the nine measurements or the subject reports an error in any repetition, additional measurements can be taken to increase result precision.
5. If the subject finishes naming the items of any task within the provided time, they should continue the test from the beginning of the sheet and count any additional items.
6. Before beginning the incongruent task, subjects should be asked to name three randomly selected items to ensure they have understood the test.

5 Conclusions

MHE represents a social, economic, and health burden, as the cognitive and motor impairment experienced by patients with this disease leads to a reduction in quality and span of life. Additionally, MHE can progress to HE with more severe symptoms. Early diagnosis of MHE could improve its treatment, enhancing the quality and life span of patients and preventing progression to HE. The current gold standard for MHE diagnosis is PHES, although as a standalone test it is insufficient to detect early neurological alterations given the proportion of patients diagnosed as without MHE by PHES, who already show cognitive and motor impairments compared to control subjects. A good approach, therefore, is to perform a combination of the PHES and other more sensitive psychometric tests in order to monitor and characterize cognitive and motor impairment and identify patients with early neurological impairments. This more exhaustive evaluation of patients allows us to measure cognitive and motor status even before MHE diagnosis is reached, which could enable us identify patients in cognitive and motor decline, track their progression meticulously, and prevent the onset of MHE and HE.

References

1. Amodio P, Montagnese S, Gatta A, Morgan MY (2004) Characteristics of minimal hepatic encephalopathy. Metab Brain Dis 19(3–4): 253–267. https://doi.org/10.1023/b:mebr.0000043975.01841.de
2. Weissenborn K, Giewekemeyer K, Heidenreich S, Bokemeyer M, Berding G, Ahl B (2005) Attention, memory, and cognitive function in hepatic encephalopathy. Metab Brain Dis 20(4):359–367. https://doi.org/10.1007/s11011-005-7919-z
3. Felipo V, Ordoño JF, Urios A, El Mlili N, Giménez-Garzó C, Aguado C, González-Lopez O, Giner-Duran R, Serra MA, Wassel A, Rodrigo JM, Salazar J, Montoliu C (2012) Patients with minimal hepatic encephalopathy show impaired mismatch negativity correlating with reduced performance in attention tests. Hepatology 55(2):530–539. https://doi.org/10.1002/hep.24704
4. Felipo V (2013) Hepatic encephalopathy: effects of liver failure on brain function. Nat Rev Neurosci 14(12):851–858. https://doi.org/10.1038/nrn3587
5. Ferenci P, Lockwood A, Mullen K, Tarter R, Weissenborn K, Blei AT (2002) Hepatic encephalopathy--definition, nomenclature, diagnosis, and quantification: final report of the working party at the 11th World Congresses of Gastroenterology, Vienna, 1998. Hepatology 35(3):716–721. https://doi.org/10.1053/jhep.2002.31250
6. Román E, Córdoba J, Torrens M, Torras X, Villanueva C, Vargas V, Guarner C, Soriano G (2011) Minimal hepatic encephalopathy is associated with falls. Am J Gastroenterol 106(3):476–482. https://doi.org/10.1038/ajg.2010.413
7. Urios A, Mangas-Losada A, Gimenez-Garzó C, González-López O, Giner-Durán R, Serra MA, Noe E, Felipo V, Montoliu C (2017) Altered postural control and stability in cirrhotic patients with minimal hepatic encephalopathy correlate with cognitive deficits. Liver Int 37(7):1013–1022. https://doi.org/10.1111/liv.13345
8. Poordad FF (2007) Review article: the burden of hepatic encephalopathy. Aliment Pharmacol Ther 25(Suppl 1):3–9. https://doi.org/10.1111/j.1746-6342.2006.03215.x

9. Ridola L, Cardinale V, Riggio O (2018) The burden of minimal hepatic encephalopathy: from diagnosis to therapeutic strategies. Ann Gastroenterol 31(2):151–164. https://doi.org/10.20524/aog.2018.0232
10. Felipo V, Urios A, Valero P, Sánchez M, Serra MA, Pareja I, Rodríguez F, Gimenez-Garzó C, Sanmartín J, Montoliu C (2013) Serum nitrotyrosine and psychometric tests as indicators of impaired fitness to drive in cirrhotic patients with minimal hepatic encephalopathy. Liver Int 33(10):1478–1489. https://doi.org/10.1111/liv.12206
11. Vilstrup H, Amodio P, Bajaj J, Cordoba J, Ferenci P, Mullen KD, Weissenborn K, Wong P (2014) Hepatic encephalopathy in chronic liver disease: 2014 Practice Guideline by the American Association for the Study of Liver Diseases and the European Association for the Study of the Liver. Hepatology 60(2): 715–735. https://doi.org/10.1002/hep.27210
12. Prakash RK, Brown TA, Mullen KD (2011) Minimal hepatic encephalopathy and driving: is the genie out of the bottle? Am J Gastroenterol 106(8):1415–1416. https://doi.org/10.1038/ajg.2011.154
13. Weissenborn K, Ennen JC, Schomerus H, Rückert N, Hecker H (2001) Neuropsychological characterization of hepatic encephalopathy. J Hepatol 34(5):768–773. https://doi.org/10.1016/s0168-8278(01)00026-5
14. Kircheis G, Wettstein M, Timmermann L, Schnitzler A, Häussinger D (2002) Critical flicker frequency for quantification of low-grade hepatic encephalopathy. Hepatology 35(2):357–366. https://doi.org/10.1053/jhep.2002.30957
15. Bajaj JS, Cordoba J, Mullen KD, Amodio P, Shawcross DL, Butterworth RF, Morgan MY (2011) International Society for Hepatic Encephalopathy and Nitrogen Metabolism (ISHEN). Review article: the design of clinical trials in hepatic encephalopathy--an International Society for Hepatic Encephalopathy and Nitrogen Metabolism (ISHEN) consensus statement. Aliment Pharmacol Ther 33(7): 739–747. https://doi.org/10.1111/j.1365-2036.2011.04590.x
16. Romero-Gómez M, Córdoba J, Jover R, del Olmo JA, Ramírez M, Rey R, de Madaria E, Montoliu C, Nuñez D, Flavia M, Compañy L, Rodrigo JM, Felipo V (2007) Value of the critical flicker frequency in patients with minimal hepatic encephalopathy. Hepatology 45(4):879–885. https://doi.org/10.1002/hep.21586
17. Stroop JR (1935) Studies of interference in serial verbal reactions. J Exp Psychol 18(6): 643–662. https://doi.org/10.1037/h0054651
18. Scarpina F, Tagini S (2017) The Stroop color and word test. Front Psychol 12(8):557. https://doi.org/10.3389/fpsyg.2017.00557
19. Bajaj JS, Thacker LR, Heuman DM, Fuchs M, Sterling RK, Sanyal AJ, Puri P, Siddiqui MS, Stravitz RT, Bouneva I, Luketic V, Noble N, White MB, Monteith P, Unser A, Wade JB (2013) The Stroop smartphone application is a short and valid method to screen for minimal hepatic encephalopathy. Hepatology 58(3): 1122–1132. https://doi.org/10.1002/hep.26309
20. Bajaj JS, Heuman DM, Sterling RK, Sanyal AJ, Siddiqui M, Matherly S, Luketic V, Stravitz RT, Fuchs M, Thacker LR, Gilles H, White MB, Unser A, Hovermale J, Gavis E, Noble NA, Wade JB (2015) Validation of EncephalApp, smartphone-based Stroop test, for the diagnosis of covert hepatic encephalopathy. Clin Gastroenterol Hepatol 13(10):1828–1835.e1. https://doi.org/10.1016/j.cgh.2014.05.011
21. Giménez-Garzó C, Garcés JJ, Urios A, Mangas-Losada A, García-García R, González-López O, Giner-Durán R, Escudero-García D, Serra MA, Soria E, Felipo V, Montoliu C (2017) The PHES battery does not detect all cirrhotic patients with early neurological deficits, which are different in different patients. PLoS One 12(2):e0171211. https://doi.org/10.1371/journal.pone.0171211
22. Brickenkamp R, Zillmer E (1998) The d2 test of attention. Hogrefe & Huber Publishers, Seattle/Washington
23. Bates ME, Lemay EP Jr (2004) The d2 test of attention: construct validity and extensions in scoring techniques. J Int Neuropsychol Soc 10(3):392–400. https://doi.org/10.1017/S135561770410307X
24. Smith A (1968) The symbol-digit modalities test: a Neuropsychologic test of learning and other cerebral disorders. In: Learning disorders, special child publications, Seattle/Washington
25. Smith A (1982) Symbol digit modalities test: manual. Los Angeles
26. Egeland J (2015) Measuring working memory with digit span and the letter-number sequencing subtests from the WAIS-IV: too low manipulation load and risk for underestimating modality effects. Appl Neuropsychol Adult 22(6):445–451. https://doi.org/10.1080/23279095.2014.992069

27. Wechsler D (1955) Manual for the Wechsler adult intelligence scale, New York
28. Yela M (1955) A test of motor speed [In Spanish]. Rev Psic Gral Apl. 33:137–148
29. Yela M, López-Ladrón LA (1955) Visuomotor coordination test [In Spanish]. Rev Psic Gral Apl 34:409–421
30. Butz M, Timmermann L, Braun M, Groiss SJ, Wojtecki L, Ostrowski S, Krause H, Pollok B, Gross J, Südmeyer M, Kircheis G, Häussinger D, Schnitzler A (2010) Motor impairment in liver cirrhosis without and with minimal hepatic encephalopathy. Acta Neurol Scand 122(1):27–35. https://doi.org/10.1111/j.1600-0404.2009.01246.x
31. Romero Gómez M, Córdoba J, Jover R, del Olmo J, Fernández A, Flavià M, Compañy L, Poveda MJ, Felipo V (2006) Red Nacional de Investigación en Encefalopatía Hepática. Normality tables in the Spanish population for psychometric tests used in the diagnosis of minimal hepatic encephalopathy [In Spanish]. Med Clin (Barc) 27(7):246–249. https://doi.org/10.1157/13091264
32. Thanapirom K, Wongwandee M, Suksawatamnuay S, Thaimai P, Siripon N, Makhasen W, Treeprasertsuk S, Komolmit P (2023) Psychometric hepatic encephalopathy score for the diagnosis of minimal hepatic encephalopathy in Thai cirrhotic patients. J Clin Med 12(2):519. https://doi.org/10.3390/jcm12020519
33. Li SW, Wang K, Yu YQ, Wang HB, Li YH, Xu JM (2013) Psychometric hepatic encephalopathy score for diagnosis of minimal hepatic encephalopathy in China. World J Gastroenterol 19(46):8745–8751. https://doi.org/10.3748/wjg.v19.i46.8745
34. Badea MA, Drug VL, Dranga M, Gavrilescu O, Stefanescu G, Popa I, Mihai C, Cijevschi-Prelipcean C (2016) Diagnosis of minimal hepatic encephalopathy in a tertiary care center from eastern Romania: validation of the psychometric hepatic encephalopathy score (PHES). Metab Brain Dis 31(6):1463–1471. https://doi.org/10.1007/s11011-016-9878-y
35. Seo YS, Yim SY, Jung JY, Kim CH, Kim JD, Keum B, An H, Yim HJ, Lee HS, Kim CD, Ryu HS, Um SH (2012) Psychometric hepatic encephalopathy score for the detection of minimal hepatic encephalopathy in Korean patients with liver cirrhosis. J Gastroenterol Hepatol 27(11):1695–1704. https://doi.org/10.1111/j.1440-1746.2012.07217

Chapter 2

Auditory P300 Event-Related Potentials as a Neurophysiological Marker for Detecting Minimal Hepatic Encephalopathy

Fátima Higuera-de-la-Tijera, Daniel Santana-Vargas, and José Luis Pérez-Hernández

Abstract

Minimal hepatic encephalopathy is frequently disregarded, primarily owing to the absence of a universally accepted diagnostic standard. The connection between minimal hepatic encephalopathy, and a dismal prognosis compromised health-related quality of life, an elevated risk of falls and accidents, and heightened mortality emphasizes the critical need for early detection. The diagnosis of minimal hepatic encephalopathy is a challenge, primarily due to the absence of a gold standard test. Existing diagnostic tools include the psychometric hepatic encephalopathy score, critical flicker fusion frequency, Stroop test, and electroencephalography. Although psychometric hepatic encephalopathy score is widely accepted, it grapples with limitations linked to age, education, and cultural factors. Acknowledging the necessity for improved diagnostic tools, this chapter reviews auditory P300 event-related potentials as a sensitive and reliable method for identifying minimal hepatic encephalopathy. Auditory P300 event-related potentials, characterized by its simplicity, cost-effectiveness, and sensitivity, emerge as a promising addition to the current diagnostic arsenal for minimal hepatic encephalopathy. Its ability to detect cognitive disturbances provides valuable insights, enhancing our understanding and management of this condition.

Key words Cirrhosis, Minimal hepatic encephalopathy, Diagnosis, Auditory P300 event-related potentials, Quality of life, Morbidity, Mortality

Abbreviations

AUROC	Area under the receiver operating curve
BBB	Blood–brain barrier
CFF	Critical flicker fusion frequency
CI	Confidence interval
CLD	Chronic liver disease
CNS	Central nervous system
CP	Child-Pugh
CSF	Cerebrospinal fluid
DOR	Diagnostic odds ratio

Tiziano Balzano (ed.), *Experimental and Clinical Methods in Hepatic Encephalopathy Research*, Neuromethods, vol. 219,
https://doi.org/10.1007/978-1-0716-4354-9_2,

EASL	European Association for the Study of Liver
EEG	Electroencephalography
ER	Evoked responses
ERP	Event-related potential(s)
HE	Hepatic encephalopathy
HR	Hazard ratio
HRQoL	Health related quality of life
MELD	Model for End-stage Liver Disease
MHE	Minimal hepatic encephalopathy
MR	Magnetic resonance
NGVU	Neurogliovascular unit
NH_3	Unionized form of ammonia
NH_4+	Ammonium
OHE	Overt hepatic encephalopathy
PHES	Psychometric hepatic encephalopathy score
sROC	Symmetrical summary receiver operating characteristic
TCA	Tricarboxylic acid
TFF	Threshold for flicker fusion

1 Introduction

Hepatic encephalopathy (HE) is a heterogeneous neuropsychiatric disorder that frequently affects patients with cirrhosis (type C HE). It represents a spectrum of neurocognitive impairments spanning from abnormal neuropsychiatric testing without evident clinical manifestations (minimal HE [MHE]) to different degrees of pronounced clinical symptoms (overt HE [OHE]) [1].

Up to 80% of patients with cirrhosis may exhibit minimal hepatic encephalopathy (MHE) [2]. Despite the high prevalence of MHE in patients with cirrhosis, it varied substantially between chronic liver disease (CLD) stages. In a study including 1,868 cirrhotic patients, the prevalence of MHE in patients with Child-Pugh (CP) A was low (25%), whereas it was high in CP B or C (42% and 52%). In patients with a Model for End-stage Liver Disease (MELD) score < 10, the prevalence of MHE was only 25%, but it was 48% in patients with a MELD score ≥ 20 [3]. MHE is considered the earliest and mildest manifestation of HE [4]. It is a condition primarily characterized by neuropsychological and neurophysiological abnormalities that are discernible only through standardized tests, even though patients may appear normal during clinical examinations [5]. There is no established gold standard for detecting MHE [4, 6], which could explain why less than 10% of clinicians routinely screen for MHE in patients with cirrhosis [7]. However, the early recognition of MHE has recently become relevant due to its association with a poor prognosis, a negative impact on health-related quality of life (HRQoL), an increased risk of falls and car accidents, and a higher mortality rate." [7–10].

The diagnosis of MHE might be easily missed as verbal functioning is typically preserved, while the cognitive deficits predominantly affect overall performance and psychomotor activities [11]. The key domains affected in patients with MHE include attention, vigilance, and integrative function; however, overt clinical manifestations are absent [11, 12]. Patients with MHE also exhibit deficiencies in visuomotor coordination, encounter psychometric slowing, face a decline in mental skills, and demonstrate early alterations in memory and learning [13].

1.1 Clinical Relevance of Detecting MHE

Patients with MHE are predisposed to falls, sleep disorders, erectile dysfunction in males, impairment in the ability to drive or operate machinery, and car accidents. Moreover, the reduction in working capacity caused by MHE may adversely impact the socioeconomic status of affected individuals. All these factors negatively affect the HRQoL of the patients [14]. Patients with MHE have a longer total reaction time compared to individuals without HE; this slowing is associated with a higher likelihood of being involved in car accidents [15]. Wein et al. reported that patients with MHE needed interventions from a driving instructor to prevent accidents at a rate ten times higher than those without MHE and controls. Furthermore, specific driving behaviors were consistently rated as poorer in patients with MHE, particularly in terms of car handling, adaptation, cautiousness, and maneuvering [16]. In a driving simulation test, Bajaj et al. demonstrated that patients with MHE had significantly more collisions and illegal turns compared to patients with cirrhosis but without MHE or controls [17].

In terms of the risk of falls, Román et al. demonstrated that patients with cirrhosis and MHE had a higher incidence of falls compared to patients with cirrhosis but without MHE or healthy controls (40%, 12.9%, and 11.6%, respectively). Additionally, in this study, falls in patients with MHE were associated with more severe injuries such as fractures, wounds, and contusions [18].

MHE also impairs the sleep quality of affected patients; this, in turn, contributes to a poorer HRQoL [19, 20]. MHE patients suffer from multiple subjective dyssomnias and changes in sleep architecture that are strongly correlated with psychological dysfunction [21]. Patients with MHE also have more daytime sleepiness and more pronounced night-time sleep disturbance than controls [22, 23]. Interestingly, Singh et al. demonstrated that after 3 months of lactulose therapy, patients experienced significant improvement in MHE, leading to improvements in sleep disturbances and HRQoL [24].

Over time, patients undiagnosed and untreated MHE may progress to overt hepatic encephalopathy (OHE), reaching an irreversible stage associated with increased morbidity and mortality [25]. In a study by Gairing et al., which included 1562 patients, the cumulative incidence of OHE at 6, 12, and 24 months of follow-up was higher in patients with MHE than in those without MHE ($p < 0.001$). In the same study, patients with MHE had lower overall survival and transplantation-free survival rates compared to those without MHE ($p < 0.001$). Upon stratifying the cohort by CP stage, MHE identified by psychometric hepatic encephalopathy score (PHES) remained consistently associated with inferior transplantation-free survival across all CP stages ($p < 0.01$) [26]. In a retrospective study with a propensity score-matching analysis, Hanai et al. also found that MHE was also associated with and increased risk of mortality (Hazard Ratio [HR], 1.92; 95% Confidence Interval [95% CI], 1.09–3.29; $p = 0.024$) [10].

1.2 Patho-physiological Mechanisms of HE

HE arises from a combination of distinct mechanisms, including hyperammonemia [27], dysbiosis [28], inflammation [29], oxidative stress [30], impaired blood–brain barrier (BBB) permeability, other neurotoxins, impaired energy metabolism of the brain [31], and central neurotransmission imbalance in favor of excess neuroinhibition mediated by gamma-aminobutyric acid (GABA) [32].

Ammonia plays a key role in the pathogenesis of HE. In fact, nowadays, the specific and effective therapy is aimed at controlling ammonia generation or promoting its metabolism and detoxification from the organism [33].

1.2.1 Overproduction of Ammonia

Ammonia is mainly produced in the large intestine, where urea from dietary protein undergoes breakdown into ammonia and carbon dioxide exclusively through urease-producing bacteria. The direct degradation of amino acids, such as glutamine, by glutaminase in enterocytes also contributes to the generation of glutamate and free ammonia. This unionized form of ammonia (NH_3) can freely traverse the intestinal epithelium and enter the portal circulation. The ionization of ammonia is influenced by the colonic luminal pH, wherein a more acidic pH results in protonation and trapping of ammonium (NH_4+) within the lumen. Non-absorbable disaccharides like lactulose not only lower colonic pH, enhancing their laxative effect, but also reduce ammonia reabsorption through acidification of luminal pH. The proximal tubular cells within the kidneys also contribute to ammonia generation, starting from glutamine, with bicarbonate as a byproduct. This renal production of ammonia is closely linked to the acid-base status and potassium balance. In cases of metabolic acidosis or hypokalemia, glutamine is utilized to assist in acid disposal or potassium recovery. Consequently, these conditions are associated with a higher risk of HE due to an increase in arterial ammonia levels. Additionally, conditions favoring intravascular hypovolemia further stimulate renal ammonia production, modulated by angiotensin II. Conversely, volume expansion improves HE [34].

1.2.2 Impaired Ammonia Catabolism

The major site of ammonia catabolism is the urea or Krebs cycle, which normally occurs within the liver. In chronic liver disease (CLD) or cirrhosis, nitrogenous waste products, such as ammonia, accumulate in the systemic circulation due to compromised or nonfunctional hepatocytes and the formation of portosystemic shunts. Consequently, ammonia bypasses the liver, reaching and crossing the BBB [35].

1.2.3 Neurogliovascular Unit (NGVU) Dysfunction

HE is characterized by multicellular dysfunction within the fundamental functional unit of the brain, known as the NGVU [36]. Astrocytes are the primary site for ammonia detoxification in the brain, actively eliminating ammonia through the synthesis of glutamine via the amidation of glutamate, a process facilitated by the enzyme glutamine synthetase. The hyperammonemia-induced accumulation of glutamine in astrocytes induces osmotic stress, causing astrocytic swelling and subsequent deterioration in neuropsychological function [37]. The pathophysiology of HE involves alterations in both astroglial and neuronal cell functions due to the accumulation not only of ammonia but also of manganese. Neuroimaging techniques reveal early deactivation of the anterior cingulate cortex and an increase in T1-weighted magnetic resonance (MR) signal hyperintensities in the basal ganglia resulting from manganese deposition, leading to Parkinsonism [38]. Moreover, the brain, with its high energy demand, primarily relies on glucose to meet its requirements. Neurovascular coupling ensures precise regulation of cerebral blood flow by astrocytes, thereby adapting the supply of nutrients and oxygen to the central nervous system (CNS) based on localized neuronal demands [39]. Under conditions of reduced oxygen availability, astrocytes support neurons nutritionally by enhancing glycolysis and generating lactate. This lactate can then function as an additional energy source for neurons [40]. Ammonia also inhibits the tricarboxylic acid (TCA) cycle, a metabolic pathway crucial for meeting brain energy needs. This deceleration leads to higher concentrations of lactate in the cerebrospinal fluid (CSF), closely linked to the severity of HE in individuals with cirrhosis [38]. Elevated cerebral lactate concentrations are an indicator of brain energy failure. [41]. Even in MHE, cerebral blood flow and oxygen metabolism are reduced due to increased cerebral vascular resistances [42, 43]. Astrocyte swelling and dysfunction lead to oxidative and nitrosative stress, resulting in persistent impairments in memory, response inhibition, and learning capacity [34]. In the long term, this condition also promotes astrocyte senescence, while ammonia additionally inhibits astrocyte proliferation [34, 36]. Microglia, as the resident immune cells in the CNS, contribute to neuroinflammation in pathological conditions. This involvement is marked by microglial activation, during which microglia transition from a ramified to an amoeboid phenotype, accompanied by an increased production of cytokines and chemokines [44]. In the dysfunction of the NGVU, impaired glymphatic clearance is equally crucial, leading to the accumulation of neurotoxic compounds [36].

1.3 Diagnostic Tools for MHE in Common Clinical Practice and Their Limitations

1.3.1 Ammonia

Despite having a central role in the pathophysiology of HE, it does not play a role in the diagnosis of MHE [45]. Even when arterial ammonia concentrations were evaluated, there was no significant difference between MHE-diagnosed patients and those without MHE (195 ± 152 mmol/L versus 148 ± 146 mmol/L; $p > 0.05$) [46].

1.3.2 PHES

In routine clinical practice, PHES is widely regarded as the gold standard for diagnosing MHE [47–49]. this tool is accessible, sensitive, and cost-effective for the diagnosis of MHE. It has standardized values and is adjusted for age and educational level [45]. As a neuropsychological test, PHES was specifically designed to assesses the main areas affected by MHE, including visual-spatial construction and orientation, concentration, attention, and motor speed and accuracy. [48–51] PHES comprises five tests: the Digit Symbol Test (DST), the Number Connection Test A (NCT-A) involving consecutive letters, the NCT-B (letters and numbers), the Serial Dotting Test (SDT), and the Line Tracing Test (LTT) [49, 52]. The limitation of PHES lies in its variation with sex, age, education, and culture, requiring the construction of normative data specific to each country [51]. In the PHES score, each patient can receive a score ranging from −15 to +3 points [45]. The cut-off point is typically defined as equal to or less than −4 points for diagnosing MHE [53]. However, this assumption cannot be universally applied, as studies in some countries such as Thailand [47], Korea [54], India [55], and Poland [56] have established −5 points or less as the standard value. A Romanian study defined the PHES score cut-off value as −3 points [48]. This variability represents a limitation of PHES, emphasizing the need for validation of values in specific populations rather than generalizing them across diverse settings.

1.3.3 Critical Flicker Fusion Frequency (CFF)

The frequency at which flickering light is perceived as continuous, serving as a measure of temporal vision processing. The upper limit of visual processing capabilities is known as the Threshold for Flicker Fusion (TFF), representing the maximum speed of flickering light perceived by the visual system. Due to its effectiveness in detecting rapid changes, it serves as an index of CNS function, specifically associated with alertness and cortical arousal in humans [57]. For MHE, a CFF value <39 Hz is diagnostic [58]. In a meta-analysis conducted by Torlot et al. to assess the performance of CFF in diagnosing MHE, nine studies with data from 622 patients were included. The sensitivity was 61% (95% CI: 55–67), specificity was 79% (95% CI: 75–83), and the diagnostic odds ratio (DOR) was 10.9 (95% CI: 4.2–28.3). The symmetrical summary receiver operating characteristic (sROC) curve yielded an area under the curve of 0.84, with a heterogeneity of 74%. Therefore, considering the language independence, simplicity in both performance and interpretation of CFF, the authors propose the use of this tool as an adjunct to psychometric testing, but not as a replacement [59].

1.3.4 Stroop Test

It assesses psychomotor speed and cognitive flexibility by measuring the Stroop effect, which refers to the delay or increased number of mistakes in processing the recognition of the color of the ink of a word describing a color in the incongruent compared to the congruent condition [60]. This test is available as a smartphone application (EncephalApp) and has demonstrated prognostic value in clinical settings [61, 62]. However, some concerns have been raised regarding the potential for overdiagnosis if used indiscriminately [63].

1.3.5 Electroencephalography (EEG)

Despite its potential to provide objectively quantifiable data that complement clinical findings and psychometric performance, guidelines do not routinely recommend its use for diagnosing MHE [45, 64, 65] due to the non-specificity of the findings, the lack of validation in the absence of a gold standard for MHE diagnosis, and challenges in accessing the required facilities [66].

1.4 Need for Better Tools

The identification and measurement of minimal hepatic encephalopathy (MHE) pose significant challenges. This condition impacts various aspects of brain function, potentially leading to varying degrees of impairment in different patients or at different times. Functional assessments of brain function, obtained through neurophysiological and neuropsychological examinations, lack specificity to MHE and may manifest in other conditions. The neuropsychological profile, characterized by deficits in psychomotor speed, sustained/divided attention, working memory, and inhibition, may also show individual variations influenced by prior cognitive training [60].

Moreover, psychometric tests, such as PHES, considered the gold standard nowadays, are susceptible to influences from variables like age, educational status, and learning effects. Failure to apply corrections for these factors may result in an overdiagnosis of MHE [67].

Compounding this complexity is the fact that recommended diagnostic tests encompass a mix of neuropsychological, neurophysiological, and psychophysiological assessments. These tests, by definition, explore diverse aspects of brain function that are only loosely interconnected [60].

The European Association for the Study of Liver (EASL) recommends considering the diagnosis of MHE based on more than one test, to be chosen depending on available local norms and expertise [64].

2 Materials

2.1 P300 Event-Related Potential (ERP) in the Diagnosis of MHE

P300 ERP have been studied and shown to be highly sensitive and reliable in the detection of cognitive disturbances in MHE [5, 68–82]. The latency of the P300 wave was significantly larger in patients with MHE compared to control subjects [81]. The auditory P300 seems to be more useful to detect MHE than visual P300 (*see* **Note 1**) [68].

2.2 Equipment to Registry P300 ERP

In our medical center, auditory and visual stimuli are delivered using STIM2 interfaced with the Grael 4K EEG System (Compumedics, Charlotte, USA). The software SCAN version 4.5 (Compumedics Neuroscan, Charlotte, USA) is used to acquire the EEG and process the P300 [68].

1. *EEG electrodes*: Electrodes are needed to record brain electrical activity. Common electrode placements for P300 include Fz, Cz, and Pz, with earlobes used for reference.
2. *EEG amplifier*: A device to amplify and record the electrical signal captured by the electrodes.
3. *Auditory or visual stimuli*: Auditory tones or visual flashes, are used to evoke the P300 response.
4. *Recording and analysis software*: Tools for recording and analyzing EEG data.

2.3 Basic Concepts to Understand the P300 ERP

The diagnosis of MHE requires new accuracy methods [68]. ERP are a simple and cost-effective tool that can be easily set up in both clinical and basic science laboratory environments. ERP are small in comparison to the large fluctuations observed in ongoing brain activity. This ongoing brain activity exhibits a rich spatiotemporal structure, indexing instantaneous cortical excitability and shaping evoked responses based on the brain state at stimulus presentation. A common approach is to divide ongoing activity into several frequency bands to extract prestimulus phase and/or amplitude information [83]. P300 is one of the most extensively studied evoked responses (ER) in EEG [84]. The P300 wave is characterized by a positive voltage deflection with a latency of about 300 ms. The analysis of the P300 (presence of the wave, latency, and amplitude) helps determine the cognitive state (*see* **Note 2** and Fig. 1) [85, 86]. The latency represents the time required for processing the stimuli and is sensitive to cognitive slowing (*see* **Note 3**)

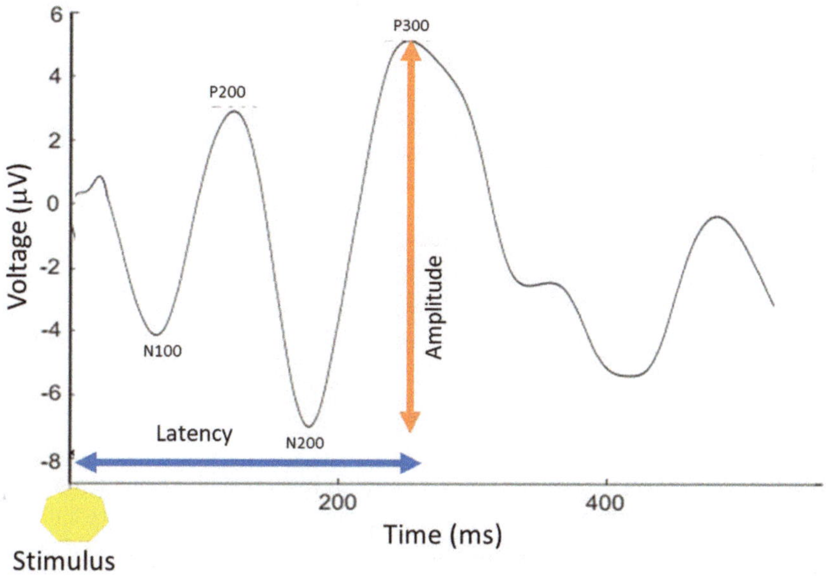

Fig. 1 Graphic representation of an event-related potential (ERP). Amplitude and latency are highlighted, as well as the most important waves (N100, P200, N200, P300). (Original image created by the author, based on ref. (86))

[68]. P300 has been hypothesized to reflect various functions, including priming, cognitive processing, memory storage, context updating, resource allocation [84], and sustained attention [68]. P300 is evaluated using the oddball paradigm (auditory or visual), where participants are tasked with detecting a target (or rare or deviant) stimulus in a sequence of standard (or frequent or non-target) stimuli (*see* **Note 4** and Fig. 2). It is common to refer to the P300 complex, which includes the earlier frontal component P3a and the later parietal component P3b (*see* Fig. 3) [84, 87].

3 Methods

3.1 Ethics and Good Clinical Practices

Auditory and visual P300 ERP are a non-invasive method (*see* **Note 5**) with minimal associated risks; however, it is recommended that the patient signs a consent form after the clinician or investigator has thoroughly explained the details of the procedure [88]. It is essential to note that conducting P300 ERP studies may require technical expertise and knowledge in neurophysiology. Additionally, environmental conditions, subject preparation, and other factors must be carefully controlled to obtain reliable results [89]. This type of study is used in clinical research and cognitive neuroscience to assess cognitive function and brain responses to specific stimuli.

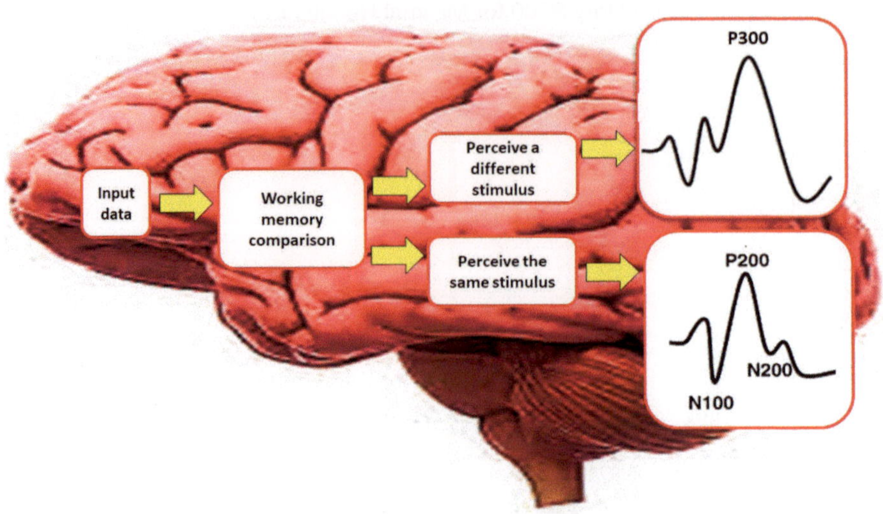

Fig. 2 Graphic representation of the context-updating model in P300 theory. Stimuli are introduced into the processing system, initiating a memory comparison process that determines whether the current stimulus matches the previous one (e.g., in the oddball task, discerning between a standard and a target stimulus). If the incoming stimulus matches, the neural model of the stimulus environment remains unchanged, and sensory evoked potentials (N100, P200, N200) are acquired through signal averaging. In cases where the incoming stimulus differs, and the subject directs attentional resources toward the target, the neural representation of the stimulus environment is modified or updated. This modification results in the generation of a P300 (P3b) potential alongside the sensory evoked potentials (Original image created by the author, based on ref. (87))

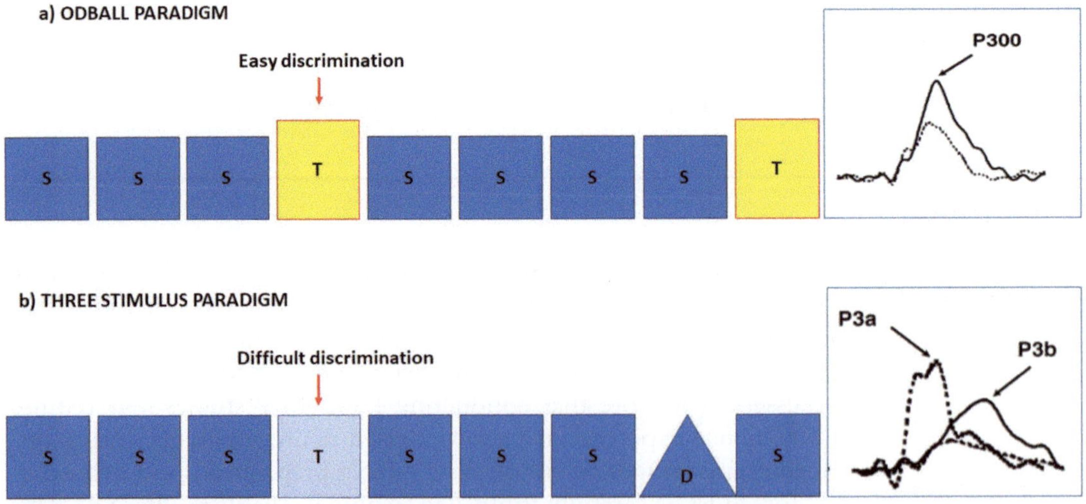

Fig. 3 P3a and P3b subcomponents of P300 ERP. Graphic representation of both (**a**) the oddball and (**b**) three-stimulus paradigms, featuring event-related potentials corresponding to each task. In the oddball task, two distinct stimuli are presented randomly, with one occurring less frequently than the other (S, standard; T, target). The subject responds exclusively to the target stimulus, triggering the P300 or P3b component. In the three-stimulus task, standards and a target are presented alongside an infrequently occurring distractor (D) stimulus, which automatically induces the P3a subcomponent in each task. The demanding discrimination task directs attention, ensuring a repeated typical stimulus elicits a robust P3a potential. (Original image created by the author, based on ref. (87))

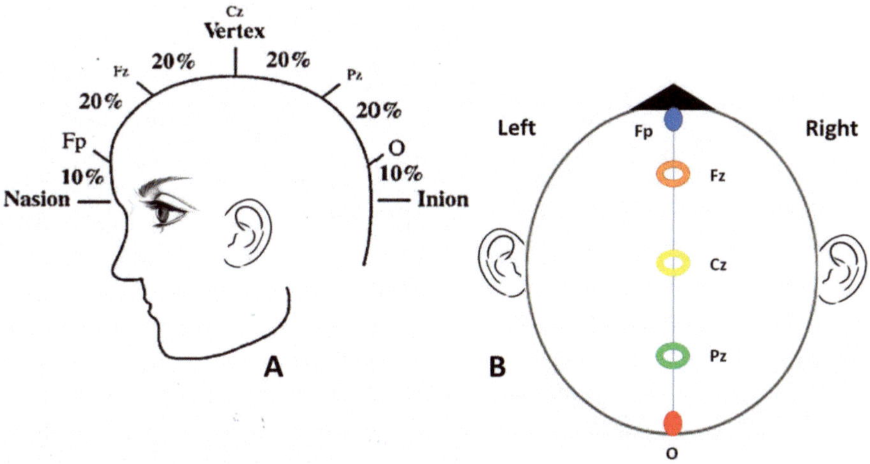

Fig. 4 Electrode positions for recording P300 ERP. (**a**) Side view. (**b**) Top view. Frontal point (Fz), Central point (Cz), Parietal point (Pz). (Original image created by the author, based on ref. (87))

3.2 Recording the Auditory and Visual P300 ERP

To record the P300, auditory or visual stimuli should be presented independently and concurrently with EEG registration. Ideally, the test must be administered by the same researcher.

1. *Patient preparation*: Place electrodes on the scalp after cleaning and preparing the skin. Head measurement may be necessary to position electrodes at specific locations. Electrodes are located at the frontal point (Fz), central point (Cz), and parietal point (Pz) positions, with the earlobes serving as reference (*see* **Note 6**, and Fig. 4).
2. *Baseline recording*: Record baseline EEG activity before stimulus presentation. EEG acquisition parameters are a sampling rate of 200 Hz, bandpass filters of 0.1–30 Hz, and a 60 Hz notch filter.
3. *Stimulus presentation and P300 recording*: Present auditory or visual stimuli to the subject in a controlled environment. For example, they might listen to tones or see flashes on a screen. For auditory stimuli, pure tones (150 ms duration, 5 ms rise/fall) are delivered in a semi-random sequence comprising 300, 1000 Hz target and 500 Hz standard tones in a proportion 1:5 (target: standard). The visual stimuli are white circles on a black background delivered in the same proportion and number as the auditory paradigm. Target and standard stimuli have diameters of 10 cm and 7 cm, respectively. Subjects must be seated in a comfortable chair 1 meter from the computer screen. The task for eliciting P300 responses are pressing a computer button each time the target is detected (reaction time). Movement and ocular artifact-free epochs (−100 to 1000 ms) are independently averaged for target and standard stimulus. P300 is identified and measured as the most

positive peak in an analysis window between 200 and 500 ms from stimuli onset.

4. *Data analysis*: Use specialized software to analyze the P300 waveform, including latency and amplitude [68].
5. *Dynamics of the EEG band power time series*: Studies on normal subjects have demonstrated a correlation between P300 amplitude and latency and EEG dynamics. These dynamics encompass the presence and strength of long-range time correlations (LRTC), providing insights into the structure of these signals. Additionally, the amplitude and latency of the P300 wave are not only linked to EEG dynamics per se but also to fluctuations in band power, particularly within the beta and gamma bands. These fluctuations, known as band power time series (BPts), have been shown to be modulated by the autonomic nervous system. The use of Detrended Fluctuation Analysis (DFA) (*see* **Note 7**) as an index to explore the structure of EEG dynamics by quantifying LRTC has been common. Furthermore, the analysis of BPts using this tool appears to be sensitive to alterations caused by neuropathies. Braveman et al. have also demonstrated its utility in detecting minimal hepatic encephalopathy (MHE).

4 Notes

1. The area under the receiver operating curve (AUROC) for auditory P300 was superior to that for visual P300 in distinguishing MHE from controls (AUROC: 0.792 vs. 0.725; $p < 0.005$ for both). The accuracy was 73.8% vs. 70.2%, sensitivity was 72.2% for both, and specificity was 74.2% vs. 69.7%, respectively. Among cirrhotic patients, only auditory P300 proved useful in detecting MHE (AUROC: 0.723; $p < 0.05$; 77.4% accuracy; 61.1% sensitivity, and 81.8% specificity) [68].
2. The first major positive peak at 250–500 ms for the rare tone is identified as the P300 auditive response.
3. Latency is measured from the stimulus point to the peak of the P300 waveform in milliseconds. P300 ERP latency is considered abnormal if it exceeds +2.5 standard deviations when compared with the mean latency measured in age-matched controls.
4. The P300 ERP wave is the most consistent wave and can be regarded as the electrophysiological counterpart of psychometric tests to diagnosis MHE, as both involve the active use of cognitive faculties. Sharma et al. agreed that the P300 ERP

response is most effectively elicited for the detection of MHE using the standard auditory oddball paradigm rather than the visual one [80].

5. Three different types of stimuli (visual, auditory, or somatosensory) can evoke the P300 response [89]; however, somatosensory stimuli have not been validated for the diagnosis of MHE.
6. The bioelectrical activity of the brain can be recorded using various electroencephalographic procedures. Typically, when studying diagnostic P300 ERP to identify the presence of MHE, electrode placement on the scalp with the use of superficial electrodes is generally adequate. To attain optimal performance at low frequencies, electrodes employed in EEG must exhibit specific characteristics. These include minimizing noise generation, minimizing polarization, reducing baseline drift (slow shift), possessing the highest specific capacitance (uF/cm^2), and minimizing contact resistance. To obtain P300 ERPs, three types of electrodes are commonly used: adhesive, contact, or cap electrodes. In our lab, we typically prefer cap electrodes because they are easier to place and more comfortable for patients. There are several electrode placement systems, but the ten-twenty system is currently the most commonly used today. In this system, the first step is to measure the distance between the nasion and the inion passing through the vertex. Ten percent above the nasion indicates the frontal polar point (Fp), while ten percent above the inion indicates the occipital point (O). Between points Fp and O, three additional points are positioned at equal intervals (20% of the nasion-inion distance between each two points). Moving from front to back, these three points are the frontal (Fz), central or vertex (Cz), and parietal (Pz) [87] (Fig. 2).
7. Detrended Fluctuation Analysis (DFA) is a statistical method used to quantify the presence and strength of long-range correlations in time series data. It is often employed in the analysis of physiological signals, including EEG data, to explore the underlying dynamics and patterns [90].

References

1. Henderson PK, Herrera JL (2015) Should we treat minimal/covert hepatic encephalopathy, and with what? Clin Liver Dis 19(3):487–495
2. Weissenborn K (2015) Diagnosis of minimal hepatic encephalopathy. J Clin Exp Hepatol 5 (Suppl 1):S54–S59
3. Gairing SJ, Mangini C, Zarantonello L et al (2020) Prevalence of minimal hepatic encephalopathy in patients with liver cirrhosis: a multicenter study. Am J Gastroenterol 118(12):2191–2200
4. Karanfilian BV, Park T, Senatore F et al (2020) Minimal hepatic encephalopathy. Clin Liver Dis 24(2):209–218
5. Amodio P, Valenti P, Del Piccolo F et al (2005) P300 latency for the diagnosis of minimal hepatic encephalopathy: evidence that spectral EEG analysis and psychometric tests are enough. Dig Liver Dis 37(11):861–868

6. Bajaj JS (2010) Review article: the modern management of hepatic encephalopathy. Aliment Pharmacol Ther 31(5):537–547
7. Flud CR, Duarte-Rojo A (2019) Prognostic implications of minimal/covert hepatic encephalopathy: large-scale validation cohort studies. J Clin Exp Hepatol 9(1):112–116
8. Moran S, López-Sánchez M, Milke-García P et al (2021) Current approach to treatment of minimal hepatic encephalopathy in patients with liver cirrhosis. World J Gastroenterol 27(22):3050–3063
9. Agrawal S, Umapathy S, Dhiman RK (2015) Minimal hepatic encephalopathy impairs quality of life. J Clin Exp Hepatol 5(Suppl 1):S42–S48
10. Hanai T, Shiraki M, Watanabe S et al (2019) Prognostic significance of minimal hepatic encephalopathy in patients with liver cirrhosis in Japan: a propensity score-matching analysis. J Gastroenterol Hepatol 34(10):1809–1816
11. Golaszewski S, Langthaler PB, Schwenker K et al (2016) Abnormal cortical synaptic plasticity in minimal hepatic encephalopathy. Brain Res Bull 125:200–204
12. Nardone R, Taylor AC, Höller Y et al (2016) Minimal hepatic encephalopathy: a review. Neurosci Res 111:1–12
13. Gupta P, Vyas S, Salan T et al (2022) Whole brain atlas-based diffusion kurtosis imaging parameters for evaluation of minimal hepatic encephalopathy. Neuroradiol J 35(1):67–76
14. Nardelli S, Gioia S, Faccioli J et al (2019) Sarcopenia and cognitive impairment in liver cirrhosis: a viewpoint on the clinical impact of minimal hepatic encephalopathy. World J Gastroenterol 25(35):5257–5265
15. Gairing SJ, Schleicher EM, Kaps L et al (2023) Development and evaluation of a virtual reality driving test for patients with cirrhosis. Hepatol Commun 7(11):e0303
16. Wein C, Koch H, Popp B et al (2004) Minimal hepatic encephalopathy impairs fitness to drive. Hepatology 39(3):739–745
17. Bajaj JS, Saeian K, Hafeezullah M et al (2008) Patients with minimal hepatic encephalopathy have poor insight into their driving skills. Clin Gastroenterol Hepatol 6(10):1135–1065
18. Román E, Córdoba J, Torrens M et al (2011) Minimal hepatic encephalopathy is associated with falls. Am J Gastroenterol 106(3):476–482
19. Cordoba J, Cabrera J, Lataif L et al (1998) High prevalence of sleep disturbance in cirrhosis. Hepatology 27(2):339–345
20. Engleman HM, Hirst WS, Douglas NJ (1997) Under reporting of sleepiness and driving impairment in patients with sleep apnoea/hypopnoea syndrome. J Sleep Res 6(4):272–275
21. Liu C, Zhou J, Yang X et al (2015) Changes in sleep architecture and quality in minimal hepatic encephalopathy patients and relationship to psychological dysfunction. Int J Clin Exp Med 8(11):21541–21548
22. Mostacci B, Ferlisi M, Baldi Antognini A et al (2008) Sleep disturbance and daytime sleepiness in patients with cirrhosis: a case control study. Neurol Sci 29(4):237–240
23. Samanta J, Dhiman RK, Khatri A et al (2013) Correlation between degree and quality of sleep disturbance and the level of neuropsychiatric impairment in patients with liver cirrhosis. Metab Brain Dis 28(2):249–259
24. Singh J, Sharma BC, Puri V et al (2017) Sleep disturbances in patients of liver cirrhosis with minimal hepatic encephalopathy before and after lactulose therapy. Metab Brain Dis 32(2):595–605
25. Ridola L, Nardelli S, Gioia S et al (2018) Quality of life in patients with minimal hepatic encephalopathy. World J Gastroenterol 24(48):5446–5453
26. Gairing SJ, Mangini C, Zarantonello L et al (2023) Minimal hepatic encephalopathy is associated with a higher risk of overt hepatic encephalopathy and poorer survival. J Intern Med. https://doi.org/10.1111/joim.13747
27. Cheon SY, Kim MY, Kim J et al (2023) Hyperammonemia induces microglial NLRP3 inflammasome activation via mitochondrial oxidative stress in hepatic encephalopathy. Biom J 46(5):100593
28. Shahbazi A, Sepehrinezhad A, Vahdani E et al (2023) Gut dysbiosis and blood-brain barrier alteration in hepatic encephalopathy: from gut to brain. Biomedicines 11(5):1272
29. Jalan R, Olde Damink SW, Hayes PC et al (2004) Pathogenesis of intracranial hypertension in acute liver failure: inflammation, ammonia and cerebral blood flow. J Hepatol 41(4):613–620
30. Bosoi CR, Rose CF (2013) Oxidative stress: a systemic factor implicated in the pathogenesis of hepatic encephalopathy. Metab Brain Dis 28(2):175–178
31. Rama Rao KV, Norenberg MD (2012) Brain energy metabolism and mitochondrial dysfunction in acute and chronic hepatic encephalopathy. Neurochem Int 60(7):697–706
32. Ahboucha S, Layrargues GP, Mamer O et al (2005) Increased brain concentrations of a

neuroinhibitory steroid in human hepatic encephalopathy. Ann Neurol 58(1):169–170

33. Kristiansen RG, Rose CF, Ytrebø LM (2016) Glycine and hyperammonemia: potential target for the treatment of hepatic encephalopathy. Metab Brain Dis 31(6):1269–1273
34. Jaffe A, Lim JK, Jakab SS (2020) Pathophysiology of hepatic encephalopathy. Clin Liver Dis 24(2):175–188
35. Weiner ID, Mitch WE, Sands JM (2015) Urea and ammonia metabolism and the control of renal nitrogen excretion. Clin J Am Soc Nephrol 10(8):1444–1458
36. Claeys W, Van Hoecke L, Lefere S et al (2021) The neurogliovascular unit in hepatic encephalopathy. JHEP Rep 3(5):100352
37. Jalan R, Shawcross D, Davies N (2003) The molecular pathogenesis of hepatic encephalopathy. Int J Biochem Cell Biol 35(8): 1175–1181
38. Butterworth RF (2019) Hepatic encephalopathy in cirrhosis: pathology and pathophysiology. Drugs 79(Suppl 1):17–21
39. Bélanger M, Allaman I, Magistretti PJ (2011) Brain energy metabolism: focus on astrocyte-neuron metabolic cooperation. Cell Metab 14(6):724–738
40. Huang L, Nakamura Y, Lo EH et al (2019) Astrocyte signaling in the neurovascular unit after central nervous system injury. Int J Mol Sci 20(2):282
41. Bosoi CR, Rose CF (2014) Elevated cerebral lactate: implications in the pathogenesis of hepatic encephalopathy. Metab Brain Dis 29(4):919–925
42. Ponziani FR, Funaro B, Lupascu A et al (2019) Minimal hepatic encephalopathy is associated with increased cerebral vascular resistance. A transcranial doppler ultrasound study. Sci Rep 9(1):15373
43. González-Regueiro JA, Higuera-de-la-Tijera MF, Moreno-Alcántar R et al (2019) Pathophysiology of hepatic encephalopathy and future treatment options. Rev Gastroenterol Mex (Engl Ed) 84(2):195–203
44. Hickman S, Izzy S, Sen P et al (2018) Microglia in neurodegeneration. Nat Neurosci 21(10):1359–1369
45. Higuera-de-la-Tijera F, Velarde-Ruiz Velasco JA, Raña-Garibay RH et al (2023) Current vision on diagnosis and comprehensive care in hepatic encephalopathy. Rev Gastroenterol Mex (Engl Ed) 88(2):155–174
46. Bragagnolo MA Jr, Teodoro V, Lucchesi LM et al (2009) Minimal hepatic encephalopathy detection by neuropsychological and neurophysiological methods and the role of ammonia for its diagnosis. Arq Gastroenterol 46(1):43–49
47. Thanapirom K, Wongwandee M, Suksawatamnuay S et al (2023) Psychometric hepatic encephalopathy score for the diagnosis of minimal hepatic encephalopathy in thai cirrhotic patients. J Clin Med 12(2):519
48. Badea MA, Drug VL, Dranga M et al (2016) Diagnosis of minimal hepatic encephalopathy in a tertiary care center from eastern Romania: validation of the psychometric hepatic encephalopathy score (PHES). Metab Brain Dis 31(6):1463–1471
49. Coubard OA, Ober KM, Gaumet M et al (2021) Standardization of the psychometric hepatic encephalopathy score in a French population. PLoS One 16(9):e0257136
50. Weissenborn K (2008) PHES: one label, different goods?! J Hepatol 49(3):308–312
51. Weissenborn K (2013) Psychometric tests for diagnosing minimal hepatic encephalopathy. Metab Brain Dis 28(2):227–229
52. Weissenborn K, Ennen JC, Schomerus H et al (2001) Neuropsychological characterization of hepatic encephalopathy. J Hepatol 34(5): 768–773
53. Weissenborn K (2019) Hepatic encephalopathy: definition, clinical grading and diagnostic principles. Drugs 79(Suppl 1):5–9
54. Seo YS, Yim SY, Jung JY et al (2012) Psychometric hepatic encephalopathy score for the detection of minimal hepatic encephalopathy in Korean patients with liver cirrhosis. J Gastroenterol Hepatol 27(11):1695–1704
55. Pawar VB, Surude RG, Sonthalia N et al (2019) Minimal hepatic encephalopathy in Indians: psychometric hepatic encephalopathy score and inhibitory control test for diagnosis and rifaximin or lactulose for its reversal. J Clin Transl Hepatol 7(4):304–312
56. Wunsch E, Koziarska D, Kotarska K et al (2013) Normalization of the psychometric hepatic encephalopathy score in polish population. A prospective, quantified electroencephalography study. Liver Int 33(9):1332–1340
57. Mankowska ND, Marcinkowska AB, Waskow M et al (2021) Critical flicker fusion frequency: a narrative review. Medicina (Kaunas) 57(10): 1096
58. Kircheis G, Wettstein M, Timmermann L et al (2002) Critical flicker frequency for quantification of low-grade hepatic encephalopathy. Hepatology 35(2):357–366
59. Torlot FJ, McPhail MJ, Taylor-Robinson SD (2013) Meta-analysis: the diagnostic accuracy

of critical flicker frequency in minimal hepatic encephalopathy. Aliment Pharmacol Ther 37(5):527–536
60. De Rui M, Montagnese S, Amodio P (2016) Recent developments in the diagnosis and treatment of covert/minimal hepatic encephalopathy. Expert Rev Gastroenterol Hepatol 10(4):443–450
61. Bajaj JS, Heuman DM, Sterling RK et al (2015) Validation of EncephalApp, smartphone-based stroop test, for the diagnosis of covert hepatic encephalopathy. Clin Gastroenterol Hepatol 13(10):1828–1835
62. Allampati S, Duarte-Rojo A, Thacker LR et al (2016) Diagnosis of minimal hepatic encephalopathy using Stroop EncephalApp: a Multicenter US-based, Norm-Based Study. Am J Gastroenterol 111(1):78–86
63. Amodio P, Cordoba J (2013) Smart applications for assessing minimal hepatic encephalopathy: novelty from the app revolution. Hepatology 58(3):844–846
64. European Association for the Study of the Liver (2022) EASL Clinical Practice Guidelines on the management of hepatic encephalopathy. J Hepatol 77(3):807–824
65. Vilstrup H, Amodio P, Bajaj J et al (2014) Hepatic encephalopathy in chronic liver disease: 2014 Practice Guideline by the American Association for the Study of Liver Diseases and the European Association for the Study of the Liver. Hepatology 60(2):715–735
66. Morgan MY, Amodio P, Cook NA et al (2016) Qualifying and quantifying minimal hepatic encephalopathy. Metab Brain Dis 31(6):1217–1229
67. Kharbanda PS, Saraswat VA, Dhiman RK (2003) Minimal hepatic encephalopathy: diagnosis by neuropsychological and neurophysiologic methods. Indian J Gastroenterol 22 (Suppl 2):S37–S41
68. Santana-Vargas ÁD, Higuera-de-la-Tijera F, Pérez-Hernández JL (2022) Auditory and visual P300 event-related potentials to detect minimal hepatic encephalopathy. Rev Esp Enferm Dig 114(2):83–88
69. Davies MG, Rowan MJ, MacMathuna P et al (1990) The auditory P300 event-related potential: an objective marker of the encephalopathy of chronic liver disease. Hepatology 12(4 Pt 1):688–694
70. Kügler CF, Lotterer E, Petter J et al (1992) Visual event-related P300 potentials in early portosystemic encephalopathy. Gastroenterology 103(1):302–310
71. Gallai V, Alberti A, Balò S et al (1995) Cognitive event-related potential in hepatic encephalopathy. Acta Neurol Scand 91(5):358–361
72. Hollerbach S, Kullmann F, Fründ R et al (1997) Auditory event-related cerebral potentials (P300) in hepatic encephalopathy--topographic distribution and correlation with clinical and psychometric assessment. Hepato-Gastroenterology 44(16):1002–1012
73. Giger-Mateeva VI, Riemslag FC, Reits D et al (2000) Visual event-related potentials in cirrhotic patients without overt encephalopathy. Metab Brain Dis 15(3):179–191
74. Saxena N, Bhatia M, Joshi YK, Garg PK, Tandon RK (2001) Auditory P300 event-related potentials and number connection test for evaluation of subclinical hepatic encephalopathy in patients with cirrhosis of the liver: a follow-up study. J Gastroenterol Hepatol 16(3):322–327. https://doi.org/10.1046/j.1440-1746.2001.02388.x
75. Jones EA, Giger-Mateeva VI, Reits D et al (2001) Visual event-related potentials in cirrhotic patients without overt encephalopathy: the effects of flumazenil. Metab Brain Dis 16(1–2):43–53
76. Saxena N, Bhatia M, Joshi YK et al (2002) Electrophysiological and neuropsychological tests for the diagnosis of subclinical hepatic encephalopathy and prediction of overt encephalopathy. Liver 22(3):190–197
77. Kramer L, Bauer E, Gendo A et al (2002) Neurophysiological evidence of cognitive impairment in patients without hepatic encephalopathy after transjugular intrahepatic portosystemic shunts. Am J Gastroenterol 97(1):162–166
78. Ciećko-Michalska I, Senderecka M, Szewczyk J et al (2006) Event-related cerebral potentials for the diagnosis of subclinical hepatic encephalopathy in patients with liver cirrhosis. Adv Med Sci 51:273–277
79. Teodoro V, Bragagnolo M Jr, Lucchesi L et al (2008) Evaluation of the event-related potential (ERP-P300) in patients with hepatic cirrhosis without encephalopathy. Arq Gastroenterol 45(1):82–86
80. Sharma P, Sharma BC (2010) Predictors of minimal hepatic encephalopathy in patients with cirrhosis. Saudi J Gastroenterol 16(3):181–187
81. Braverman-Jaiven D, Vargas DS, Bojorges-Valdez E et al (2022) Exploration of the severity of hepatic encephalopathy deterioration process through dynamics of the EEG band power time series. Annu Int Conf IEEE Eng Med Biol Soc 2022:3603–3606

82. Torres DS, Abrantes J, Brandão-Mello CE (2020) Cognitive and neurophysiological assessment of patients with minimal hepatic encephalopathy in Brazil. Sci Rep 10(1):8610
83. Helfrich RF, Knight RT (2019) Cognitive neurophysiology: event-related potentials. Handb Clin Neurol 160:543–558
84. Studenova A, Forster C, Engemann DA et al (2023) Event-related modulation of alpha rhythm explains the auditory P300-evoked response in EEG. elife 12:RP88367
85. de la Calzada MD, Poca MA, Sahuquillo J et al (2010) Cognitive event-related brain potentials (P300) in patients with normal pressure hydrocephalus. Results of a prospective study. Neurologia 25(1):32–39
86. Wall LG, Davidson SA, Dalebout SD (1991) Determining latency and amplitude for multiple peaked P300 waveforms. J Am Acad Audiol 2(3):189–194
87. Polich J (2004) Clinical application of the P300 event-related brain potential. Phys Med Rehabil Clin N Am 15(1):133–161
88. Otte A, Maier-Lenz H, Dierckx RA (2005) Good clinical practice: historical background and key aspects. Nucl Med Commun 26(7): 563–574
89. Figueroa-Saavedra C, Otzen T, Romero-Mejías F et al (2020) Morfología de los potenciales evocados auditivos de latencia tardía P300 y su rol en la evaluación de la conducta suicida. Int J Morphol 38(2):356–362
90. Bashan A, Bartsch R, Kantelhardt JW et al (2008) Comparison of detrending methods for fluctuation analysis. Phys A Stat Mech Appl 387(21):5080–5090

Chapter 3

Magnetic Resonance Imaging Quantification of Steatosis, Iron, and Fibrosis in Chronic Liver Disease

David Martí-Aguado and Luis Martí-Bonmatí

Abstract

A comprehensive overview is provided on how Magnetic Resonance (MR) imaging stands as a pivotal non-invasive diagnostic tool in evaluating Chronic Liver Disease (CLD). The advancement in multiparametric MR protocols offers a simultaneous, accurate, and quantitative assessment of the key pathological features of CLD: steatosis, iron content, and fibrosis. The automated organ segmentation process allows for a comprehensive and accurate evaluation of the whole liver volume, surpassing the limitations of manual regional analyses and reducing sampling errors inherent in biopsy-based assessments. This chapter details specific MR protocols including Multiecho Chemical Shift Imaging (MECSE) with Proton Density Fat Fraction (PDFF) and R2* dephasing calculations for quantification of steatosis and iron load, as well as Magnetic Resonance Elastography (MRE) for liver fibrosis assessment. The use of hepatobiliary contrast agents is also highlighted in assessing liver function, providing insights into the functional reserve and regional functionality. These MR techniques, enhanced by convolutional neural networks (CNNs) for automated whole liver segmentation, reduce variability and increase the precision of CLD assessment, facilitating better clinical decision-making and longitudinal monitoring of the disease.

Key words Magnetic resonance imaging, Steatosis, Iron overload, Fibrosis, Deep learning, Whole liver segmentation

Abbreviations

ALD	Alcohol-related liver disease
CLD	Chronic liver disease
CNN	Convolutional neural network
DL	Deep learning
FLIS	Functional Liver Imaging Score
GANs	Generative Adversarial Networks
HBP	Hepatobiliary phase
HCV	Hepatitis C virus
LIC	Liver iron concentration
LSM	Liver stiffness measurement
MASLD	Metabolic dysfunction-associated steatotic liver disease.
MECSE	Multi-echo chemical-shift-encoded

Tiziano Balzano (ed.), *Experimental and Clinical Methods in Hepatic Encephalopathy Research*, Neuromethods, vol. 219,
https://doi.org/10.1007/978-1-0716-4354-9_3,

MR	Magnetic Resonance
MRE	Magnetic Resonance Elastography
RLE	Relative liver enhancement
ROI	Region of interest
PDFF	Proton density fat fraction
TE	Time of echoes
WLS	Whole liver segmentation

1 Introduction: Where Does MR Stand in the Evaluation of CLD?

Chronic liver disease (CLD) is a major cause of morbidity, mortality, and health care resource use, mainly related to the growing epidemic of obesity and metabolic dysfunction-associated steatotic liver disease (MASLD) [1]. Steatosis is the histopathologic hallmark of MASLD and is also a common feature in other liver diseases. Iron overload is often depicted in patients with diffuse liver diseases and metabolic disorders. Iron deposition can be both a cause and effect of CLD, and when combined with MASLD has a synergistic impact on the severity of the disease [2]. Chronic liver injury triggers the development of liver fibrosis that may progress to cirrhosis. Thus, steatosis, iron, and fibrosis often co-exist as key features in the complex pathophysiology of CLD.

Despite the high prevalence of CLD, histologic evaluation is still considered the reference standard to assess disease severity. However, its high sampling variability and invasive nature makes it inadequate for serial analyses and monitorization of liver disease. Non-invasive imaging techniques are increasing in usefulness in the diagnosis and monitoring of CLD. Advances in Magnetic Resonance (MR) imaging sequences have led to the development of multiparametric MR techniques that provide accurate, quantitative, and simultaneous evaluation of the degree and distribution of liver steatosis, iron, and fibrosis. Abbreviated MR protocols have positioned as the standard of care for the prompt evaluation of CLD to guide clinical decision-making, enable risk stratification, and facilitate longitudinal disease monitoring and treatment response assessment. More recently, convolutional neural network (CNN) solutions have been developed to obtain automated whole liver segmentation that facilitate image analysis and reduces variability errors [3].

This chapter will review the aforementioned MR strategies and will present the standard-of-care image protocols available to clinicians for diagnosing and monitoring CLD. A concise overview will be provided for the sequences that allow a triple-screen of liver disease: steatosis, iron, and fibrosis. Additionally, insights into the assessment of liver function through MR images after the

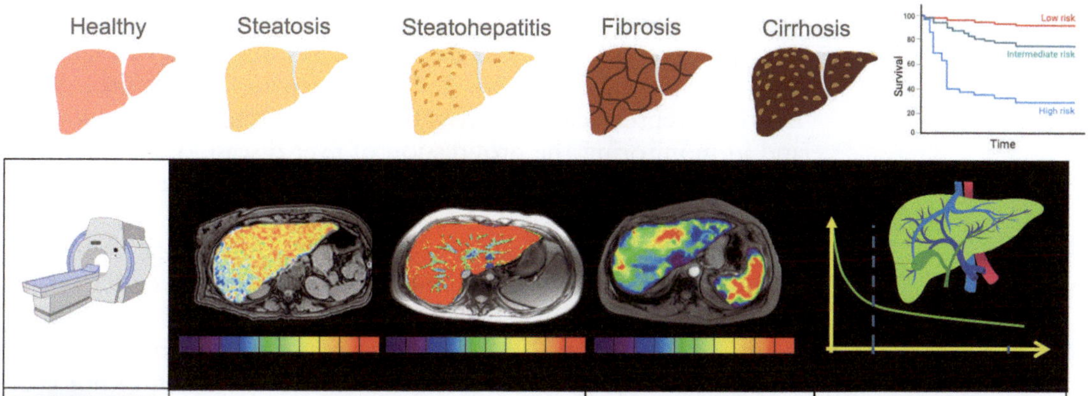

	MECSE: PDFF, R2* steatosis, steatohepatitis, iron overload	**MRE:** elastography fibrosis, cirrhosis	**T1 - HBP:** RLE, FLIS liver function
• Applicability	High, independent of MR field-strength & vendor	Requires specific device	Requires iv contrast agent
• Diagnosis	Excellent steatosis and iron; good steatohepatitis	Excellent for fibrosis	Limited to cirrhosis
• Prognosis	Liver disease progression	Liver-related events	Transplant, risk-stratification
• Monitoring	PDFF decrease >30%, chelation therapy	Cirrhosis development	HCV regression

Fig. 1 Graphical abstract of the Magnetic Resonance Imaging sequences used in clinical practice, their role in diagnosis, prognosis, and monitoring liver disease

administration of hepatobiliary contrast agents will be complementary explored. All these sequences can be combined in a single abbreviated MR protocol to offer a complete imaging profile and risk-stratification of CLD (Fig. 1).

2 Material: Automated Whole Liver Volume Segmentation

The rapid advancement in medical imaging techniques, particularly MR and CT scans, has significantly enhanced the diagnostic process for various liver diseases. The delineation and evaluation of the whole liver segmentation (WLS) volume, as opposed to the analysis of Regions of Interest (ROI), offers several advantages in the context of CLD screening of fibrosis, fat, and iron quantification. WLS provides a comprehensive assessment, allowing for the measurement and distribution changes across the entire organ. This is crucial because liver fat, iron, and fibrosis contents can vary significantly across different segments and regions, and focusing on limited ROIs might not capture the full extent and dispersion heterogeneity of the disease [4]. Discordance between biopsy- and MR-based fibrosis staging is associated with heterogeneity in LSM variability, as depicted with MR elastography (MRE) [5]. WLS offers an increased accuracy limiting the potential for sampling error compared to when only few ROIs are considered [6]. This WLS approach offers a more accurate quantification of hepatic fibrosis, steatosis, and iron, which is essential not only for diagnosing and staging, but also for monitoring the diseases, avoiding variability related to ROIs placement. Therefore, WLS improves reproducibility across time points in longitudinal studies and in monitoring the progression of liver disease or the response to treatment, where consistency in measurement techniques is crucial.

Manual segmentation of liver volume is time-consuming and prone to inter-observer variability. Recent studies have focused on the development of automated algorithms for whole liver volume segmentation, promising to revolutionize this field by improving accuracy, consistency, and efficiency. All this information not only underscores the accuracy and reliability of automated segmentation but also its potential in reducing the time burden on radiologists. WLS facilitates the automated evaluation and streamline the analysis process, reducing the time burden on radiologists and potentially leading to more widespread use in clinical practice [7]. Comparative analyses highlight the quantitative advantages of automatic WLS over manual sampling in MRI strategies, particularly in diagnosing MASLD.

Automated WLS using MR and CT images employs sophisticated algorithms, often based on deep learning (DL) models, to delineate the liver's boundary accurately from surrounding tissues [8, 9]. For instance, U-Net is an encoder–decoder model that improves the performance by applying a skip connection to compensate for missing data due to convolution. Recently, state-of-the-art approaches that apply an attention-mechanism, deep supervision, and nested architecture techniques, have improved segmentation performance. In abdominal CT images, deep 3D attention using convolutional long short-term memory (CLSTM) U-Net models provides accurate and quantitative liver segmentation, as an aid in estimating liver volume for living-donor liver transplantation [10]. Similarly, other studies have demonstrated the utility of automated MR image prescriptions and segmentation in assessing liver conditions, thereby facilitating the diagnosis, and staging of CLDs [3]. For automatic WLS, some of the most effective AI techniques are grounded in DL, particularly those involving convolutional neural networks (CNNs) due to their prowess in handling image data. These techniques have demonstrated significant success in achieving accurate, reliable, and efficient segmentation of the liver from medical imaging scans such as CT and MRI. Some of the leading AI techniques used in this domain are:

- U-Net and its Variants, such as 3D U-Net and Attention U-Net; V-Net, an extension of U-Net designed specifically for volumetric image segmentation [10].

- Dense Convolutional Networks (DenseNet) facilitate the training of deep networks by ensuring maximum information flow between layers [11].
- Generative Adversarial Networks (GANs) where one network generates segmentations, and another evaluates them, helping to refine the segmentation performance iteratively [12].
- Transfer Learning, where a model pretrained on a large dataset is fine-tuned on liver-specific images, has proven to be an effective strategy.

These AI techniques are at the forefront of innovations in medical imaging, continually evolving to offer more precise, efficient, and robust solutions for WLS. It is important to note that the approach requires sophisticated analysis techniques, which might not be readily available in all clinical settings. To solve this issue, cloud computing solutions are starting to be use, offering accurate segmentation solutions.

As the field advances, we can expect to see further improvements in these models and the emergence of new techniques that will enhance their performance and applicability in clinical settings. Despite the promising advancements, remaining challenges include algorithmic transparency, standardization across different imaging modalities, and the integration of these tools into clinical workflows. Moreover, ethical considerations regarding data privacy and the reliance on automated systems for critical health decisions need careful deliberation.

3 Methods: MR Screening Protocol

In this section, the main techniques used for quantification of fat, iron, fibrosis, and evaluation of liver functional status will be addressed. Figure 1 illustrates the main MR standard-of-care sequences that will be discussed, their applicability and role in diagnosis, prognosis, and monitoring CLD.

3.1 MECSE—Steatosis and Iron Overload

3.1.1 Concepts and Technique

The quantification of liver fat and iron using MR techniques has become increasingly important in the management of diffuse CLD such as MASLD and alcohol-related liver disease (ALD) [13]. MR offers a non-invasive, radiation-free modality for the accurate, fast, and precise assessment of liver fat and iron content. This discussion will explore the most important methods for fat and iron quantification, referencing the MR protocol and recent advancements in the field.

Proton Density Fat Fraction (PDFF) is a quantitative MR technique that measures the proportion of fat signal in the liver compared to the total proton signal. This technique has been validated against histology and is considered a gold standard for non-invasively quantifying liver fat content for both clinical practice and clinical trials [14].

Chemical Shift Imaging (CSI), including Dixon techniques, separates water and fat signals based on their different resonance frequencies. Recent updates have improved its accuracy and robustness, making it a reliable tool for liver fat quantification [15]. In addition, DL models have shown efficacy in inferring hepatic PDFF from high-resolution T1-weighted in-phase and opposed-phase MRI, enhancing the automation and accuracy of liver fat quantification to other image sequences [11].

Multiecho chemical shift magnetic resonance (MECSE-MR) imaging is a highly effective technique for predicting and quantifying the liver PDFF. This method leverages the differences in precession frequencies between water and fat protons, due to their chemical environment, to separate fat and water signals, enabling accurate quantification of fat content. Multiecho acquisitions at different echo times (TEs), within a single breath-hold, result in a series of images where the signal intensity of fat and water varies periodically, allowing for their separation and quantification (Fig. 2). By fitting the signal intensity variations across the echoes to a model that accounts for the known differences in precession frequencies between water and fat, the algorithm can accurately quantify the relative contributions of fat and water in each voxel.

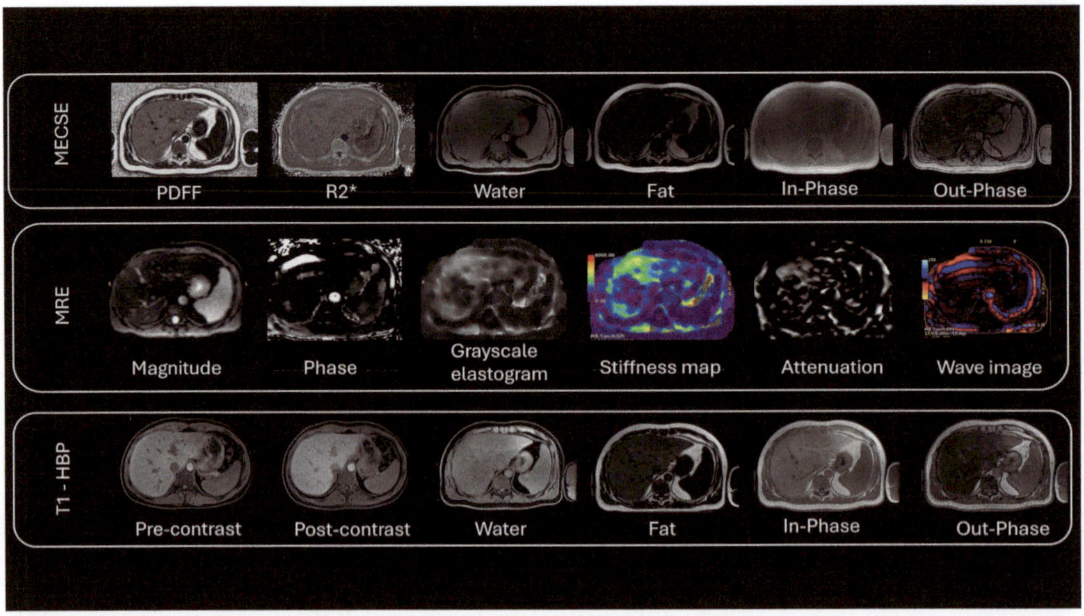

Fig. 2 Axial MR images obtained with the protocol examination of multi-echo chemical-shift-encoded (MECSE), Magnetic Resonance Elastography (MRE) and T1-weighted images before and after injection of contrast in the hepatobiliary phase (T1—HBP)

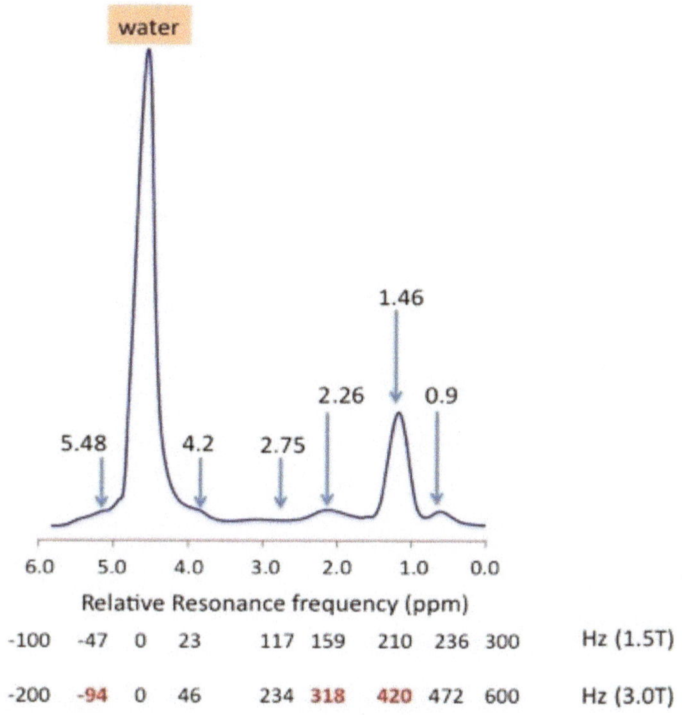

Fig. 3 Illustration of the liver multipeak spectrum composition of fat. Fat peaks have different frequencies (Hz) for 1.5 and 3 T magnets, as frequency is related to the main field strength (B0)

Reliable MECSE analysis requires the standard six peaks model, as simpler spectral complexity models introduce a low but clinically significant bias in fat quantification and steatosis grading [16]. The six principal distinct spectral peaks at different resonance frequencies are 5.48, 4.20, 2.75, 2.26, 1.46, and 0.90 ppm, and their relative amplitudes are 4.7, 3.9, 0.6, 12, 70, and 8.8%, respectively [17] (Fig. 3). Then, PDFF is calculated as the ratio of the signal from fat to the total signal (fat + water) in each voxel, expressed as a percentage that is independent of scanner and patient-specific variables. Table 1 summarized the protocol details to obtain a reliable MECSE sequence.

Signal intensity ratios, R2 (transverse relaxation rate) and R2* (effective transverse relaxation rate) are MR techniques used to accurately quantify liver iron concentration (LIC) [18]. Iron particles create magnetic field inhomogeneities, accelerating the dephasing of proton spins, quantifiable through the relaxation rates measurements. Gradient echo T2*-weighted imaging is fast and sensitive to evaluate R2* magnetic susceptibility effects caused by iron deposition in the liver and is preferred over the signal intensity ratios and spin-echo R2 measurements. The R2* value is inversely related to the T2* relaxation time, meaning that higher

Table 1
Magnetic resonance image protocol for standard series in a MR exam 3.0 T Philips Healthcare

MR sequence	MECSE	MRE	T1-HBP
Biomarker	PDFF, R2*	Elastography	RLE, FLIS
Slice thickness (mm)	6.0	10.0	1.7
TR (ms)	9.3	50	2.67
TE (ms)	0.9–8.6	20.2	0.92
Number of echoes	12	1	1
Flip angel (°)	10	30	13
FOV (mm)	380	450	430
Matrix	192 × 192	256 × 64	288 × 216
Image segmentation	Automatic	Manual	Manual
Liver segmentation	Whole liver volume	ROIs on elastograms	Whole liver volume
Acquisition time	15 s	2 min	16 s

Abbreviations: *HBP* The hepatobiliary phase, *FOV* field of view, *MECSE* Multiecho Chemical Shift Imaging, *MRE* magnetic resonance elastography, *PDFF* proton density fat fraction, *RLE* Relative liver enhancement, *ROI* region of interest, *TR* repetition time, *TE* time of echoes

concentrations of iron in the liver lead to faster signal decay and higher R2* values. This direct relationship between R2* values and liver iron content, makes R2* relaxometry a reliable and sensitive method for detecting and quantifying liver iron overload, playing a crucial role in the diagnosis and management of conditions related to iron metabolism disorders [19].

Despite great reproducibility, the mathematical calibration formulas between R2* and LIC (evaluated with atomic absorption spectrometry of the liver biopsy) are diverse among different studies (Table 2). This can be explained by differences in sequence parameters and postprocessing fitting algorithms, as well as by sampling bias related to liver biopsy and different post-biopsy sample processing. Available confounder-corrected R2*-based LIC quantification at both 1.5 T and 3 T is the most practical method with the strongest level of evidence for accurate and reproducible quantification of LIC [19].

3.1.2 Diagnostic Performance: Accuracy and Precision

Several studies have demonstrated a high accuracy of WLS and quantification of PDFF and R2* on MR images for assessing liver fat and iron content [20–22]. The relationship between PDFF and histology is predominantly linear ($r = 0.85$–0.90) [19, 20]. Established cut-offs to differentiate among grades of steatosis with PDFF are: <5.75% for mild steatosis (≥S1), 15.50–21.35% for moderate steatosis (≥S2), and >21.35% for severe steatosis

Table 2
Conversion formulas used in the different published methods [24–27] to obtain R2* and liver iron content (LIC)

Study and population	Conversion formula for: LIC in mg/g and μmol/g to R2* (s^{-1})	Cut-of value for: LIC > 2 mg/g or 36 μmol/g
Wood JC, 2005 [24] Thalassemia and sickle cell disease	Fe [mg/g] = 0.0254 × R2* + 0.202 Fe [μmol/g] = R2*/2.18 + 3.6	71 s^{-1}
Garbowski MW, 2014 [25] Transfusional hemosiderosis	Fe [mg/g] = 0.032 × R2* − 0.14 Fe [μmol/g] = R2*/1.74–2.5	67 s^{-1}
Henninger B, 2015 [26] Hemochromatosis	Fe [mg/g] = 0.024 × R2* + 0.277 Fe [μmol/g] = R2*/2.31 + 4.8	70 s^{-1}
Hankins JS, 2009 [27] Thalassemia and sickle cell disease	Fe [mg/g] = 0.028 × R2* × 0.454 Fe [μmol/g] = R2*/1.98–8.1	88 s^{-1}

Abbreviations: *Fe* iron, *LIC* liver iron concentration

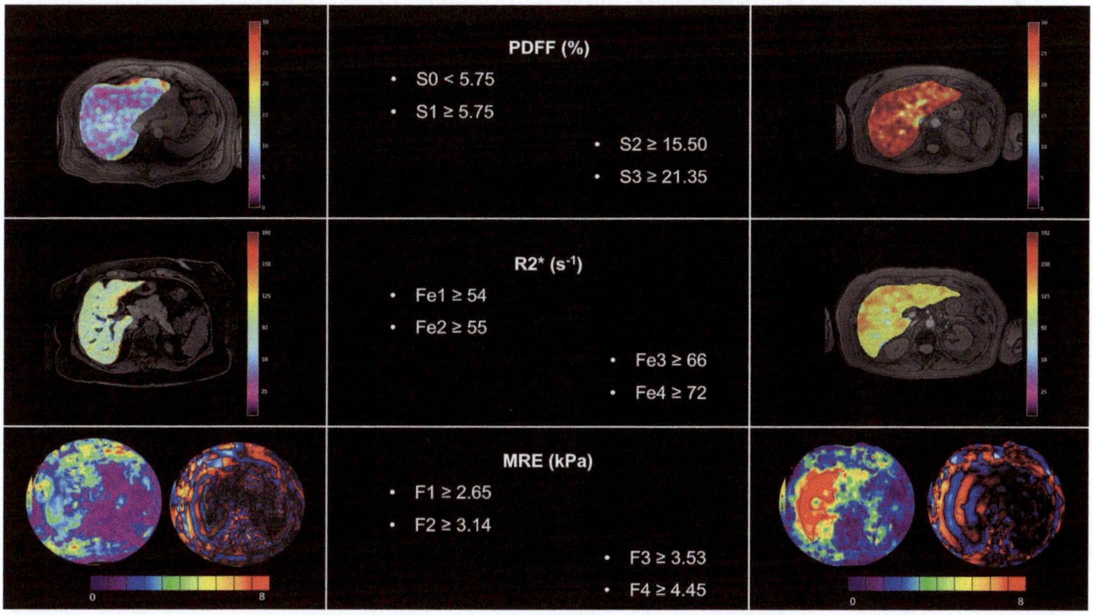

Fig. 4 Thresholds proposed for grading steatosis with proton density fat fraction (PDFF), iron deposits with effective transverse relaxation rate (R2*), and staging liver fibrosis with Magnetic Resonance Elastography (MRE)

(S3) (Fig. 4). Importantly, above 5% and towards higher levels of steatosis, absolute values of the percentage of lipid droplets assessed by pathologists and PDFF tend to diverge, with histology exceeding MR. On average, 100% histological steatosis corresponded to a PDFF of 33% [22]. Possible reasons for these numerical differences are (1) sampling error as MR evaluates the whole liver volume and only 1/50000 of the whole liver tissue is sampled during a liver biopsy, and (2) while pathologist visually estimate liver fat as the fraction of macrovesicular lipid droplet-containing hepatocytes, PDFF considers the spectral complexity of fat, including the smaller fat peaks relative to that of water. Other non-hepatocyte cell types as well as components of the extracellular matrix can influence the MR-visible water-to-fat ratio and might contribute to the differences between PDFF and histological steatosis. Regarding monitorization, a 1.5% change for MRI PDFF should be used as the threshold to discern measurement error from longitudinal real change in CLD patients [23].

Regarding iron measurement, several studies among patients with transfusional hemosiderosis demonstrated an excellent linear relationship between R2* and LIC ($r = 0.94$) [24–27]. More recent studies with digital pathology among CLD and metabolic disorders have shown a moderate linear relationship ($r = 0.65$) [21]. Differences can be due to the following reasons: (1) spectrum bias as some studies evaluate conditions with very high levels of iron overload (hereditary hemochromatosis, hemolytic anemias, or secondary hemosiderosis associated with chronic blood transfusions) and others evaluate diseases with mainly low levels of iron deposits (MASLD or ALD), and (2) use of different relaxometry R2*/R2 methods (MECSE, FerriScan). It is important to highlight, however, that the differences between relaxometry R2* methods do not appear to be clinically significant, as long as validated methods are utilized [19]. Established cut-offs to differentiate among grades of iron overload with R2* in non-hemochromatosis CLD are $<55\ s^{-1}$ for mild iron deposits (≥Fe1-2), and $> 70\ s^{-1}$ for moderate-to-severe iron overload (≥Fe3-4) [21]. For clinical purposes and considering all the spectrum of primary and secondary liver diseases, a cut-off $70\ s^{-1}$ can be used as reference for iron overload that should ponder in the management of patients.

3.1.3 Biases: Confounding Factors and Limitations

MECSE technique is robust to several confounding factors that can affect fat quantification, such as T1 bias, T2* decay, multipeak far frequencies, and noise. The dual influence of iron and fat on liver R2* values underscores the need to also account for confounding factors in iron quantification [28].

Despite the advancements in MRI techniques for liver fat and iron quantification, several challenges remain. The simultaneous presence of both fat and iron within the liver can interfere with the accuracy of MRI measurements, as both elements affect the MRI signal in complex ways. It has even been suggested in MASLD, that hepatic R2* can be more strongly associated with PDFF than histologic liver iron [29]. Addressing the simultaneous presence of fat and iron in the liver, which can confound the quantification of each component due to their overlapping effects on MR signals, is crucial and highlights the importance of advanced imaging protocols that can dissect each individual contributions [20, 21]. This advancement is particularly relevant in conditions such as MASLD and ALD, where both steatosis and iron overload usually coexist, requiring a precise quantification for appropriate management. Regarding R2* relaxometry, measurements must be performed with validated gradient echo MR sequence parameters (Table 1), which should be acquired before the administration of intravenous gadolinium chelate, especially in the case of hepatocyte-specific contrast agents, as these contrast agents may interfere with R2* measurements.

3.2 MRE—Fibrosis

3.2.1 Concepts and Technique

Many liver diseases cause profound changes in the mechanical properties of tissues (elasticity and stiffness). In patients with CLD, destruction of normal hepatic parenchyma, progressive accumulation of excessive extracellular matrix and distortion of the parenchymal architecture leads to increased liver stiffness.

MRE is an MRI-based technique for quantitatively assessing tissue stiffness, which correlates with the extent of hepatic fibrosis. MRE is based on the physical principle that the propagation characteristics of mechanical waves within the liver tissue is determined by its elasticity and stiffness. MRE is available as a software and device (external driver) that can be easily integrated into standard clinical liver MR scanners. The technique consists of three steps:

- An acoustic passive driver (flat disk-shaped vibration source) is placed with an elastic band in the right upper cuadrant of the patient and connected by a flexible plastic tube to an acoustic active driver outside the MR room. Continuous vibrations at 60 Hz are generated by the active driver and delivered by the tube to the passive driver, which then transmits the vibrations into the body, producing shear waves in the liver.
- Imaging the propagating mechanical waves is performed with a phase contrast two-dimensional gradient recalled echo (2D-GRE-MRE) pulse sequence while the vibrations are transmitted. Usually, 4 axial slices of 10-mm thick are acquired in a 12–15 s breath-hold through the widest transverse dimension of the liver. Table 1 summarizes 2D-GRE-MRE sequence data. The acoustic waves generated by the driver produce microscopic shear displacement of tissues, which when imaged continuously by the motion sensitive MRE sequence, synchronized to the mechanical waves, produces images of propagating shear waves.
- Processing the information to calculate liver stiffness measurement (LSM) is done with an inversion algorithm in the magnitude and phase images, to generate gray scale and colored stiffness maps (elastograms). Then, ROIs are manually drawn within the confidence map over the liver, avoiding liver edge, artifacts, fissures, and wave interference, to obtain a reliable

LSM. MRE evaluates a larger portion of the liver, offering the option of choosing the best ROI, which can be performed in conjunction with conventional MRI.

The mechanical property measured with MRE is the magnitude of the complex shear modulus expressed in kilopascals (0–8 kPa range).

3.2.2 Diagnostic Performance: Accuracy and Precision

Liver fibrosis is the most important predictor of the occurrence of liver-related events and overall mortality in patients with CLD [30]. Therefore, the accurate detection and staging of liver fibrosis is a key issue for the management of patients with liver diseases. Among imaging mechanical biomarkers of stiffness, MRE plays a major role.

MRE has high sensitivity and specificity for predicting the stage of hepatic fibrosis, showing a higher diagnostic performance than transient elastography and shear wave elastography [31]. An individual patient data meta-analyses including almost 800 MASLD patients demonstrated that MRE achieves an excellent diagnostic accuracy for staging liver fibrosis. Established cut-offs are 2.65 kPa for mild fibrosis (≥F1), 3.14 kPa for significant fibrosis (≥F2), 3.53 kPa for advanced fibrosis (≥F3), and 4.45 kPa for cirrhosis (F4) (Fig. 4) [32]. Although liver shear stiffness increases with increased severity of the fibrotic process, the relationship between fibrosis and stiffness is non-linear, and predicting fibrosis stage may be challenging in early stages [33].

Multiple studies have assessed the test-retest repeatability of MRE in liver imaging [23]. A difference in MRE-derived LSM greater than 20% or a threshold of 0.75 kPa difference represent meaningful longitudinal changes. This information can be used to discern measurement error from real change in patients with CLD.

In summary, the utility of MRE has been demonstrated in a variety of CLD and this technique is currently regarded as the most accurate non-invasive image biomarker for evaluation of liver fibrosis.

3.2.3 Biases: Confounding Factors and Limitations

Liver stiffness measurements are subject to some biological confounding variables [34].

Elastography measurements should be acquired in the fasting state (4 h) as postprandial state may increase liver stiffness. Elevated inflammatory activity and increased GGT level may affect the diagnostic accuracy of MRE, leading to overestimation of liver fibrosis in early stage [32]. This information should be considered for early-stage fibrosis cohorts to optimize the diagnostic accuracy of MRE in staging liver fibrosis. Extrahepatic cholestasis, diffuse infiltrative disorders (such as liver amyloidosis or malignant cellular infiltrates), decompensated right heart failure, and hepatic venous congestion can also increase liver rigidity [35]. Although liver fibrosis is the predominant factor that causes increased stiffness in CLD, other pathologic processes often coexisting with fibrosis. Differentiating these processes from liver fibrosis is challenging and research work and efforts are being made in this direction with evaluation of other mechanical parameters obtained with MRE. Importantly, assessment of fibrosis with MRE is not confounded by the presence of steatosis or metabolic disorders such as obesity and diabetes.

The most common reason for technical failure of MRE is hepatic iron overload, which is common in patients with CLD. With conventional 2D-GRE-MRE sequence, high liver iron content causes a low signal intensity of the liver parenchyma, limiting the visualization of the mechanical waves and resulting in a failure rate around 5%. The newly-introduced Spin-echo Echo-planar imaging (SE-EPI-MRE) sequences are much less sensitive to iron overload, making these technical failures much less common. Liver MRE can be performed before or after gadolinium-based contrast material administration, as studies have shown that MR contrast materials do not affect LSM by MRE [36].

3.3 T1-HBP—Liver Function

3.3.1 Concepts and Technique

Evaluating liver function through MR images after hepatobiliary contrast agents' administration, such as gadoxetic acid (Gd-EOB-DTPA) and gadobenate dimeglumine (Gd-BOPTA), represents a significant advancement. This technique not only provides detailed anatomical information but also insights into the functional reserve and regional function of the liver, crucial for managing patients with liver diseases.

Hepatobiliary contrast agents are uptake by functioning hepatocytes. A standard dose of Gd-EOB-DTPA is 0.025 mmol/kg (Primovist in Europe and Eovist in the United States; Bayer Healthcare, Germany), while for Gd-BOPTA is 0.05 mmol/Kg (MultiHance, Bracco, Italy). Between 4% (Gd-BOPTA) and 50% (Gd-EOB-DTPA) of the injected contrast agent is taken up by normal hepatocytes through an active transport mechanism involving organic anion-transporting polypeptides (OATPs). Afterwards, consecutive biliary excretion is produced via multidrug resistance protein 2 (MRP2) at the canalicular membrane of hepatocytes. The hepatobiliary phase (HBP) provides a maximum liver enhancement about 120 min (Gd-BOPTA) or 20 min (Gd-EOB-DTPA) after injection and can last up to a few hours. During this HBP, the liver parenchyma shows increased signal intensity on T1-weighted images due to the highest concentration of the contrast agent within functioning hepatocytes. The degree of relative liver enhancement (RLE) is directly related to the functioning liver cell mass and the efficiency of the hepatocytes in up-taking the contrast agent [37, 38]. Areas of the liver with reduced function or damaged hepatocytes will show less enhancement. This relative or differential enhancement allows for the assessment of liver function. Hepatobiliary-enhanced MR and subsequent signal intensity changes in the HBP has been increasingly integrated into clinical practice, offering a non-invasive method to assess liver function [39].

The contrast-enhanced standard protocol includes axial 3D Gradient-echo T1-weighted, volume-interpolated breath-hold examination sequence obtained before and after contrast injection (details of the HBP are summarized in Table 1). Additionally, axial Gradient-echo in- and opposed-phase T1-weighted images can also be acquired (Fig. 2). MR sequences should cover the entire liver both before the contrast agent administration and at the HBP, which is obtained after 20 min if Gd-EOB-DTPA is used and 120 min if Gd-BOPTA is administered. Furthermore, automated WLS tools have extended their utility beyond volume quantification to assess liver function through innovative metrics like the MELIF score derived from Gd-EOB-DTPA-enhanced MR images [40]. Such comprehensive applications represent a big step towards integrated diagnostic platforms that could offer holistic assessments of liver health. Advanced techniques, including radiomics, DL, and machine learning, can analyze the complex imaging data to extract quantitative biomarkers that correlate with liver function [41–43]. These methods can stratify hepatic functional reserve, predict outcomes, and guide clinical decision-making in patients with different CLDs.

3.3.2 Diagnostic Performance: Accuracy and Precision

As a reflection of liver function, the hepatic hepatobiliary agent uptake correlates with CLD severity, mainly identifying cases with late-stage cirrhosis [39]. Although HBP enhancement using T1 relaxation times (R1) correlates with fibrosis stage ($r = 0.51$), its Area Under the Curve (AUC) for diagnosing advanced fibrosis is significantly lower than using MRE (0.82 vs. 0.92) [32, 44]. Given that MRE can be performed without the injection of *iv* contrast and with a shorter acquisition time, it has displaced the use of hepatobiliary contrast agents for the evaluation of fibrosis in clinical practice (Table 1). The current use of hepatobiliary-enhanced MRI is restricted to characterize liver lesions and to confer prognostic information. Recently, Bastati et al. developed the functional liver imaging score (FLIS) based on three simple visual features on Gd-EOB-DTPA-enhanced MRI: RLE; biliary excretion; and portal vein signal intensity [45]. This score has been associated with graft survival after liver transplantation and can stratify the risk of decompensation in patients with cirrhosis. This finding suggests that MR-based scores could complement traditional blood-based scores (like MELD or Child-Pugh), providing a more comprehensive assessment of liver outcomes. In this regard, other studies have assessed the potential of Gd-EOB-DTPA-enhanced MRI in monitoring liver function changes post-intervention, which is vital for treatment planning and follow-up [42, 46].

3.3.3 Biases: Confounding Factors and Limitations

Visual analysis of the contrast enhanced HBP-MR images provides the simplest estimate of liver function and has become an integral part of the radiology report [39]. The main limitations of this sequence are (1) a 20 to 120-min delay between dynamic and HBP imaging is needed to be wasted depending on the type of contrast used; and (2) the presence of transient severe motion artifacts which may degrade arterial-phase images in nearly 20% of patients [47]. Derived quantitative parameters from contrast-enhanced MR remain limited by complex mathematical modeling, long examination times, and respiratory artifacts. There is no standardization across MR scanners, and no established threshold values that define normal and abnormal measures, limiting these methods to research purposes. The diagnostic accuracy of T1-HBP relaxation time is low for necroinflammatory activity [44].

Moreover, the HBP-derived RLE is not constant over serial examinations. Signal intensity-based measurements after contrast agents administration provide relative parameters on an arbitrary scale which can neither be compared between repeated examinations of the same patient, nor between different patients. Regarding magnetic field strength, 3.0 T MRI gives a higher signal-to-noise ratios than 1.5 T MRI for the liver parenchyma on contrast-enhanced MRI. Finally, clinical applicability is also challenged due to the non-linear relationship between the gadolinium concentration and MR-signal intensity [37]. Most studies evaluating this sequence include a selection bias of patients, because contrast enhanced MRI is broadly used for the diagnostic work-up of liver nodules in patients with advanced CLD.

4 Clinical Impact of MR Multiparametric Data

Improvements in MR multiparametric imaging technology provide quantitative biomarkers for the diagnosis and follow-up of patients with CLD. Currently, imaging biomarkers are being used in clinical practice for risk-stratification in referral pathways, screening strategy to identify candidates for clinical trials, and treatment monitorization. Compared with liver biopsy, multiparametric MRI has several advantages, such as quantitative assessment of the whole liver, low sampling variability, high reproducibility, and the possibility of serial examinations.

In patients with chronic hepatitis C virus (HCV) and cirrhosis, the regression of fibrosis and improvement in liver-related outcomes have been reported after viral eradication [48]. However, available non-invasive tests (serum scores or transient elastography) cannot capture fibrosis regression and the residual risk of liver-related complications. As a possible solution for this scenario, contrast enhanced MRI studies have shown that an increase in RLE after HCV eradication can discriminate between individuals at low and high risk of hepatic decompensation [49]. This could allow for individualized surveillance strategies after sustained virological response is obtained in HCV patients.

Multiparametric MRI offers an attractive option for the evaluation of MASLD. In the algorithm of cirrhosis prevention in MASLD, persons with an indeterminate-high risk of liver fibrosis based on LSM by transient elastography merit referral to the liver specialist and a multidisciplinary approach to management [50]. Specific liver tissue quantification of fat, iron, and fibrosis allows a triple screen of clinically meaningful endpoints:

- *Steatohepatitis*: participants enrolled in MASLD clinical trials are required to have steatohepatitis and significant fibrosis (≥F2). There is a major need to develop biomarkers that will non-invasively predict either steatohepatitis resolution or > 1 stage improvement in fibrosis stage. Given that treatments can have different mechanisms of action, steatohepatitis and fibrosis components need to be evaluated independently to avoid confusion with the endpoints of interest. Beyond steatosis, PDFF can identify steatohepatitis component when combined with other parameters like waist circumference (MR-MASH score) or corrected T1 (Fig. 5) [51, 52]. These studies showed that quantitative MRI-derived biomarkers are suitable for the differentiation between steatosis and steatohepatitis, which have prognostic implications. Therefore, PDFF has important clinical utility to help guide the appropriate use of interventions in MASLD clinical care pathways. Moreover, in longitudinal studies using MRI-PDFF, an increase >15% is associated with

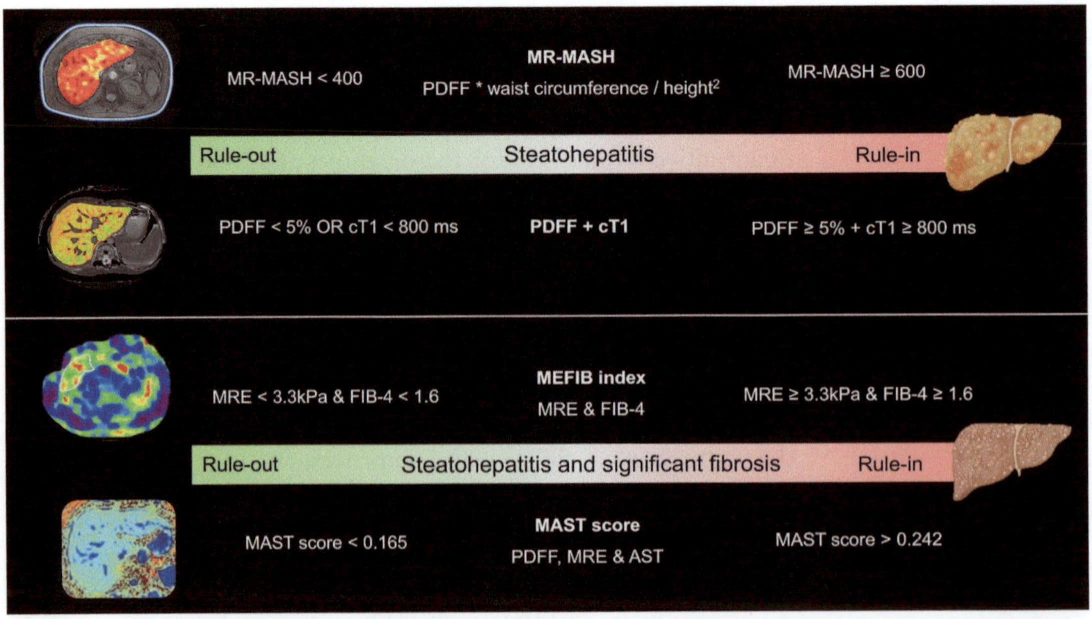

Fig. 5 Multiparametric MR scores developed and validated for the diagnosis of steatohepatitis and significant fibrosis as relevant endpoints in clinical trials

fibrosis progression (odds ratio = 6.7), while a reduction >30% is associated with steatohepatitis resolution (odds ratio = 5.5), and fibrosis regression (odds ratio = 6.5) [53, 54].

- *Iron overload*: noninvasive quantitative methods are needed to detect and grade iron overload, but also to initiate and monitor treatment (phlebotomy or chelation). Therefore, R2* is crucial not only for diagnosis of iron overload, but also to guide iron-reducing therapies [55]. Despite the good correlation between serum ferritin and R2* relaxometry in monitoring iron overload, treatment response may be misinterpreted when only serum ferritin is considered. Thus, MRI should be performed for monitorization, especially in case under chelation therapy [56].
- *Fibrosis*: Advanced fibrosis and cirrhosis are associated with increased risks of liver-related complications and death. New onset of hepatic decompensation is rare, being encephalopathy the most common event with an incidence per 100 person-years that increases with fibrosis stage as follows: 0.02 for F0-F2 vs. 0.75 for F3 vs. 2.39 for F4 [30]. Using LSM by MRE, the MEFIB index and the MAST score have been developed for the non-invasive identification of candidates for pharmacological therapy (Fig. 5) [57, 58]. MRE has excellent negative predictive value for hepatic decompensation and a 15% increase with follow-up is associated with fibrosis progression [59, 60]. This evidence, together with the overwhelming grey zones that restricts the clinical use of other MR techniques like diffusion weighted images, suggests that MRE in combination with MECSE can offer a complete metabolic imaging longitudinal profile of CLD [61].

5 Conclusion

Early detection and management of patients with CLD is constrained by the use of biopsy as the gold standard. Recent developments in multiparametric MR imaging offer an accurate diagnosis and monitoring, with implications for the clinical pathway in MASLD, HCV, or hemochromatosis. The most relevant MR techniques with clinical applications in the risk stratification and treatment monitorization of CLD are MECSE (derived biomarkers of PDFF and R2*) and MRE. The combination of these techniques allows clinicians to perform a triple screen for the evaluation of steatosis (steatohepatitis), iron overload, and fibrosis. Other image advances include the use of hepatobiliary contrast agents to assess liver function for prognostic information, and the use of DL approaches for the automatic whole liver segmentation, to reduce variability and increase the precision of CLD assessment. Technical concepts and biases must be considered to facilitate the use of multiparametric MR for clinical decision-making and longitudinal monitoring of the liver disease.

References

1. Younossi ZM, Golabi P, Paik JM, Henry A, Van Dongen C, Henry L (2023) The global epidemiology of nonalcoholic fatty liver disease (NAFLD) and nonalcoholic steatohepatitis (NASH): a systematic review. Hepatology (Baltimore, Md.) 77(4):1335–1347. https://doi.org/10.1097/HEP.0000000000000004
2. Nelson JE, Wilson L, Brunt EM, Yeh MM, Kleiner DE, Unalp-Arida A, Kowdley KV, Nonalcoholic Steatohepatitis Clinical Research Network (2011) Relationship between the pattern of hepatic iron deposition and histological severity in nonalcoholic fatty liver disease. Hepatology (Baltimore, Md.) *53*(2):448–457. https://doi.org/10.1002/hep.24038
3. Jimenez-Pastor A, Alberich-Bayarri A, Lopez-Gonzalez R, Marti-Aguado D, França M, Bachmann RSM, Mazzucco J, Marti-Bonmati L (2021) Precise whole liver automatic segmentation and quantification of PDFF and R2* on MR images. Eur Radiol 31(10): 7876–7887. https://doi.org/10.1007/s00330-021-07838-5
4. Bonekamp S, Tang A, Mashhood A et al (2014) Spatial distribution of MRI-determined hepatic proton density fat fraction in adults with nonalcoholic fatty liver disease. J Magn Reson Imaging 39(6): 1525–1532. https://doi.org/10.1002/jmri.24321
5. Kawamura N, Imajo K, Kalutkiewicz KJ et al (2022) Influence of liver stiffness heterogeneity on staging fibrosis in patients with nonalcoholic fatty liver disease. Hepatology 76(1): 186–195. https://doi.org/10.1002/hep.32302
6. Vu KN, Gilbert G, Chalut M, Chagnon M, Chartrand G, Tang A (2016) MRI-determined liver proton density fat fraction, with MRS validation: comparison of regions of interest sampling methods in patients with type 2 diabetes. J Magn Reson Imaging 43(5):1090–1099. https://doi.org/10.1002/jmri.25083
7. Cao D, Yang Y, Li M et al (2024) Quantitative comparison of liver volume, proton density fat fraction, and time burden between automatic whole liver segmentation and manual sampling MRI strategies for diagnosing metabolic dysfunction-associated Steatotic liver disease in obese patients. Curr Med Imaging 20. https://doi.org/10.2174/0115734056282249231206060136
8. Geng R, Buelo CJ, Sundaresan M et al (2023) Automated MR image prescription of the liver using deep learning: development, evaluation, and prospective implementation. J Mag Resona Imaging 58(2):429–441. https://doi.org/10.1002/jmri.28564
9. Lee S, Elton DC, Yang AH et al (2022) Fully automated and explainable liver segmental volume ratio and spleen segmentation at CT for diagnosing cirrhosis. Radiol Artif Intell 4(5): e210268. https://doi.org/10.1148/ryai.210268
10. Jeong JG, Choi S, Kim YJ, Lee WS, Kim KG (2022) Deep 3D attention CLSTM U-Net based automated liver segmentation and volumetry for the liver transplantation in abdominal CT volumes. Sci Rep 12(1):6370. https://doi.org/10.1038/s41598-022-09978-0
11. Wang K, Cunha GM, Hasenstab K et al (2023) Deep learning for inference of hepatic proton density fat fraction from T1-weighted in-phase and opposed-phase MRI: retrospective analysis of population-based trial data. AJR Am J Roentgenol 221(5):620–631. https://doi.org/10.2214/AJR.23.29607
12. Hagen F, Mair A, Bitzer M, Bösmüller H, Horger M (2021) Fully automated whole-liver volume quantification on CT-image data: comparison with manual volumetry using enhanced and unenhanced images as well as two different radiation dose levels and two reconstruction kernels. PLoS One 16(8): e0255374. https://doi.org/10.1371/journal.pone.0255374
13. Martí-Aguado D, Martí-Bonmatí L (2023) Iron deposits in alcoholic and metabolic liver disease. An RANM 140(03):277–283. https://doi.org/10.32440/ar.2023.140.03.doc01
14. Moura Cunha G, Navin PJ, Fowler KJ, Venkatesh SK, Ehman RL, Sirlin CB (2021) Quantitative magnetic resonance imaging for chronic liver disease. Br J Radiol 94(1121):20201377. https://doi.org/10.1259/bjr.20201377
15. Zerunian M, Pucciarelli F, Masci B et al (2022) Updates on quantitative MRI of diffuse liver disease: a narrative review. Biomed Res Int 2022:1147111. https://doi.org/10.1155/2022/1147111
16. Martí-Aguado D, Alberich-Bayarri Á, Martín-Rodríguez JL et al (2020) Differences in multi-echo chemical shift encoded MRI proton density fat fraction estimation based on multifrequency fat peaks selection in non-alcoholic fatty liver disease patients. Clin Radiol 75(11): 880.e5–880.e12. https://doi.org/10.1016/j.crad.2020.07.031

17. Kukuk GM, Hittatiya K, Sprinkart AM et al (2015) Comparison between modified Dixon MRI techniques, MR spectroscopic relaxometry, and different histologic quantification methods in the assessment of hepatic steatosis. Eur Radiol 25(10):2869–2879. https://doi.org/10.1007/s00330-015-3703-6
18. França M, Carvalho JG (2020) MR imaging assessment and quantification of liver iron. Abdom Radiol (New York) 45(11): 3400–3412. https://doi.org/10.1007/s00261-020-02574-8
19. Reeder SB, Yokoo T, França M et al (2023) Quantification of liver iron overload with MRI: review and guidelines from the ESGAR and SAR. Radiology 307(1):e221856. https://doi.org/10.1148/radiol.221856
20. França M, Alberich-Bayarri Á, Martí-Bonmatí L et al (2017) Accurate simultaneous quantification of liver steatosis and iron overload in diffuse liver diseases with MRI. Abdom Radiol (New York) 42(5):1434–1443. https://doi.org/10.1007/s00261-017-1048-0
21. Martí-Aguado D, Jiménez-Pastor A, Alberich-Bayarri Á et al (2022) Automated whole-liver MRI segmentation to assess steatosis and iron quantification in chronic liver disease. Radiology 302(2):345–354. https://doi.org/10.1148/radiol.2021211027
22. Qadri S, Vartiainen E, Lahelma M et al (2023) Marked difference in liver fat measured by histology *vs.* magnetic resonance-proton density fat fraction: a meta-analysis. JHEP Rep Innov Hepatol 6(1):100928. https://doi.org/10.1016/j.jhepr.2023.100928
23. Fowler KJ, Venkatesh SK, Obuchowski N et al (2023) Repeatability of MRI biomarkers in nonalcoholic fatty liver disease: the NIMBLE consortium. Radiology 309(1):e231092. https://doi.org/10.1148/radiol.231092
24. Wood JC, Enriquez C, Ghugre N, Tyzka JM, Carson S, Nelson MD, Coates TD (2005) MRI R2 and R2* mapping accurately estimates hepatic iron concentration in transfusion-dependent thalassemia and sickle cell disease patients. Blood 106(4):1460–1465. https://doi.org/10.1182/blood-2004-10-3982
25. Garbowski MW, Carpenter JP, Smith G, ak. (2014) Biopsy-based calibration of T2* magnetic resonance for estimation of liver iron concentration and comparison with R2 Ferriscan. J Cardiovasc Magn Reson 16(1):40. https://doi.org/10.1186/1532-429X-16-40
26. Henninger B, Zoller H, Rauch S, Finkenstedt A, Schocke M, Jaschke W, Kremser C (2015) R2* relaxometry for the quantification of hepatic iron overload: biopsy-based calibration and comparison with the literature. RoFo 187(6):472–479. https://doi.org/10.1055/s-0034-1399318
27. Hankins JS, McCarville MB, Loeffler RB et al (2009) R2* magnetic resonance imaging of the liver in patients with iron overload. Blood 113(20):4853–4855. https://doi.org/10.1182/blood-2008-12-191643
28. Karlsson M, Ekstedt M, Dahlström N et al (2019) Liver R2* is affected by both iron and fat: a dual biopsy-validated study of chronic liver disease. J Mag Reson Imaging 50(1):325–333. https://doi.org/10.1002/jmri.26601
29. Bashir MR, Wolfson T, Gamst AC et al (2019) Hepatic R2* is more strongly associated with proton density fat fraction than histologic liver iron scores in patients with nonalcoholic fatty liver disease. J Mag Reson Imaging 49(5): 1456–1466. https://doi.org/10.1002/jmri.26312
30. Sanyal AJ, Van Natta ML, Clark J et al (2021) Prospective study of outcomes in adults with nonalcoholic fatty liver disease. N Engl J Med 385(17):1559–1569. https://doi.org/10.1056/NEJMoa2029349
31. Selvaraj EA, Mózes FE, Jayaswal ANA et al (2021) Diagnostic accuracy of elastography and magnetic resonance imaging in patients with NAFLD: a systematic review and meta-analysis. J Hepatol 75(4):770–785. https://doi.org/10.1016/j.jhep.2021.04.044
32. Liang JX, Ampuero J, Niu H et al (2023) An individual patient data meta-analysis to determine cut-offs for and confounders of NAFLD-fibrosis staging with magnetic resonance elastography. J Hepatol 79(3):592–604. https://doi.org/10.1016/j.jhep.2023.04.025
33. Loomba R, Wolfson T, Ang B et al (2014) Magnetic resonance elastography predicts advanced fibrosis in patients with nonalcoholic fatty liver disease: a prospective study. Hepatology (Baltimore, Md) 60(6):1920–1928. https://doi.org/10.1002/hep.27362
34. Pepin KM, Welle CL, Guglielmo FF, Dillman JR, Venkatesh SK (2022) Magnetic resonance elastography of the liver: everything you need to know to get started. Abdom Radiol (New York) 47(1):94–114. https://doi.org/10.1007/s00261-021-03324-0
35. Venkatesh SK, Wells ML, Miller FH, Jhaveri KS, Silva AC, Taouli B, Ehman RL (2018) Magnetic resonance elastography: beyond

liver fibrosis-a case-based pictorial review. Abdom Radiol (New York) 43(7): 1590–1611. https://doi.org/10.1007/s00261-017-1383-1

36. Plaikner M, Kremser C, Zoller H, Steurer M, Glodny B, Jaschke W, Henninger B (2019) Does gadoxetate disodium affect MRE measurements in the delayed hepatobiliary phase? Eur Radiol 29(2):829–837. https://doi.org/10.1007/s00330-018-5616-7
37. Haimerl M, Verloh N, Zeman F et al (2017) Gd-EOB-DTPA-enhanced MRI for evaluation of liver function: comparison between signal-intensity-based indices and T1 relaxometry. Sci Rep 7:43347. https://doi.org/10.1038/srep43347
38. Eiras-Araújo AL, Parente DB, da Silva AC et al (2023) Relative enhancement index can be used to quantify liver function in cirrhotic patients that undergo gadoxetic acid-enhanced MRI. Eur Radiol 33(7):5142–5149. https://doi.org/10.1007/s00330-023-09402-9
39. Poetter-Lang S, Bastati N, Messner A, Kristic A, Herold A, Hodge JC, Ba-Ssalamah A (2020) Quantification of liver function using gadoxetic acid-enhanced MRI. Abdom Radiol (New York) 45(11):3532–3544. https://doi.org/10.1007/s00261-020-02779-x
40. Río Bártulos C, Senk K, Bade R et al (2022) MELIF, a fully automated liver function score calculated from Gd-EOB-DTPA-enhanced MR images: diagnostic performance vs. the MELD score. Diagnostics (Basel, Switzerland) 12(7):1750. https://doi.org/10.3390/diagnostics12071750
41. Hu J, Wang X, Prince M et al (2024) Gd-EOB-DTPA enhanced MRI based radiomics combined with clinical variables in stratifying hepatic functional reserve in HBV infected patients. Abdom Radiol (New York) 49(4): 1051–1062. https://doi.org/10.1007/s00261-023-04176-6
42. Zhu L, Wang F, Chen X, Dong Q, Xia N, Chen J, Li Z, Zhu C (2023) Machine learning-based radiomics analysis of preoperative functional liver reserve with MRI and CT image. BMC Med Imaging 23(1):94. https://doi.org/10.1186/s12880-023-01050-1
43. Park HJ, Yoon JS, Lee SS, Suk HI, Park B, Sung YS, Hong SB, Ryu H (2022) Deep learning-based assessment of functional liver capacity using Gadoxetic acid-enhanced hepatobiliary phase MRI. Korean J Radiol 23(7): 720–731. https://doi.org/10.3348/kjr.2021.0892
44. Ding Y, Rao SX, Zhu T, Chen CZ, Li RC, Zeng MS (2015) Liver fibrosis staging using T1 mapping on gadoxetic acid-enhanced MRI compared with DW imaging. Clin Radiol 70(10):1096–1103. https://doi.org/10.1016/j.crad.2015.04.014
45. Bastati N, Beer L, Ba-Ssalamah A et al (2022) Gadoxetic acid-enhanced MRI-derived functional liver imaging score (FLIS) and spleen diameter predict outcomes in ACLD. J Hepatol 77(4):1005–1013. https://doi.org/10.1016/j.jhep.2022.04.032
46. Fabritius MP, Garlipp B, Öcal O et al (2024) Assessing regional hepatic function changes after hypertrophy induction by radioembolisation: comparison of gadoxetic acid-enhanced MRI and ^{99m}Tc-mebrofenin hepatobiliary scintigraphy. Eur Radiol Exp 8(1):15. https://doi.org/10.1186/s41747-023-00409-x
47. Davenport MS, Viglianti BL, Al-Hawary MM et al (2013) Comparison of acute transient dyspnea after intravenous administration of gadoxetate disodium and gadobenate dimeglumine: effect on arterial phase image quality. Radiology 266(2):452–461. https://doi.org/10.1148/radiol.12120826
48. D'Ambrosio R, Degasperi E, Anolli MP et al (2022) Incidence of liver- and non-liver-related outcomes in patients with HCV-cirrhosis after SVR. J Hepatol 76(2):302–310. https://doi.org/10.1016/j.jhep.2021.09.013
49. Haider L, Mandorfer M, Güngören Z et al (2018) Noninvasive monitoring of liver disease regression after hepatitis C eradication using Gadoxetic acid-enhanced MRI. Contrast Media Mol Imaging 2018:8489709. https://doi.org/10.1155/2018/8489709
50. Cusi K, Isaacs S, Barb D et al (2022) American Association of Clinical Endocrinology Clinical Practice Guideline for the diagnosis and Management of Nonalcoholic Fatty Liver Disease in primary care and endocrinology clinical settings: co-sponsored by the American Association for the Study of Liver Diseases (AASLD). Endocr Pract 28(5):528–562. https://doi.org/10.1016/j.eprac.2022.03.010
51. Marti-Aguado D, Arnouk J, Liang JX et al (2024) Development and validation of an image biomarker to identify metabolic dysfunction associated steatohepatitis: MR-MASH score. Liver Int 44(1):202–213. https://doi.org/10.1111/liv.15766
52. Andersson A, Kelly M, Imajo K et al (2022) Clinical utility of magnetic resonance imaging biomarkers for identifying nonalcoholic Steatohepatitis patients at high risk of progression: a multicenter pooled data and meta-analysis. Clin Gastroenterol Hepatol 20(11): 2451–2461.e3. https://doi.org/10.1016/j.cgh.2021.09.041

53. Stine JG, Munaganuru N, Barnard A et al (2021) Change in MRI-PDFF and histologic response in patients with nonalcoholic Steatohepatitis: a systematic review and meta-analysis. Clin Gastroenterol Hepatol 19(11): 2274–2283.e5. https://doi.org/10.1016/j.cgh.2020.08.061
54. Ajmera V, Park CC, Caussy C et al (2018) Magnetic resonance imaging proton density fat fraction associates with progression of fibrosis in patients with nonalcoholic fatty liver disease. Gastroenterology 155(2):307–310.e2. https://doi.org/10.1053/j.gastro.2018.04.014
55. Wood JC, Zhang P, Rienhoff H, Abi-Saab W, Neufeld EJ (2015) Liver MRI is more precise than liver biopsy for assessing total body iron balance: a comparison of MRI relaxometry with simulated liver biopsy results. Magn Reson Imaging 33(6):761–767. https://doi.org/10.1016/j.mri.2015.02.016
56. Plaikner M, Kremser C, Zoller H, Jaschke W, Henninger B (2018) Monitoring iron overload: relationship between R2* Relaxometry of the liver and serum ferritin under different therapies. J Clin Imaging Sci 8:40. https://doi.org/10.4103/jcis.JCIS_30_18
57. Noureddin M, Truong E, Gornbein JA et al (2022) MRI-based (MAST) score accurately identifies patients with NASH and significant fibrosis. J Hepatol 76(4):781–787. https://doi.org/10.1016/j.jhep.2021.11.012
58. Jung J, Loomba RR, Imajo K et al (2021) MRE combined with FIB-4 (MEFIB) index in detection of candidates for pharmacological treatment of NASH-related fibrosis. Gut 70(10):1946–1953. https://doi.org/10.1136/gutjnl-2020-322976
59. Ajmera V, Kim BK, Yang K et al (2022) Liver stiffness on magnetic resonance Elastography and the MEFIB index and liver-related outcomes in nonalcoholic fatty liver disease: a systematic review and meta-analysis of individual participants. Gastroenterology 163(4): 1079–1089.e5. https://doi.org/10.1053/j.gastro.2022.06.073
60. Ajmera VH, Liu A, Singh S et al (2020) Clinical utility of an increase in magnetic resonance Elastography in predicting fibrosis progression in nonalcoholic fatty liver disease. Hepatology (Baltimore, Md.) 71(3):849–860. https://doi.org/10.1002/hep.30974
61. Marti-Aguado D, Rodríguez-Ortega A, Alberich-Bayarri A, Marti-Bonmati L (2020) Magnetic Resonance imaging analysis of liver fibrosis and inflammation: overwhelming gray zones restrict clinical use. Abdom Radiol (New York) 45(11):3557–3568. https://doi.org/10.1007/s00261-020-02713-1

Chapter 4

Neuroimaging Applications in Hepatic Encephalopathy: Acquisition Protocols and Basic Preprocessing Techniques

Miguel López-Aguirre and Jorge El-Khatib-Núñez

Abstract

Hepatic encephalopathy (HE) is a potentially reversible metabolic syndrome resulting from liver dysfunction. Typically, diagnosis of HE relies on a combination of clinical findings, laboratory tests to assess liver function, and evaluation of portal-systemic circulation. However, in certain cases, these methods may not be good enough. Hence, additional auxiliary tools may be necessary for an accurate differential diagnosis. Neuroimaging has emerged as a powerful asset in this regard, enabling identification of prevalent features of HE such as manganese accumulation, brain edema, and brain atrophy. Moreover, neuroimaging helps exclusion of other neurological conditions with similar clinical signs to HE, thereby increasing diagnostic accuracy. Consequently, neuroimaging is gaining relevance in clinical practice for the differential diagnosis of HE and other multiple neurological conditions. In this chapter, we will outline diverse neuroimaging procedures suitable for identifying pathological features of HE, describing the protocols and specific features observable with each modality. Finally, we will explore some basic preprocessing techniques designed to enhance neuroimaging quality and increase their reliability and sensitivity to pathological alterations.

Key words Neuroimaging, MRI, Hepatic encephalopathy, Protocols, Preprocessing

Acronyms

ADC	Apparent diffusion coefficient
ASL	Arterial Spin Labeling
BOLD	Blood oxygenation level dependence
CNS	Central nervous system
CT	Computed Tomography
DMN	Default mode network
DKI	Diffusion kurtosis imaging
DTI	Diffusion tensor imaging
DWI	Diffusion-weighted imaging
FA	Fractional anisotropy
FC	Functional connectivity
FLAIR	Fluid-attenuated inversion recovery
fMRI	Functional magnetic resonance imaging

Tiziano Balzano (ed.), *Experimental and Clinical Methods in Hepatic Encephalopathy Research*, Neuromethods, vol. 219, https://doi.org/10.1007/978-1-0716-4354-9_4,

FoV	Field of view
FWHM	Full width at half maximum
HE	Hepatic encephalopathy
MD	Mean diffusivity
MRI	Magnetic resonance imaging
MRS	Magnetic resonance spectroscopy
NODDI	Neurite orientation dispersion and density imaging
PET	Positron emission tomography
rs-fMRI	Resting-state functional magnetic resonance imaging
ROI	Region of interest
SNR	Signal-to-noise ratio
SPECT	Single-photon emission computed tomography
SWI	Susceptibility-weighted imaging
T1	Longitudinal relaxation time
T1w	T1-weighted
T2	Transverse relaxation time
T2w	T2-weighted
tb-fMRI	Task-based functional magnetic resonance imaging
WM	White matter

1 Introduction

Hepatic encephalopathy (HE) represents a multifaceted and potentially reversible neuropsychiatric syndrome arising from liver dysfunction. Despite hepatic failure can stem from a broad spectrum of factors such as cirrhosis, portal hypertension, or portosystemic shunts [1], it typically results in the accumulation of toxins within the bloodstream (e.g., ammonia, inflammatory cytokines, and benzodiazepine-like compounds [2]). These toxins eventually reach the brain, causing diverse pathological alterations that potentially disrupt normal brain function. Thus, they may trigger diverse neuropsychiatric disturbances and motor impairment [1]. The diagnosis of HE is typically straightforward, relying on a combination of clinical findings, laboratory tests to assess liver function, and evaluation of portal-systemic circulation [2]. However, challenges may arise when laboratory tests yield inconclusive results, when patients exhibit atypical neurological or motor signs, or when they display significant comorbidities. In such instances, supplementary neuroimaging data may constitute an asset in clinical practice. These tools allow for non-invasive observation of both brain structure and activity, thus offering a powerful resource for the identification of alterations in the central nervous system (CNS). Indeed, neuroimaging has demonstrated its efficacy in identifying prevalent features of HE, such as manganese accumulation, brain edema, and brain atrophy, among others [1]. Furthermore, they can assist in excluding other neurological disorders with compatible clinical signs (e.g., Wernicke–Korsakoff's syndrome, alcohol-withdrawal delirium, etc.). Consequently, they are gaining clinical relevance for the differential diagnosis of HE and other multiple neurological conditions.

In this chapter, we will outline conventional and widely available neuroimaging procedures suitable for defining a standard clinical protocol for HE, alongside complementary neuroimaging techniques that may become of clinical interest in the future. In addition, we will delve into some advanced sequences that enable the investigation of specific pathophysiological features in this clinical condition, thus constituting a valuable asset to understand HE pathogenesis and characterizing its pathophysiology. Finally, we will introduce some preprocessing techniques designed to enhance image quality and sensitivity to pathological alterations.

2 Materials

Patients showing clinical signs compatible with HE are frequently evaluated with CT (computed tomography) or MRI (magnetic resonance imaging) scans, mainly to exclude emergent processes such as hemorrhage, ischemia, or other neurological processes. However, despite CT is frequently used in HE and is highly beneficial in the clinical setting, its application is limited and more often based on routine rather than evidence [3]. Overall, brain CT results ineffective in determining the cause of altered mental status in HE without a history of trauma or focal neurological signs [3]. Indeed, except for findings like frontal cortical atrophy and mild cerebral edema, brain CT scans are typically normal in HE [3, 4]. Therefore, CT should be reserved for specific cases and not used routinely in HE [3, 4]. On the other hand, molecular imaging techniques like PET (positron emission tomography) and SPECT (single-photon emission computed tomography), while potentially useful, are frequently contraindicated in HE since patients commonly exhibit renal insufficiency [5]. Consequently, MRI is the most deployed neuroimaging technique in HE, providing critical insights beyond basic diagnostic confirmation.

2.1 Magnetic Resonance Imaging System

While many HE features can be detected with a 1.5T MRI scanner, a 3T scanner is recommended owing to its superior performance. This scanner should be equipped with a multi-channel phased array head coil for parallel imaging. Along this chapter we will use a 3T Siemens Magnetom Vida as a reference to describe the acquisition protocols.

2.2 Research Workstations

Although neuroimaging preprocessing techniques are not generally deemed essential in the clinical setting, they are commonly used by researchers to exploit the full potential of medical images. These steps aim to refine image quality (e.g., artifact removal, signal-to-noise ratio (SNR) enhancement, bias field correction, etc.) to pave the way for more robust and reliable analyses. In this case, additional tools are required to execute these tasks effectively:

1. *Computer hardware:* Most neuroimaging tools are developed for PC and Mac platforms. Accordingly, we suggest working on these environments. Required computer specifications may vary depending on the type of processing or analysis applied. In any case, it is generally recommended, but not mandatory, to operate on a workstation equipped with at least an 8-core, 16-thread CPU, together with 32 GB of RAM. While not essential, having access to a dedicated GPU with CUDA support (*https://developer.nvidia.com/cuda-toolkit*) for accelerated processing in some computationally demanding steps is also recommended.
2. *Operative system:* Considering compatibility with neuroimaging software, GNU Linux-based operative systems are highly suggested, with Ubuntu (*https://ubuntu.com*) or CentOS (*https://www.centos.org*) among the most compatible and well-supported distributions. MacOS is also generally supported, although incompatibilities may exist with some tools. Windows-based operating systems are not indicated owing to their high degree of incompatibility with most neuroimaging software. Linux virtualization can be used to bypass this problem, but this approach is not recommended given its inferior performance compared to native Linux environments.
3. *Software:* Nowadays, there is a broad spectrum of neuroimaging software, with numerous tools offering similar functionalities. The selection of these tools may vary depending on the study demands and your personal preferences. Therefore, in order to provide a basic understanding of common preprocessing steps, along this chapter, we will only focus on some functionalities provided by dcm2niix [6], FSL [7], ANTs [8], 3D Slicer [9], MRtrix3 [10], and fmriprep [11].

3 Methods

3.1 Patients Positioning Within the MRI Scanner

All participants must be scanned in a comfortable and appropriate position to minimize motion during brain scans:

1. Participants should be positioned supine in the head coil with back and knee supports to enhance comfort.

2. Ensure the head is straight and the neck is neither hyper-extended nor retracted.
3. Center the participant's head in the head coil using the nasion as an anatomical landmark.
4. Attach available head immobilization devices.
5. Use the positioning lasers to align the nasion with the magnet's isocenter.

3.2 MRI Modalities

HE is conventionally differentiated into four main categories based on the duration and characteristics of clinical manifestations, with each category displaying specific radiologic features [12]:

1. *Acute HE*: The primary finding in acute HE is diffuse cerebral edema, appearing as thickening and hyperintensities within the cerebral gyri in T2-weighted or FLAIR acquisitions [13]. In the most severe cases, these features may also be extended to the cingulate gyrus and insular cortex [13]. Involvement of the basal ganglia and the thalamus is also frequent but variable. Acute HE is also characterized by abnormalities in the periventricular white matter (WM), thalami and basal ganglia in patients with cirrhosis and HE [14], as well as a characteristic glutamate/glutamine spike in magnetic resonance spectroscopy and water diffusion restriction [15].
2. *Chronic HE*: Classically, the most recognizable finding in patients with chronic liver disease is a bilateral hyperintensity in the globus pallidus in T1-weighted scans caused by manganese overaccumulation [16]. This signal alteration in the striatopallidal system may decline or even be completely reversed after normalization of liver function [17, 18]. Abnormalities in WM in or around the corticospinal tracts have also been documented in T2-weighted and FLAIR MRI scans in cirrhotic patients [19]. These signal alterations may also be normalized after liver transplantation [20]. Additionally, cirrhotic patients with long-standing clinical HE and high-flow portosystemic shunts may exhibit brain tissue loss, most notably in those with a long history of HE-associated neurological symptoms [21]. This condition is known as acquired hepatocerebral degeneration and presents with multiple manifestations [22].
3. *Acute-on-chronic liver failure*: It refers to cirrhotic patients experiencing acute hepatic deterioration, typically triggered by inflammatory or infectious processes. Often, signs of both acute and chronic HE overlap in cases of acute worsening of liver function in individuals with preexisting chronic liver disease. This condition has a high short-term mortality rate and can be triggered by bacterial and fungal infections. These patients exhibit a combination of changes due to chronic liver disease and acute liver failure, secondary to sudden-onset

hyperammonemia with cortical edema. In cases of alcoholic liver disease, Wernicke's encephalopathy may also be present [12].

4. *Minimal HE*: Also known as subclinical or latent HE, it is characterized by the absence of obvious neurological abnormalities on clinical examination in cirrhotic patients. It includes patients with cirrhosis or portosystemic shunts who have a preserved level of consciousness but mildly abnormal neurological or cognitive function. This can negatively impact quality of life and the ability to perform complex tasks, making additional diagnostic techniques important for detection. Changes in cerebral blood flow, brain metabolism, and structural abnormalities can be identified using some neuroimaging techniques. However, there is no consensus on the cost-effectiveness of these different methods.

3.2.1 T1-Weighted Imaging

Description T1-weighted (T1w) imaging is one of the most elementary MRI techniques, as it relies on one of the foundations of MRI, the longitudinal relaxation time (T1). This metric quantifies the time required for water protons to realign with the main magnetic field after excitation with a radiofrequency pulse, offering an MRI contrast that is useful for distinguishing different soft tissue types, such as muscle, fat, and biofluid spaces. Thus, T1w images provide detailed scans that can be used for anatomical reference, among many other applications.

Background It has been consistently demonstrated that HE patients suffering from liver cirrhosis or portosystemic shunts exhibit bilateral and symmetrical T1w signal hyperintensities within the globus pallidus (Fig. 1a), the substantia nigra (Fig. 1b), and to a lesser extend in the other regions of the midbrain [20, 23, 24]. These hyperintensities are observed in up to 90% of cirrhotic patients [25] and can also affect the pituitary gland [22] (Fig. 1c).

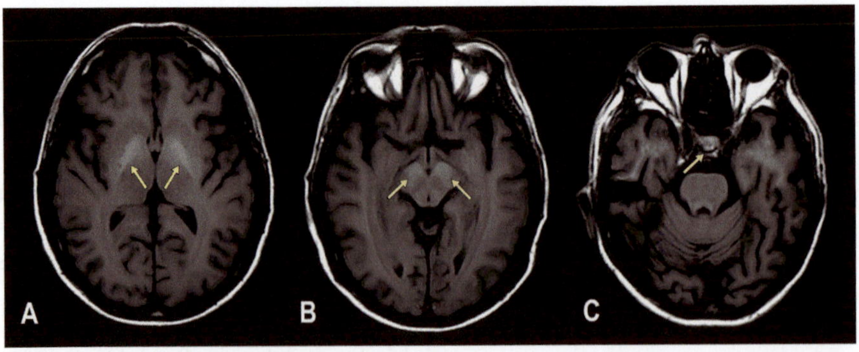

Fig. 1 T1-weighted scans displaying signal hyperintensities caused by manganese deposition in a cirrhotic patient. Axial scans highlight these signal alterations in the globus pallidus (**a**), anterior midbrain (**b**), and pituitary gland (**c**)

These signal alterations are generally accepted to be caused by an abnormal accumulation of manganese [16], a paramagnetic substance excreted by the biliary tract that is deposited in the CNS [26]. Its paramagnetic properties lead to shortened T1, resulting in this hyperintensity on T1w sequences. The manganese hypothesis is grounded in the elevated manganese concentrations found in the blood and cerebrospinal fluid of cirrhotic patients, along with the increased manganese levels in the globus pallidus observed in autopsies of HE patients [16]. However, although pallidal T1w signal hyperintensities are present in most cirrhotic patients, they do not directly correlate with the clinical stage [27, 28]. Indeed, this radiologic feature has also been described in other entities not related to elevated brain manganese levels, such as nonketotic hyperglycemic episodes, hypoxic-ischemic encephalopathy, basal ganglia calcification, neurofibromatosis type I, and Japanese encephalitis [29]. Additionally, this signal pattern can be observed in both inherited and acquired conditions causing hypermanganesemia. It also manifests in non-cirrhotic patients with chronic manganese poisoning from parenteral nutrition [30], with occupational exposures (such as professional welders exposed to manganese), and patients with congenital portosystemic shunts lacking underlying hepatocellular disease that restricts biliary elimination. Therefore, although it is a highly distinctive feature of chronic HE, this signal alteration is not specific to the disease and does not reliably measure tissue manganese levels quantitatively. Nonetheless, there seems to be some correlation with liver function, as this signal change can manifest following TIPS placement and can be reversed after resolving the underlying cause and reducing manganese levels, typically through liver transplantation [20].

Accumulation of manganese in the basal ganglia is also believed to interfere with normal dopaminergic function, potentially triggering the onset of extrapyramidal parkinsonian signs observed in some HE patients. Accordingly, the occurrence of parkinsonism may be related to the hyperintensity in the substantia nigra, thus constituting a distinctive radiologic hallmark in patients with cirrhotic-related parkinsonism [27, 31]. In any case, T1w signal hyperintensities are expected to return to normal levels after normalization of liver function, or after occlusion of congenital portosystemic shunts [17, 18] (Fig. 2a).

Standard clinical protocol T1w TSE SAG (TR/TE: 946/9.3 ms; flip angle: 90°; FoV: 256 mm; in-plane matrix: 320 × 256; in-plane resolution: 0.4×0.4 mm^2; slice thickness: 5 mm; 30 sagittal slices) and T1w TSE AX (TR/TE: 531/6.7 ms; flip angle: 90°; FoV: 220 mm; in-plane matrix: 320 × 256; in-plane resolution: 0.3×0.3 mm^2; slice thickness: 3 mm; 42 axial slices).

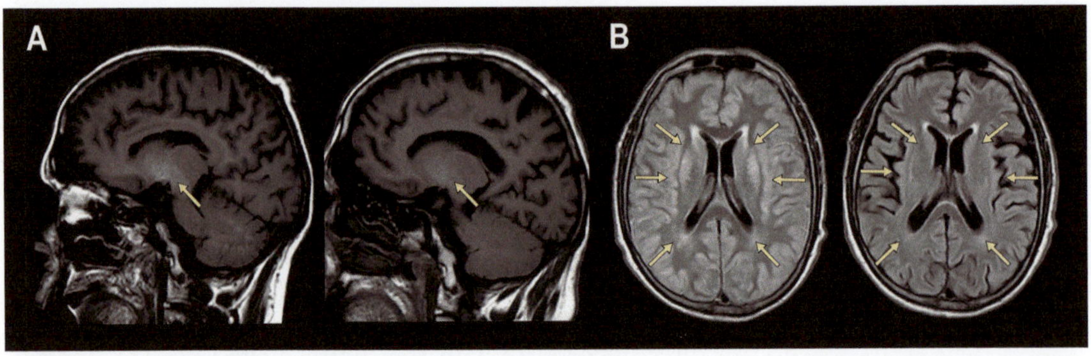

Fig. 2 Signal normalization one year after liver transplantation. (**a**) Sagittal T1-weighted scans showing signal recovery in the globus pallidus. (**b**) FLAIR scans displaying normalization in the putamen, periventricular white matter, and multiple cortical regions, especially within the insular cortex

Recommended research protocol 3D T1w GRE SAG (TR/TE: 2300/2.98 ms; flip angle: 9°; FoV: 256 mm; in-plane matrix: 256 × 256; in-plane resolution: 1 × 1 mm^2; slice thickness: 1 mm; 192 sagittal slices).

3.2.2 T2-Weighted Imaging

Description T2-weighted (T2w) imaging is the other basis for MRI, as it is derived from the transverse relaxation time (T2). This feature determines the time necessary for protons to decay after excitation with a radiofrequency pulse. This imaging modality is useful for highlighting differences in tissue water content and is commonly employed in the diagnosis of edema, inflammation, and certain tumors.

Background Patients with liver failure or portosystemic shunts frequently display elevated levels of ammonium, a pivotal neurotoxin linked to HE that is mainly produced by colonizing bacteria in the colon [32]. The gastrointestinal tract is the primary source of ammonia, which enters the circulation through the portal vein and is subsequently cleared mostly by the liver, preventing it from reaching the systemic circulation. Hepatic dysfunction induces a state of hyperammonemia. This condition allows ammonium to permeate the blood-brain barrier, resulting in brain-blood ammonia concentration ratios up to four times higher than normal [33]. This hyperammonemia induces one of the neuropathological hallmarks of HE, the presence of Alzheimer type II astrocytes, which eventually lead to astrocyte edema [34] (Fig. 3). Acute HE is characterized by this type of edema, visible as a cortical signal hyperintensity on T2w MRI scans, primarily affecting the insular border and cingulate gyrus, but sparing the occipital and prerolandic cortex [13]. This involvement is typically displayed symmetrically but may also manifest asymmetrically [13]. In severe cases, cortical edema may appear diffusely [15]. Astrocyte edema is

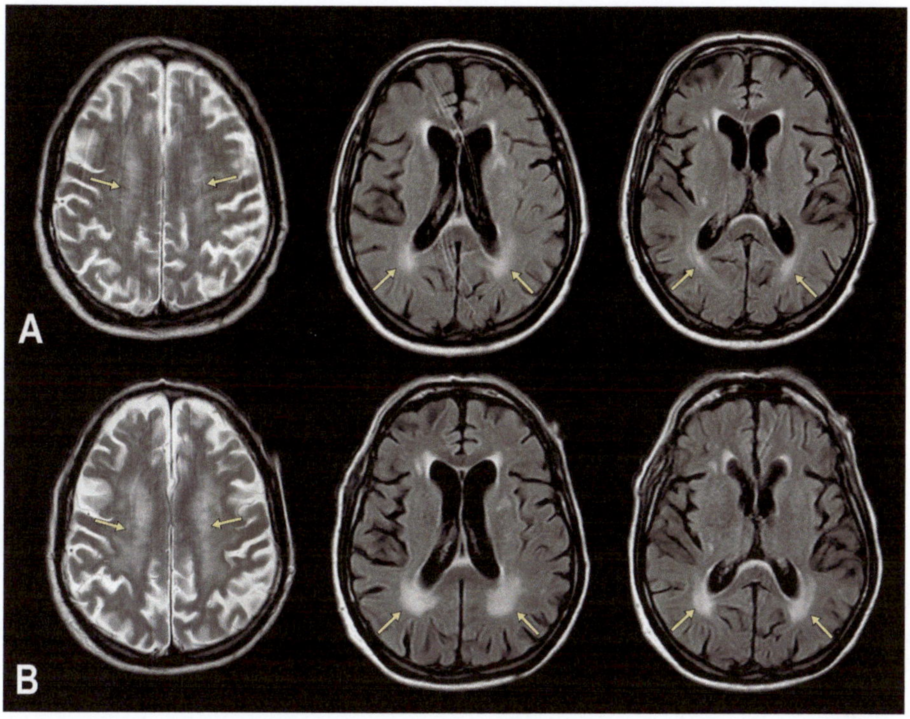

Fig. 3 Axial T2-weighted (left) and FLAIR (middle and right) brain images of a patient with chronic liver failure displaying the transition from asymptomatic to acute hepatic encephalopathy. (**a**) Baseline brain MRI conducted during an asymptomatic period. In here, multiple confluent hyperintensities can be observed in the periventricular white matter in both FLAIR and T2-weighted images. (**b**) Brain MRI during severe acute hepatic encephalopathy. In these scans, we can notice the increased hyperintensity of the periventricular white matter compared to the previous MRI, indicating rapid changes over a short period

thought to result from excessive glutamine production in astrocytes, a byproduct of ammonium detoxification via glutamine synthetase [35]. While increased glutamine is present in both acute and chronic HE, cerebral edema and death from intracranial hypertension are rare in chronic liver failure, occurring mainly in cases of sudden onset encephalopathy [15]. T2w signal alterations return to normal levels after normalization of liver function post-liver transplant [20].

Ammonium has multiple detrimental effects on several neurotransmitter systems. Firstly, ammonium promotes the conversion of glutamate to glutamine in astrocytes, thereby reducing the availability of glutamate, a key excitatory neurotransmitter in the brain [35, 36]. Concurrently, ammonium elevates GABA production, an inhibitory neurotransmitter [37]. The alteration of both neurotransmitter systems impacts normal synaptic transmission, thus disrupting the excitatory-inhibitory synaptic balance in the brain. In addition, excess ammonium can interfere with dopamine synthesis and metabolism, potentially leading to extrapyramidal symptoms, such as the parkinsonism observed in some patients with liver dysfunction [31]. Furthermore, ammonium can alter tryptophan levels, the precursor to serotonin, thereby altering serotonin synthesis [38]. This can influence mood and behavior, contributing to the neuropsychiatric symptoms [31].

Standard clinical protocol T2w TSE AX (TR/TE: 3120/79 ms; flip angle: 150°; FoV: 220 mm; in-plane matrix: 320 × 240; in-plane resolution: 0.3 × 0.3 mm^2; slice thickness: 4 mm; 30 axial slices) and T2 TSE COR (TR/TE: 8280/80 ms; flip angle: 150°; FoV: 230 mm; in-plane matrix: 528 × 528; in-plane resolution: 0.2 × 0.2 mm^2; slice thickness: 3 mm; 40 coronal slices).

Recommended research protocol 3D T2w SE SAG (TR/TE: 3120/418 ms; flip angle: 120°; FoV: 256 mm; in-plane matrix: 256 × 256; in-plane resolution: 1 × 1 mm^2; slice thickness: 1 mm; 160 sagittal slices).

3.2.3 Fluid-Attenuated Inversion Recovery Imaging

Description Commonly known as T2-FLAIR or just FLAIR, this MRI technique was developed to suppress signals coming from fluids, unveiling potential lesions that might otherwise be obscured, especially within cortical and periventricular areas. Moreover, as they do not nullify the signal coming from blood, FLAIR images are also very useful for detecting microhemorrhages, edema and inflammation.

Background Patients with chronic liver failure often show increased signal in the hemispheric WM or around the corticospinal tracts in FLAIR sequences [19]. This elevated signal is likely attributed to mild cerebral edema [39]. The gradual normalization of signal following liver transplantation supports this theory (Fig. 4), as demyelination and axonal loss are irreversible processes

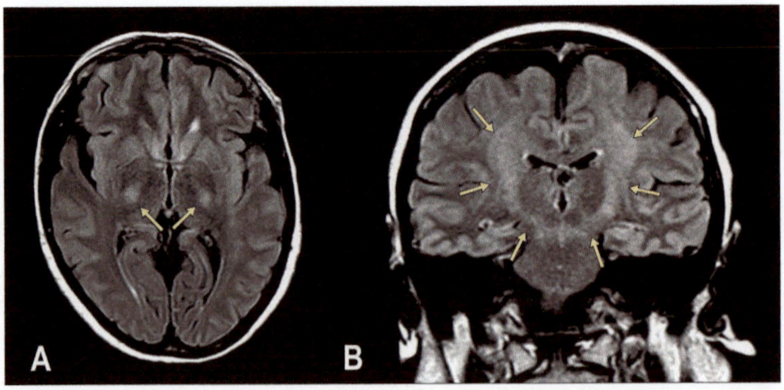

Fig. 4 Axial (**a**) and coronal (**b**) FLAIR images displaying signal hyperintensities along the corticospinal tracts

[40]. Cirrhotic patients also exhibit focal WM lesions, with or without evident HE [41]. These lesions resemble those observed in several forms of small-vessel cerebrovascular diseases (e.g., arteriosclerosis, amyloid angiopathy, or CADASIL) that are commonly seen in elderly population without liver disease. Given their small size, these lesions can sometimes be overlooked or mistaken for normal age-related changes or chronic ischemic lesions [42]. However, unlike lesions caused by microvascular disease, these lesions tend to decrease following improvements in HE or after liver transplantation [41] (Fig. 2b). Importantly, the reduction in these lesions closely correlates with improvements in neuropsychological function [43].

Standard clinical protocol FLAIR AX (TR/TE: 9000/90 ms; flip angle: 150°; FoV: 220 mm; in-plane matrix: 304 × 258; in-plane resolution: 0.4 × 0.4 mm^2; slice thickness: 3 mm; 40 axial slices).

Recommended research protocol 3D FLAIR SAG (TR/TE: 7000/388 ms; flip angle: 120°; FoV: 250 mm; in-plane matrix: 256 × 230; in-plane resolution: 1 × 1 mm^2; slice thickness: 1 mm; 176 sagittal slices).

3.2.4 Diffusion-Weighted Imaging

Description Diffusion-weighted imaging (DWI) is an MRI technique sensitive to water motion. This can be evaluated by measuring the signal intensity attenuation in relation to a diffusion sensitization factor called b-value. This parameter is influenced by the gyromagnetic ratio of water protons and the strength and timing of the diffusion-sensitizing gradients [44]. Thus, water motion can be captured along the axis of these gradients. Given that water motion is inherently non-isotropic within living tissues (it is hindered by biological barriers), the distance a water molecule travels in a particular direction varies depending on the local microstructure. Accordingly, water diffusion can be locally characterized through the application of multiple magnetic gradient directions and mathematically modeled to extract different parameters that can be exploited to assess microstructural integrity.

Background In cases of acute HE and acute-on-chronic hepatic failure, it is observed a decrease in the apparent diffusion coefficient (ADC), which is indicative of cytotoxic edema [45] (Fig. 5). In contrast, chronic HE patients exhibit an increase in ADC values, suggesting water accumulation in the extracellular space, among other alterations [46]. These findings challenges the hypothesis that astrocyte swelling is the primary cause of diffuse cerebral

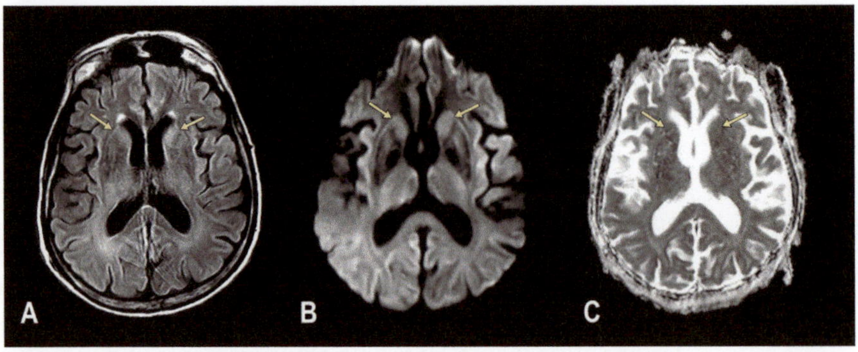

Fig. 5 Brain MRI scans of a patient with liver cirrhosis exhibiting bradyphrenia and disorientation. (**a**) Axial T2-weighted FLAIR image showing a bilateral and symmetric signal abnormality in the head of the caudate nuclei. (**b**) Diffusion-weighted image and (**c**) apparent diffusion coefficient showing a hyperintense signal in the caudate heads

edema in these patients [46]. Incremented ADC values have been predominantly documented in the thalamus, basal ganglia, and periventricular WM [14]. The precise mechanism underlying this increase is uncertain, but two of the most plausible explanations includes an increase in the permeability of the blood-brain barrier [15] or a decrease in the expression of genes encoding astrocyte proteins, such as GFAP (glial fibrillary acidic protein), owing to brain exposure to ammonium [45]. These proteins are an integral part of mature astrocytes, so decreased expression of GFAP induces changes in astrocyte morphology, favoring diffusivity in the extracellular space [46]. These increased ADC values correlate, as do alterations in FLAIR scans, with neuropsychological impairment and serum ammonium levels [45].

This kind of pathological alterations have also been described using Diffusion Tensor Imaging (DTI), a slightly more advanced diffusion MRI technique that takes into account water diffusion directionality [47]. Thus, DTI enables estimation of several biomarkers with greater sensitivity to pathological changes than ADC, like the fractional anisotropy (FA) or the mean diffusivity (MD), which have already been used to characterize WM alterations within multiple brain areas either in minimal and chronic HE [48–50]. Reduced FA and increased MD have been linked to low-grade brain edema [49, 50], elevated ammonium levels [51], and neurocognitive impairment [49]. More recently, advanced diffusion models like DKI (diffusion kurtosis imaging) [52] or NODDI (neurite orientation dispersion and density imaging) [53] have also been used to explore HE alterations. In contrast to classical DWI and DTI, DKI quantifies the degree of non-Gaussian water diffusion, allowing detection of subtle microstructural alterations in minimal HE [54–56]. DKI-based parameters have shown widespread microstructural abnormalities in multiple brain regions associated with cognitive function [56–58] and have also identified mild brain edema before neurocognitive decline, suggesting its potential for differentiating subjects with and without minimal HE [57]. On the other hand, NODDI is a multi-compartment diffusion model that differentiates between three water compartments: intracellular, extracellular, and free-water. Although not yet widely applied in HE, this model has proven useful for detecting white matter alterations in patients with minimal HE, showing an association between frontal lobe alterations and cognitive dysfunction [59].

Standard clinical protocol DWI single-shell (b-value = 1000 s/mm^2; gradient directions: 1; phase-encoding direction: AP; TR/TE: 5700/97 ms; flip angle: 90°; FoV: 220 mm; in-plane matrix: 192 × 192; in-plane resolution: 0.6 × 0.6 mm^2; 4 slice thickness: mm; 27 slices) with a b_0 scan (b-value = 0 s/mm^2) at the beginning of the sequence.

Recommended research protocol for DTI DWI single-shell (b-value = 1000 s/mm^2; gradient directions: 64; phase-encoding direction: AP; TR/TE: 10,000/102 ms; flip angle: 90°; FoV: 256 mm; in-plane matrix: 128 × 128; in-plane resolution: 2 × 2 mm^2; slice thickness: 2 mm; 24 slices) with 4 b_0 scans (b-value = 0 s/mm^2) at the beginning of the sequence. Four additional b_0 scans with reverse phase-encoding direction (PA) are recommended in order to correct EPI distortions.

Recommended research protocol for DKI or NODDI DWI multi-shell (b-values = 400/1000/2000 s/mm^2; gradient directions: 20/64/64; phase-encoding direction: AP; TR/TE: 10,000/102 ms; flip angle: 90°; FoV: 256 mm; in-plane matrix: 128 × 128; in-plane resolution: 2 × 2 mm^2; slice thickness: 2 mm; 24 slices) with 4 b_0 scans at the beginning of the sequence. Four additional b_0 scans with reverse phase-encoding direction (PA) are recommended in order to correct EPI distortions.

3.2.5 Susceptibility-Weighted Imaging

Description Susceptibility-weighted imaging (SWI) is an advanced MRI technique that enhances the visualization of structures with different magnetic properties. This modality incorporates phase information from a T2* signal to derive supplementary data that allow estimation of the phase shift caused by local tissues. Given that paramagnetic and diamagnetic substances respectively induce positive and negative phase shifts, phase images can be used to extract a mask to enhance the contrast of the corresponding magnitude image, weighting it according to tissue susceptibilities. SWI is suitable for detecting small veins, microhemorrhages,

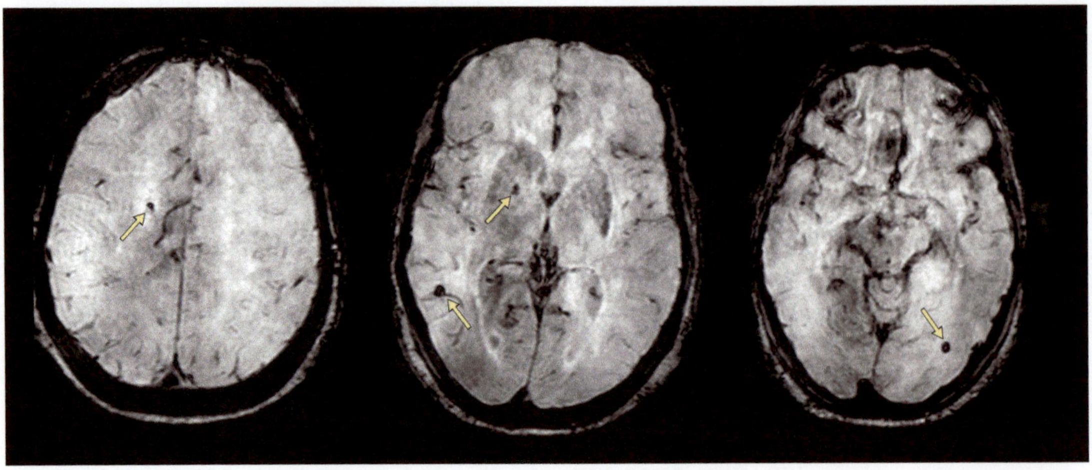

Fig. 6 Axial susceptibility-weighted scans showing bilateral microbleeds in a patient with chronic hepatic encephalopathy. These alterations are shown as punctate foci regions of low signal in bilateral periventricular white matter and right basal ganglia

calcifications, and the accumulation of paramagnetic or diamagnetic elements.

Background Hypointense punctate foci have been observed in approximately half of the patients with acute HE, probably corresponding to microbleeds [60]. These foci are more commonly found in the periventricular WM, cortical gray matter, and subcortical WM [60] (Fig. 6), although they do not consistently correspond with signal alterations in DWI or FLAIR acquisitions [60], nor with clinical severity [60]. It is hypothesized that these microbleeds may arise from an underlying coagulopathy, vasculopathy, or dysfunction of the blood-brain barrier due to endothelial damage, possibly exacerbated by disrupted permeability in hyperammonemic states [61]. Elevated ferritin levels, typically observed in acute liver failure, could also contribute to microbleed pathogenesis [60, 62].

Standard clinical protocol SWI AX (TR/TE: 28/20 ms; flip angle: 15°; FoV: 220 mm; in-plane matrix: 384 × 292; in-plane resolution: 0.3 × 0.3 mm^2; slice thickness: 2.5 mm; 56 axial slices).

Recommended research protocol 3D SWI AX (TR/TE: 57.4/23.2 ms; flip angle: 30°; FoV: 220 mm; in-plane matrix: 512 × 512; in-plane resolution: 0.43 × 0.43 mm^2; slice thickness: 2 mm; 136 axial slices).

3.2.6 Magnetic Resonance Spectroscopy

Description Magnetic Resonance spectroscopy (MRS) is a non-invasive technique applied to assess metabolic and biochemical alterations in the brain. Grounded in the principle of chemical shift, MRS generates a spectral plot wherein frequencies correspond to specific metabolites, and the area under the curve reflects their density. Since this spectrum is relative, a reference point is essential for calibration. Then, a distinct peak associated with a well-known metabolite is typically selected for this purpose. Expressed in parts per million (ppm), this reference enables differentiation of signals from other metabolites, facilitating the assessment of metabolite concentrations within a selected region. While MRS can analyze diverse chemical elements, in humans is primarily focused on hydrogen due to its abundance in the body. This allows evaluation of key metabolites like lactate, choline, glutamine/glutamate, or N-acetylaspartate. Therefore, by measuring changes in these metabolite concentrations, MRS facilitates detection of alterations in cellular metabolism, neuronal integrity, and biochemical processes linked to different neurological and psychiatric conditions.

Background Since MRS studies in HE have consistently reported a normal N-acetylaspartate peak, a marker for neuronal integrity, it is usually used as a reference, setting it at 2.02 ppm [63, 64]. HE patients typically exhibit the following metabolite signature (Fig. 7):

1. Increased glutamine/glutamate signal: These metabolites resonate at 2.2–2.4 ppm and 3.6–3.8 ppm [65]. Elevated levels of glutamine and glutamate are indicative of increased brain ammonium levels [66]. Glutamine accumulates within astrocytes as part of the process of ammonium detoxification mediated by glutamine synthetase [67]. This feature is observed in both acute and chronic HE [15].

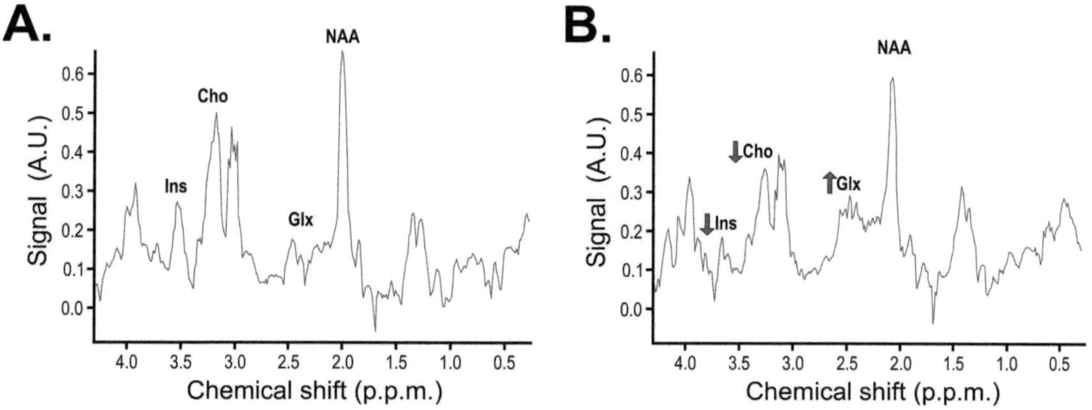

Fig. 7 Illustrative example of the differences in the magnetic resonance spectroscopy spectra within the thalamus between a normal control (**a**) and a cirrhotic patient (**b**). In here can be observed a reduction in the myoinositol (Ins) and choline (Cho) signal, together with an increased glutamine/glutamate

2. Decreased myoinositol signal: Typically resonates at 3.61 ppm [65]. Myoinositol functions as an intracellular osmolyte in the brain [68]. A decrease in myoinositol levels is interpreted as a compensatory response to the increased osmolarity within astrocytes [69].
3. Decreased choline signal: Resonates at 3.2 ppm [65]. Choline is a component of membrane phospholipids and is also found in myelin degradation products. Changes in choline levels can reflect alterations in membrane turnover and myelin sheath integrity [70].

This triad is considered a good indicator of neuropsychiatric dysfunction in HE patients, showing a strong correlation with the severity of HE [71]. However, similar MRS alterations have also been observed in cirrhotic patients who do not exhibit apparent signs of brain dysfunction [66, 72, 73]. Spectroscopic changes resolve following liver transplantation or shunt occlusion, preceding signal normalization within globus pallidus in T1w scans [74, 75]. The choline peak normalizes first [74, 75], followed by the glutamine/glutamate peak which usually reaches a normal value within 1–2 months [74, 75]. The myoinositol peak normalizes more slowly and may take 3–7 months to reach normal values [74, 75]. This normalization of the MRS signal correlates with neurological improvement of HE manifestations [76, 77]. For this reason, it has been postulated as an indicator of treatment response.

Standard clinical protocol Non-standard acquisition but recommended.

Recommended research protocol Single-voxel point-resolved ^{1}H-MRS with water suppression (TR/TE: 2000/30 ms; flip angle: 90°; FoV: mm; in-plane matrix: 1 × 1; in-plane resolution: 20 × 20 mm^2; slice thickness: 20 mm; 1 slice) guided with a T1w TSE SAG (TR/TE: 450/7.3 ms; flip angle: 90°; FoV: 256 mm; in-plane matrix: 256 × 256; in-plane resolution: 1 × 1 mm^2; slice thickness: 1 mm; 192 sagittal slices).

3.2.7 Perfusion Imaging

Description Perfusion imaging refers to a variety of techniques designed to quantitatively estimate blood flow or its perfusion to tissues. Numerous methods are currently available to assess cerebral blood flow such as contrast-enhanced CT, molecular imaging (like SPECT or PET), and MRI (with or without contrast enhancement). However, given that many HE patients may exhibit renal insufficiency [5], perfusion techniques are not usually indicated since they typically involve injections of radioactive tracers or contrast agents. Nonetheless, alternatives such as arterial spin labeling (ASL) MRI emerge as promising options. ASL employs radiofrequency pulses to magnetically label inflowing arterial blood water, using it as endogenous contrast. Thus, ASL is a Gd-free MRI technique that does not require exogenous contrast agents, making it safer for assessing hemodynamics in patients with HE.

Background Not many studies have evaluated HE using perfusion techniques and reports are heterogeneous, probably due to the variability of used techniques and degree in HE [78]. Focusing on ASL, an increase in cerebral blood flow in minimal HE patients has been reported within the basal ganglia and thalami compared to normal controls [79, 80], feature that has been correlated with reduced performance in neuropsychological tests [80]. Unfortunately, there is limited evidence, so further studies are still needed.

Standard clinical protocol Non-standard acquisition.

Recommended research protocol Still to be defined. Example protocol obtained from [79]: Pulsed ASL (TR/TE: 2500/11 ms; TI1/TI2 = 700/1800 ms; flip angle: 90°; FoV: 256 mm; in-plane matrix: 64 × 64; in-plane resolution: 4 × 4 mm^2; slice thickness: 8 mm; 11 slices).

3.2.8 Functional Magnetic Resonance Imaging

Description Functional magnetic resonance imaging (fMRI) utilizes oxygenated blood as an endogenous marker to indirectly assess metabolic activity. This method is based on two principles: (1) When neurons activate, they raise their metabolic requirements, thus demanding more oxygen (2). Oxygen-rich arterial blood exhibits different electromagnetic properties compared to oxygen-poor venous blood. Consequently, an increased flow of oxygenated blood induces small fluctuations in the electromagnetic signal, referred as BOLD (Blood Oxygenation Level Dependence) signal, that can be captured using MRI. Taking this into account, there are two main approaches for fMRI studies: resting-state (rs-fMRI) and task-based (tb-fMRI). During rs-fMRI studies, subjects remain still without performing any task or receiving any stimulus. This facilitates the analysis of potential alterations or abnormal patterns in brain connectivity, i.e., the loss of synchrony between functionally connected brain regions that normally exhibit highly correlated BOLD signal activity. On the other hand, during tb-fMRI studies, the subject performs specific actions or is exposed to diverse stimuli during event-related or block-design paradigms. This allows identification of those brain regions associated with the performed task or the detection of any dysfunction or abnormal pattern in brain activation.

Background fMRI is widely available and can be utilized to examine changes in brain function across numerous neurological and psychiatric disorders, including HE. Owing to its primary role on numerous basic brain functionalities [81], most rs-fMRI studies have focused on the default-mode network (DMN). Among them, the most widely used metric is the functional connectivity (FC), which assesses the temporal dependencies between remote brain regions. This approach has revealed a significant disruption in multiple brain nodes associated with the DMN [82–84], some of which have already been linked to the hyperammonemia [85] or the neurocognitive deficit typically exhibited by HE patients [86–88]. Moreover, since FC progressively decreases as minimal HE begins to develop [83, 89], it has been proposed as a potential biomarker to differentiate minimal HE from non-HE patients [90]. Although less studied, other neural networks exhibiting functional alterations in HE are the salience network [84, 87], the central executive network [87], and the basal ganglia-thalamocortical circuits [84, 86, 91, 92]. On the other hand, tb-fMRI is much less widespread in HE. Nevertheless, this kind of studies allow identification of neurocognitive alterations like the impaired coupling between visual judgment areas [93] or changes in spatial working memory [94]. Finally, both rs-fMRI and tb-fMRI have been applied to assess the recovery of brain activation after liver transplantation [95, 96] or the administration of therapeutic drugs [97, 98].

Standard clinical protocol Non-standard acquisition.

Recommended research protocol for rs-fMRI 2D T2* single-shot EPI (TR/TE: 2400/30 ms; flip angle: 75–90° (need to be optimized for the BOLD signal); FoV: 192 mm; in-plane matrix: 64 × 64; in-plane resolution: 3 × 3 mm^2; slice thickness: 3 mm; 40 slices; 250 volumes) and an auxiliar fieldmap extracted from two 2D GRE scans (TR: 455 ms, TE1/TE2: 4.92/7.38 ms; flip angle: 60°; FoV: 192 mm; in-plane matrix: 64 × 64; in-plane resolution: 3 × 3 mm^2; slice thickness: 3 mm; 40 slices).

Recommend research protocol for tb-fMRI Acquisition protocols in tb-fMRI are highly dependent on the task being studied. While the basic setup is similar to that used in rs-fMRI, the TRs must be optimized for the task and typically require synchronization with the task triggers. The number of volumes also strongly varies according to the duration of the study.

3.3 Preprocessing Methods

Image preprocessing is a crucial stage for researchers and practitioners in order to refine imaging data for more accurate and reliable analysis. Preprocessing workflow may vary according to image modality, the quality of the scans, and the intended analysis, but overall, preprocessing aims to enhance imaging contrast, improve SNR, mitigate acquisition artifacts, and standardize images across datasets. These efforts collectively boost the performance of analytical methodologies in medical imaging studies, improving their reliability and sensibility to pathological alterations. In this section, we will examine several preprocessing procedures commonly utilized in neuroimaging. Before moving on, please check **Notes 1** and **2**.

3.3.1 Quality Control

Before any kind of preprocessing, it is advisable to routinely inspect your imaging data as soon as possible, ideally immediately after acquisition. Prioritizing this practice is cautious, as it allows early identification of potential problems or anomalies in your imaging studies. In case of data abnormalities, it is more advantageous to allocate time for immediate correction or re-acquisition as needed, rather than dealing with data inconsistencies post-analysis that may force you to dismiss your data without opportunity for reacquisition.

Quality control stages aim to ensure that images have enough quality, exhibit good contrast, and are free from significant artifacts. This task is usually performed manually by reviewing images individually using an ortho-viewer (e.g., fsleyes, mricron, mrview, mango...). However, this may not be feasible when large cohorts are studied. If this is your case, consider using packages such as *eddyqc* or *MRIQC* to generate reports that can be reviewed much faster.

3.3.2 Database Structure

The adoption of BIDS (Brain Imaging Data Structure) or BIDS-like standards is highly recommended when organizing your imaging database. This approach guarantees that all your data is stored in a rigidly structured format, thereby simplifying its localization. In addition, it also streamlines data exchange between researchers due to its intuitive nature and easy adoption. Given that some neuroimaging software has embraced this standard, its implementation may be mandatory. For more information, visit *https://bids.neuroimaging.io*.

3.3.3 General Preprocessing for Structural MRI

Before moving forward, please note that all commands described in this section are presumed to be executed in a Linux or MacOS terminal window. If you have doubts about how to execute any of them, check **Note 3**. For illustrative purposes and because it is a terminal scripting feature (it relies on the Bash programming language), the symbol "\" will be used as a line continuation character. This symbol allows you to split a long command into multiple lines for better readability without affecting the command's execution.

Transformation from DICOM to NIFTI Format

Most medical imaging devices export imaging data using the DICOM (Digital Imaging and Communication In Medicine) standard. However, working with this format can be cumbersome, as it often fragments volumetric information into multiple files. Consequently, it is common practice within the neuroimaging community to utilize the NIFTI (Neuroimaging Informatics Technology Initiative) standard. This format condenses volumetric information into a single file, making it more manageable. There are several packages available to perform this transformation (e.g., *dcm2nii, mrconvert* or *mri_convert*), but our preferred choice is *dcm2niix* [6]. Aside from the image file (*.nii*), this package also generates a JSON text file containing metadata that is usually lost during format conversion with other packages. An example illustrating its application would be:

```
dcm2niix -o output_directoty -z o input_directoty
```

where "*output_directory*" should be an existing directory and the option "*-z o*" denotes that the output file should be compressed in gzip format to save storage space. If the flag "*-o output_directory*" is not provided, output files will be generated within the input directory.

Image Reorientation

Sometimes images may not be correctly oriented, either due to difficulties during patient positioning or because of data corruption. In such scenarios, the *fslreorient2std* package can be applied to reorient the image. This tool, integrated in FSL, aligns the image according to the standard MNI152 by applying rotations of 90, 180 or 270°:

```
fslreorient2std image_in.nii.gz image_reoriented.nii.gz
```

Optimize Field of View (Optional)

Frequently, MRI images are acquired with an excessively wide field of view (FoV), covering irrelevant background regions. In this situation, the *robustfov* utility from FSL can be employed to automatically crop the FoV, restricting it just to the brain:

```
robustfov -i image_in.nii.gz -r image_cropped.nii.gz
```

If the FoV remains too large, it can be further narrowed using *fslroi*:

```
fslroi image_in.nii.gz image_cropped.nii.gz \
xmin xsize ymin ysize zmin zsize
```

where "*xmin*," "*ymin*," and "*zmin*" respectively denote the coordinates along the x, y, and z axes from which the image will be cropped, and "*xsize*," "*ysize*," and "*zsize*" indicate the dimensions of the cropping along these axes. These coordinates can be manually checked by opening the image with *fsleyes* or any other ortho-viewer.

Denoising (Optional)

While structural brain images are not typically denoised, including this stage could be considered in some situations: (1) If the image contains significant noise, denoising may improve image quality and clarity. However, if the SNR is too low, it may remove a significant fraction of the signal along with noise (2). Denoising may also be used to enhance visualization when noise obscures important anatomical or pathological details, making subtle features more discernible (3). Finally, denoising could help mitigate small artifacts derived from motion or scanner-related distortions. In any case, denoising is contraindicated when intensity values are quantified within small regions of interest (ROIs) or when lesion boundaries need to be delimited. A good noise estimation can be computed with *DenoiseImage* [99] (Fig. 8), included in ANTs:

```
DenoiseImage -d 3 -i image_in.nii.gz \
-o [image_corrected.nii.gz, noise_estimation.nii.gz]
```

Bias Field Correction

Bias field is a low-frequency non-uniformity that is present in all MRI images, introducing intensity gradients that are unrelated to the tissue being scanned. Several factors can potentially introduce it, such as non-uniformities in the coil sensitivity, coil defects, or magnetic field inhomogeneities. Although this signal may be so subtle as to be visually imperceptible, it can substantially alter any quantitative metric or automatic segmentation extracted from the image. Bias field correction is therefore an essential preprocessing step in almost every MRI acquisition. This field can be subtracted using the *N4BiasFieldCorrection* [100] command included in ANTs (Fig. 8):

```
N4BiasFieldCorrection -d 3 -i image_in.nii.gz \
-o [image_corrected.nii.gz, biasfield_estimation.nii.gz]
```

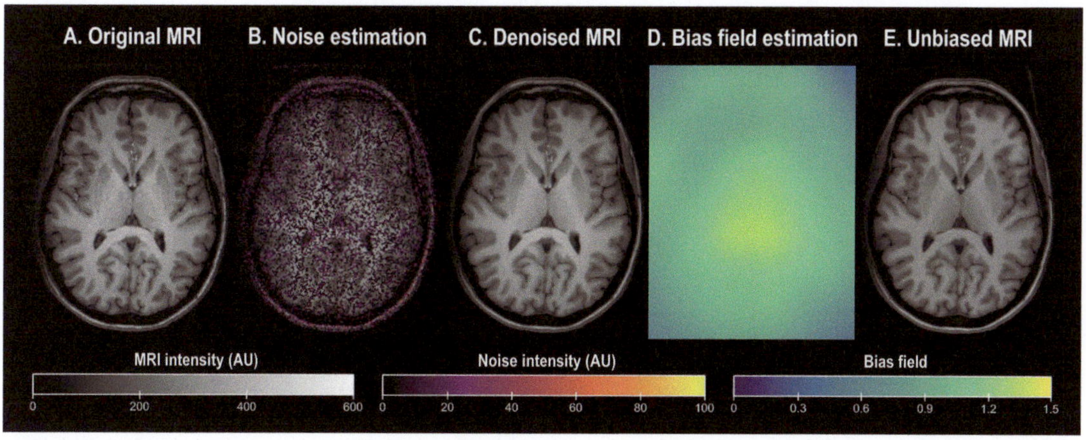

Fig. 8 Main preprocessing workflow for structural MRI images. In the same axial plane, from left to right are displayed: (**a**) The original MRI. (**b**) Noise estimation with "DenoiseImage" (ANTs). (**c**) The original MRI after denoising. (**d**) Bias field estimation using the "N4BiasFieldCorrection" utility from ANTs. (**e**) Structural MRI after denoising and bias field correction. AU Arbitrary units

Non-brain Tissue Removal

The elimination of non-brain tissue is a common step in MRI analyses. Visually, it helps focus attention on brain tissue and speeds up visualization through ortho-viewers, avoiding manually adjusting intensity clipping ranges by eliminating artifacts that may be present within extracerebral regions. In addition, by reducing the number of non-zero voxels, it accelerates subsequent processing stages. Brain extraction can be easily carried out in FSL with *BET* [101]:

```
bet image_in.nii.gz brain_extracted.nii.gz -f thresh -m
```

where the option "*-m*" generates a binary mask (Fig. 9) of the brain named "*brain_extraded_mask.nii.gz*" and "*thresh*" is the fractional intensity threshold, ranging from 0 to 1. The lower this value, the larger the estimated brain outlines will be. Even though keeping only isolated brain images may save disk space, we recommend also preserving whole head images (see subheading "Affine Registration").

3.3.4 Co-registration Between Image Spaces

Co-registration between medical images is a key step for multiple image analyses, providing a spatial matching among images. This is particularly useful for:

1. Monitoring patient progression: Aligning images from different imaging sessions allows for easier comparison of the progression of lesions or pathological features over time.
2. Establishing a spatial correlation between functional features and the anatomy. Functional imaging modalities, like PET, SPECT, or fMRI lack anatomical references. Therefore, aligning functional images with high-quality structural images assists in establishing the spatial correspondence between functional features and brain structure.

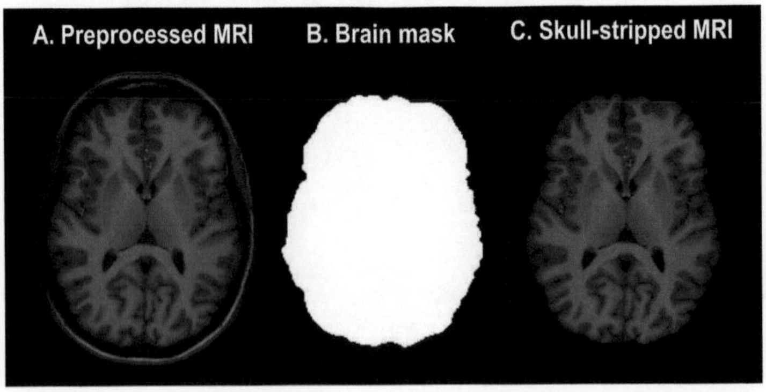

Fig. 9 Brain extraction stage. (**a**) Preprocessed image. (**b**) Estimated brain mask with "bet" (FSL). (**c**) Skull-stripped volume

3. Spatial correlation among pathological features. Each imaging modality is sensitive to specific features, making imaging alignment essential for establishing correlations among pathological features and/or lesions.
4. Transport imaging segmentations from one space to another. Some features or structures may be more evident in certain modalities than in others. Thus, defining ROI segmentations in one modality and subsequently transforming them to others may be beneficial for evaluation purposes.

In this section, we will illustrate registration steps using ANTs, which are applicable for all neuroimaging modalities.

Affine Registration

This linear technique is designed to align two images by a combination of translation, rotation, scaling, and shearing operations. This technique does not apply any kind of local deformation, maintaining the overall shape and orientation of the structures. Affine registrations are computationally efficient and commonly used to align images acquired from the same subject. When several modalities and/or sessions are studied, it is recommended to register all images to a common reference image, such as a baseline T1w scan, which acts as a common link for all transformations. Affine coregistration between images can be achieved using *antsRegistration* [102]:

```
antsRegistration --dimensionality 3 --interpolation Linear \
--metric MI[ref_image.nii.gz,moving_image.nii.gz, \
 1,32,Regular,0.15] \
--output [transf_matrix,transformed_image.nii.gz] \
--transformation Rigid[0.1]
```

This illustrates the computation of a three-dimensional rigid-body transformation, i.e., with no scaling or shearing, using a mutual information strategy. Here, "*ref_image.nii.gz*" denotes the reference image, "*moving_image.nii.gz*" the image undergoing transformation, "*transformed_image.nii.gz*" the output following transformation, and "*transf_matrix*" the file containing the affine transformation matrix required to align both images in the reference image space. This file should be defined only as a filename (e.g., "outname_"). The output file will be named as "outname_0GenericAffine.mat." Once coregistration has been computed, you must verify its accuracy. Successful registration is not guaranteed, especially when significantly different images are trying to be matched. If coregistration fails, please check **Note 4**.

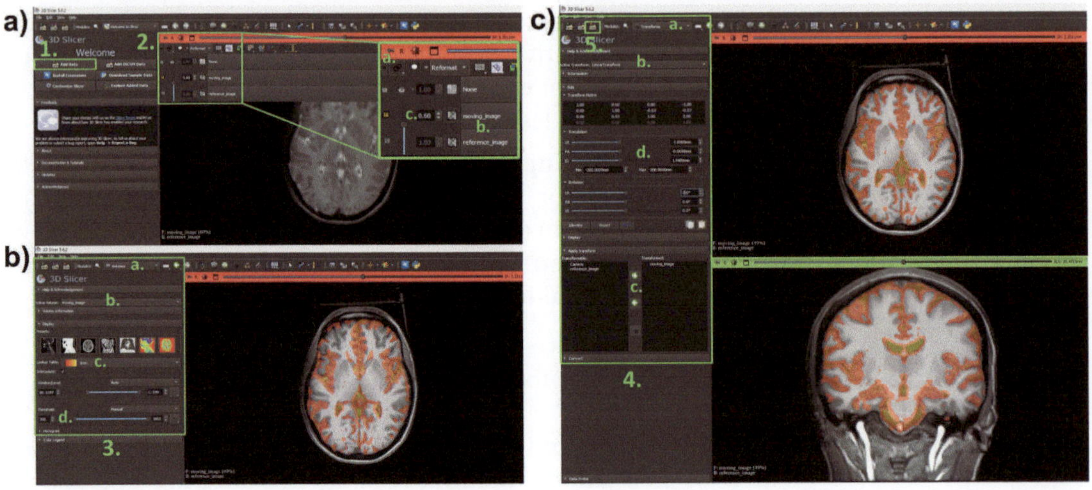

Fig. 10 Snapshots illustrating the different steps required to complete a manual coregistration between two images. (**a**) Stages 1 and 2; (**b**) Stage 3; (**c**) Stages 4 and 5

Coregistration can also be performed manually as a last resort when automatic methods fail. This can be done following the next steps in 3DSlicer [9] (Fig. 10):

1. Open both reference and moving images: "Add Data → Choose File(s) to Add."
2. Display both images:
   (a) Link slice views: "Thumbtack icon → Two-ring icon."
   (b) Select your reference image in the first dropdown menu (below).
   (c) Select your moving image in the second dropdown menu (middle) and turn the opacity to 0.6.
3. Adjust the moving image contrast and colormap in order to visualize both images at the same time:
   (a) Select "Volumes" module.
   (b) Change active volume to the moving image.
   (c) Change moving image colormap (Active volume: Moving image; Lookup Table: Iron).
   (d) Adjust the intensity threshold to display both images by removing the lowest values.
4. Register both images:
   (a) Select "Transforms" module.
   (b) Create a new linear transform (Active Transform: Create new LinearTransform).

   (c) Set your moving image from Transformable to Transformed (Apply transform).
   (d) Modify "Translation" and "Rotation" options until both images are registered. You can adjust contrast settings at any time to better visualize the coregistration process. Move the cursor to the corresponding slice viewer and scroll the mouse wheel to move the display across the selected axis.
5. Store your transformation matrix:
   (a) Name the transformation matrix.
   (b) Change the output file to .mat format.
   (c) Select your storing directory.
   (d) Press "Save."

Diffeomorphic Registration

Diffeomorphic registrations are nonlinear methods that integrate affine registrations along with local deformations to fully establish spatial correspondence between transformed and reference images. These transformations are commonly employed to register images to standard spaces like the MNI152 template [103] (Fig. 11), facilitating population-level analyses in the same space (e.g., voxel-based morphometry). This type of transformation can be done with the command *antsRegistrationSyN.sh* [104] (check **Note 5** if you want to boost computing performance):

```
antsRegistrationSyN.sh -d 3 -f ref_image.nii.gz \
-m oving_image.nii.gz -o outname_
```

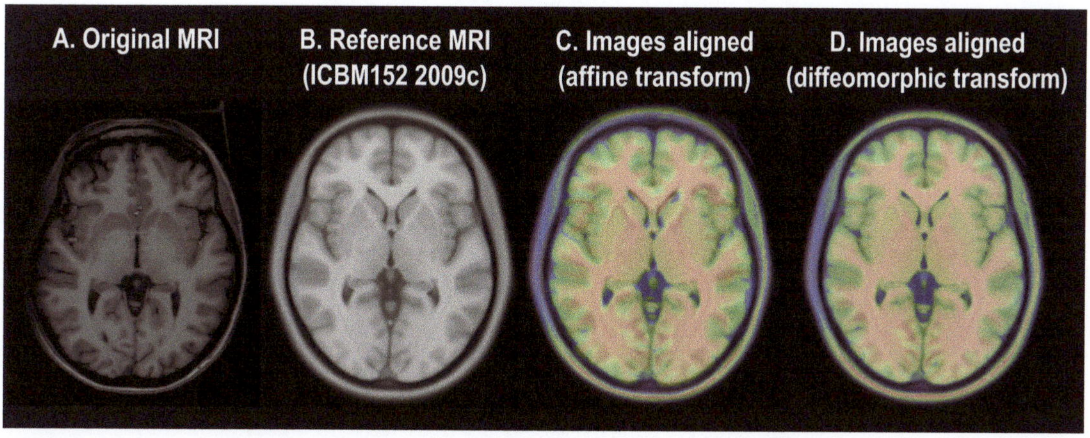

Fig. 11 Differences between affine and diffeomorphic registration strategies. (**a**) Original moving image. (**b**) Fixed reference image. (**c**) Image overlay after alignment using an affine registration strategy. The mismatch between both images is clearly observed in CSF (blue in the moving image and black in the reference one) and GM (green and gray respectively) regions. (**d**) Image overlay after alignment using a diffeomorphic registration strategy. No mismatch between images is observed in this case

This command computes the transformation in two stages: first, it applies an affine transformation to line up both images and second, it warps the moving image to fully match the reference one. Several output files are then generated:

1. "*outname_0GenericAffine.mat*": Affine transformation to ref_-image.nii.gz.
2. "*outname_1Warp.nii.gz*": Warp applied to match ref_image.nii.gz once the affine transformation has been applied.
3. "*outname_1InverseWarp.nii.gz*": Inverse warp transformation of the previous one.
4. "*outname_1Warped.nii.gz*": Moving image transformed into reference image space.
5. "*outname_1InverseWarped.nii.gz*": Reference image transformed into moving image space.

Applying Previously Computed Transformations

Once all registrations have been computed, any image can be transferred to other spaces by sequentially applying a few transformation matrices with *antsApplyTransforms*. Please note that all transformations should be concatenated in reverse order to keep algebraic consistency. Below are some illustrative examples for clarity:

Example 1: Transform a FLAIR Image into MNI152 Template Space:

```
antsApplyTransforms -d 3 -n Linear -i FLAIR.nii.gz \
-r T1wtemplate_MNI152.nii.gz -o FLAIR_toMNI152.nii.gz \
-t warp_fromT1wBL_toMNI152.nii.gz \
-t affine_fromT1wBL_toMNI152.nii.gz \
-t affine_fromFLAIR_toT1wBL.nii.gz
```

Here, "*-i*" is the input image, "*-r*" the reference one, "*-o*" the output, and "*-t*" the transformation to be applied. Additionally, "*-n*" denotes the interpolation strategy used to resample our FLAIR image after transforming it to the reference space. The rationale behind these transformations would be as follows: (1) Transformation from FLAIR native space to T1wBL space (our nexus); (2) Affine transformation from T1wBL space to MNI152 space; (3) Warp from T1wBL space to MNI152 space.

Example 2: Transform It Back to the Original Space:

```
antsApplyTransforms -d 3 -n Linear -i FLAIR_toMNI152.nii.gz \
-r FLAIR.nii.gz -o FLAIRback_toFLAIR.nii.gz \
-t [affine_fromFLAIR_toT1wBL.nii.gz,1] \
-t [affine_fromT1wBL_toMNI152.nii.gz,1] \
-t invwarp_fromT1wBL_toMNI152.nii.gz
```

Here, "*-t [transform_sel,1]*" denotes that the transformation to be applied will be the inverse. This applies only to affine transformations as inverse warps should be applied as a direct transformation. The rationale behind this example is (1) inverse warp from MNI152 space to T1wBL space; (2) inverse affine transformation from MNI152 space to T1wBL space; and (3) inverse affine transformation from T1wBL space to FLAIR space.

Example 3: Transform This FLAIR Image to Match a Second FLAIR Acquisition Obtained During a Different Imaging Session

```
antsApplyTransforms -d 3 -n Linear -i FLAIR_session1.nii.gz \
-r FLAIR_session2.nii.gz \
-o FLAIR_session1_toFLAIRsession2.nii.gz \
-t affine_fromFLAIRsession1_toT1wBL.nii.gz \
-t [affine_fromFLAIRsession2_toT1wBL.nii.gz,1]
```

Here the logic is simple: (1) Transformation from FLAIR native space to T1wBL space; (2) Inverse transformation from T1wBL space to the second FLAIR space.

3.3.5 Segmentation of Regions of Interest

Segmentations play a pivotal role in many imaging studies, especially those focused on ROIs, like lesions, tumors, or brain structures. Depending on segmentation's nature, it can be executed through different methodologies:

Manual Segmentation

This is the most basic methodology. In this approach, the researcher delimits the ROI voxel by voxel on one or several aligned images. This process can be carried out with different packages (fsleyes, 3DSlicer, ITK-SNAP, . . .), although the underlying workflow remains consistent across all of them. Below we outline the pipeline for manual segmentations using *fsleyes* (Fig. 12):

1. Open your reference image with *fsleyes.* In the terminal, run "*fsleyes image_in.nii.gz.*"
2. Activate edit mode: "Tools → Edit Mode." This can also be done by pressing "Alt+E".
3. Create a new image: "File → New Image." Alternatively, "Ctrl + Shift + N." Press then Ok. The new image will acquire the same metadata as the one used to define the segmentation.
4. Set drawing configuration:
   (a) Select a 2D (square) or a 3D (cuboid) drawing box.
   (b) Define the size, in voxels, of this drawing box (Selection size).

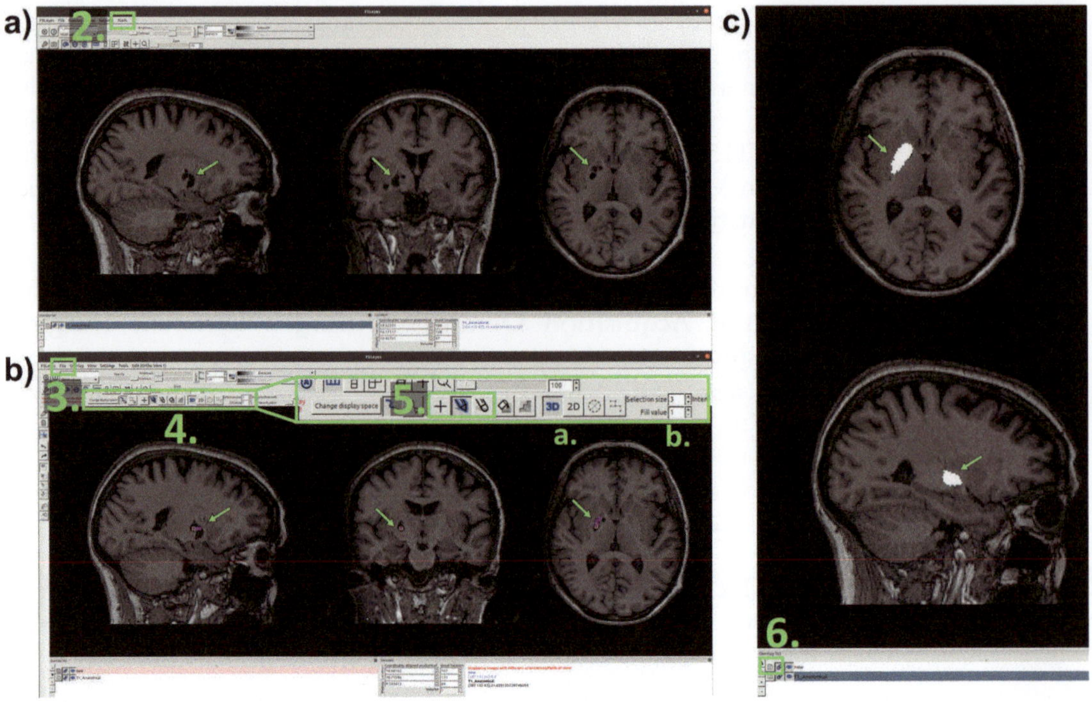

Fig. 12 Manual segmentation workflow with fsleyes. This subject exhibits a lesion in the globus pallidus, which makes automatic segmentation of this structure with FIRST impossible. (**a**) Illustration of the lesion in the globus pallidus (green arrows) and stage 2 of the segmentation process. (**b**) Stages 3, 4, and 5 of the segmentation pipeline. (**c**) Final segmentation of the globus pallidus and location of the icon to save the segmentation (stage 6)

5. Use "pencil" or "erase" modes to manually define your segmentation and "navigate" mode (crosshair icon) to move between imaging planes.
6. Store your segmentation: Press the floppy disk icon near to "new" in the "overlay list" (lower left corner). Then, select the directory and the name under which your manual segmentation will be stored.

FIRST-Based Segmentation

FSL incorporates a useful package that enables automatic segmentation of several deep brain nuclei of interest, including the accumbens nucleus, amygdala, caudate, globus pallidus, hippocampus, putamen, and thalamus. If you want to study any of these structures, FIRST-based segmentations are optimal, as they are suited to the morphology of the brain under investigation. These segmentations can be derived from any structural T1w image using the following command (Fig. 13):

```
run_first_all -i T1w.nii.gz -o out_directory/output_name
```

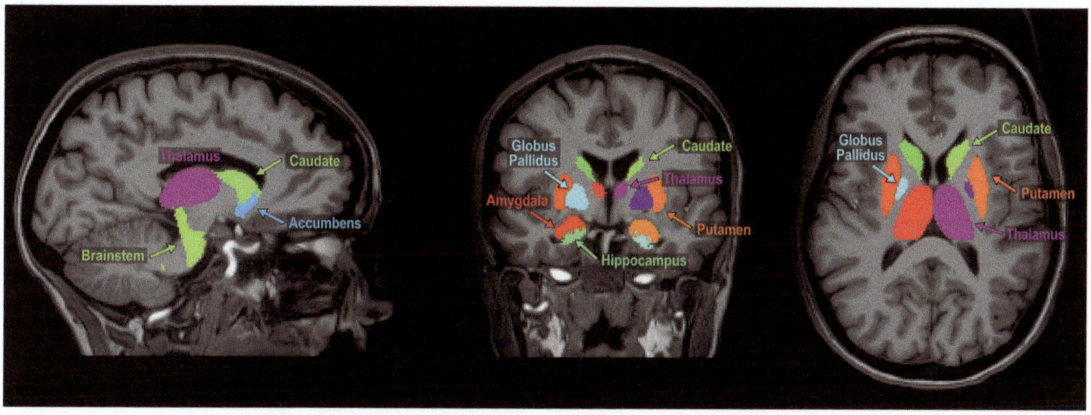

Fig. 13 Automated segmentation of subcortical structures from a T1w image using "run_first_all' (FSL)

Table 1
Brain regions that can be segmented with FIRST

	Assigned value		Code name	
ROI	**L**	**R**	**L**	**R**
Accumbens	26	58	L_Accu	R_Accu
Amygdala	18	54	L_Amyg	R_Amyg
Brainstem	16		BrStem	
Caudate	11	50	L_Caud	R_Caud
Globus Pallidus	13	52	L_Pall	R_Pall
Hippocampus	17	53	L_Hipp	R_Hipp
Putamen	12	51	L_Puta	R_Puta
Thalamus	10	49	L_Thal	R_Thal

L Left hemisphere; *R* Right hemisphere

The main output file containing all segmentations will be allocated as "*output_directory/output_name_all_fast_firstseg.nii.gz.*" Within this file, every segmentation will be assigned with an index that can be used to locate and isolate each individual structure. These indexes are listed in Table 1. In addition, although of lesser interest, the output directory will also contain numerous .vtk files containing the meshes of the segmented nuclei, and several .bvars files with the mode parameters and the model used for the segmentation. Structures can also be individually segmented using the "-*s struct_name*" option (input structure names can be found in Table 1). The output file will then be named "*output_struct_name_corr.nii.gz.*"

Sometimes, either because of poor data quality, artifacts or image anomalies, automatic segmentation of these nuclei may fail. If this happens, check **Note 6**. Structures that cannot be segmented using this method can alternatively be extracted manually, as previously described, or by using an atlas-based approach (see below).

Atlas-Based Segmentation

Creating customized segmentations of brain structures for the subject under examination is the optimal scenario, but not always feasible. In such instances, we can take advantage of different MRI-based atlases that are publicly available. Most of these atlases are defined in MNI152 standard space, allowing a straightforward transformation to our image spaces (see subheading "Applying Previously Computed Transformations"). While these segmentations might not perfectly match the morphology of the ROIs once they are transformed into subject's space, they provide a robust and reliable inter-subject allocation for your population-based analyses. A large compilation of MRI-based atlases can be found on the Lead-DBS project [105] website:

https://www.lead-dbs.org/helpsupport/knowledge-base/atlasesresources/

3.3.6 Preprocessing for Diffusion-Weighted Imaging

DWI preprocessing can be conducted using FSL and ANTs functionalities. However, this entails multiple steps that may prove challenging to set up, especially for non-proficient users. Hence, we recommend MRtrix3 [10], which offers a more user-friendly configuration for this kind of preprocessing, along with some additional functionalities not available in FSL and ANTs.

Transformation from DICOM to .mif Format

As mentioned above (subheading "Transformation from DICOM to NIFTI Format"), MRI scans are typically exported in DICOM format, which may be inconvenient to work with. To manage DWI data effectively, MRtrix3 utilizes the .mif file format, consolidating all diffusion data into a single file. DICOM to .mif conversion can be easily performed with *mrconvert*:

```
mrconvert dicom_dwi_directory output_directory/dwi.mif
```

Data already stored in NIFTI format can also be converted to . mif extension using *mrconvert*:

```
mrconvert dwi.nii.gz -fslgrad dwi.bvec dwi.bval /
-json_import dwi.json dwi.mif
```

Besides the usual .nii.gz and .json files, it's important to note that DWI files in NIFTI format also include two additional text files with .bval and .bvec extensions. These files respectively represent the acquisition b-values and magnetic gradient directions.

If you are dealing with multi-shell data, an additional step is necessary. Typically, these data are acquired or exported as multiple single-shell acquisitions with different b-value, resulting in multiple directories. In this case, just individually transform each DICOM directory to .mif format as indicated above, and then concatenate your .mif files with the *dwicat* command:

```
dwicat dwi_b1.mif dwi_b2.mif ... dwi_bn.mif dwi_merged.mif
```

Preparation of the b0 Pair (Optional)

Beyond the primary DWI acquisition, typically acquired with a fixed phase-encoding direction, additional image acquisition with reverse phase-encoding is advisable (e.g., primary DWI in anterior-posterior direction and auxiliary image in posterior-anterior). This auxiliary image proves highly beneficial during the distortion correction stage (see below), enabling cancellation of the warping effects normally observed in DWI acquisitions (Fig. 14). In order to perform these corrections, a minimum of two images with b-value = 0 s/mm^2 (b0) and opposite phase-encoding direction is required. The first one, with parallel phase-encoding, can be extracted from the main DWI image by applying:

```
dwiextract dwi.mif -bzero b0_parallel.mif
```

Reverse phase-encoding image should be extracted from the secondary protocol and converted to the .mif extension, thus enabling creation of the b0 pair necessary for the distortion correction stage:

```
mrcat b0_parallel.mif b0_reverse.mif b0_pair.mif
```

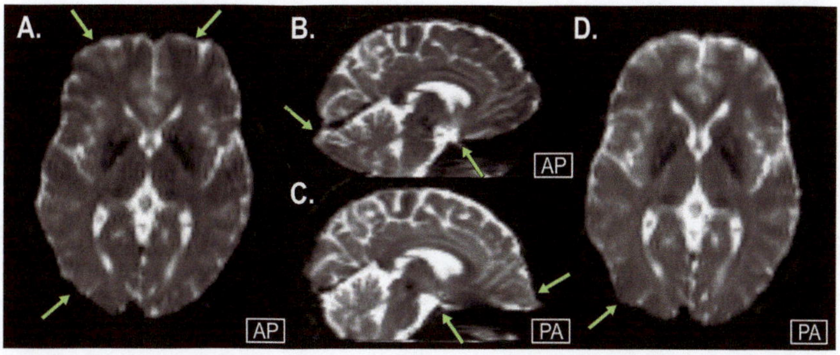

Fig. 14 Illustrative example of two b = 0 s/mm^2 images acquired with opposite phase-encoding direction (b0 pair). (**a**, **b**) depict geometric- and susceptibility-induced distortions typically observed in DWI images acquired with anterior to posterior (AP) phase encoding direction. (**c**, **d**) show the analogous in DWI images with opposite phase encoding, i.e., posterior to anterior (PA). The location of some of these distortions is indicated by green arrows

Sometimes more than one b0 image is collected during either the main or secondary protocols. In such instance, b0 images should be averaged before creating the b0 pair file:

```
mrmath b0_dir.mif mean b0_dir_mean.mif -axis 3
```

Denoising

DWI are inherently noisy, exhibiting low SNR resulting from a low signal amplitude and thermal noise [106]. This feature distorts DWI analysis and introduces biases into the estimation of quantitative diffusion parameters. Therefore, denoising becomes almost imperative before conducting any quantification (e.g., DTI [47], FWE-DTI [107] or NODDI [53] models) or advanced post-processing (e.g., tractography or tract-based statistics). Denoising can be performed with *dwidenoise* [106] (Fig. 15):

```
dwidenoise dwi_in.mif dwi_denoised.mif -noise noise_map.mif
```

Gibbs Artifacts Removal (Optional)

Gibbs ringing artifacts are a recognizable type of artifacts in MRI, characterized by a series of alternating raise and drop ringing patterns in image intensity. To correct them, the *mrdegibbs* command can be applied (Fig. 15):

```
mrdegibbs dwi_denoised.mif dwi_deringed.mif
```

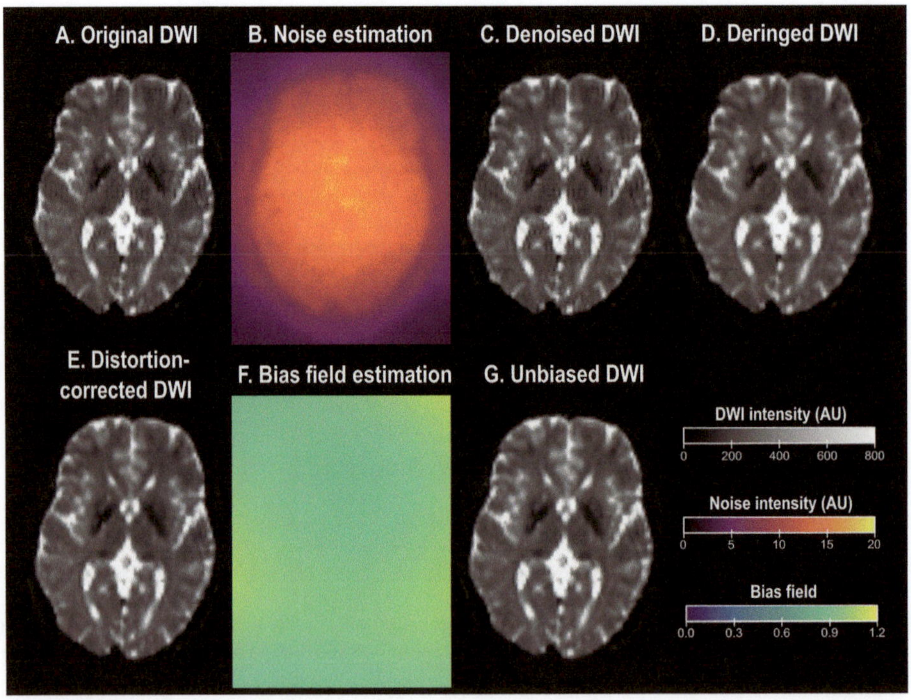

Fig. 15 Main preprocessing workflow for DWI scans. In the same axial plane: (**a**) Original DWI. (**b**) Noise estimation. (**c**) Denoised DWI. (**d**) Deringed DWI after Gibbs ringing correction. (**e**) Distortion-corrected DWI after eddy and topup corrections. (**f**) Bias field estimation. (**g**) Unbiased DWI after undergoing the whole preprocessing workflow

While this algorithm effectively removes such artifacts, it also introduces some smoothing effects that may be detrimental if you are studying subtle tissue alterations. Keep this in mind before introducing this step in your DWI preprocessing workflow.

Distortion Correction

Driven by the inherent features of DWI acquisition, the following distortions are commonly observed:

1. Geometric distortions owing to inhomogeneities in the main magnetic field. These distortions typically manifest in regions with strong magnetic gradients, such as near air-tissue (e.g., paranasal sinuses) or air-bone interfaces. These artifacts are exhibited along the encoding direction of the acquisition.
2. Susceptibility artifacts arising from significant magnetic susceptibility gradients between neighboring tissues. These artifacts are more prominent in regions near tissue interfaces, such as the frontal sinus or the olfactory bulb.
3. Eddy current-induced distortions, generated in conductive materials of the MRI system when fast-switching gradient magnetic fields are applied during DWI acquisition.
4. Distortions caused by patient motion.

All these distortions can be fixed (Fig. 15) by a couple of algorithms implemented in FSL: *topup* and *eddy*. However, they are hard to configure, making *dwifslpreproc* a preferable choice due to its user-friendly setup. Moreover, this algorithm automatically checks for CUDA availability, adjusting it accordingly to the available version (eddy, eddy_cuda8.0, eddy_cuda9.1, or eddy_cuda10.2). This can significantly accelerate the computationally intensive eddy correction stage:

```
dwifslpreproc dwi_deringed.mif dwi_preproc.mif /
 -pe_dir phase_dir -rpe_pair -se_epi b0_pair.mif /
-align_seepi -readout_time read_time
```

where "*phase_dir*" is the phase-encoding direction of the main DWI protocol (e.g., AP, LR, IS ...) and "read_time" denotes the readout time of the acquisition in seconds. This value can be checked in your .mif file with the command *mrinfo* (field "*TotalReadOutTime*"). Adding the option "*-nocleanup*" to preserve all intermediate files generated during this process is usually recommended to review topup and eddy performance. Also, consider adding the "*-nthreads CPUthreads*" option to speed up the computation (where "*CPUthreads*" is the number of CPU threads to be enabled).

Please note that this process assumes your data were acquired with a single phase-encoding direction, that you have at least one b0 image available with opposite phase encoding, and that all images were acquired with the same readout time. Your data may not meet these requirements. For example, DWI data are often extracted from clinical protocols in which images with opposite phase-encoding directions are not typically acquired. In that situation, distortion correction by *topup* is not achievable and the option "*-rpe_pair -se_epi b0_pair.mif*" needs to be replaced by "*-rpe_none*." Alternatively, your data may be multi-shelled. In this case, *dwifslpreproc* configuration need to be adjusted:

```
dwifslpreproc dwi_deringed.mif dwi_preproc.mif /
 -pe_dir phase_dir -rpe_pair -se_epi b0_pair.mif /
-align_seepi -readout_time read_time /
-eddy_options "--data_is_shelled"
```

Please refer to the *dwifslpreproc* documentation for detailed information on how to adjust the setup to suit your data.

Bias Field Correction

As previously explained, all MRI images exhibit some bias that must be removed. This bias can be estimated in DWI using *dwibiascorrect* (Fig. 15), which internally executes the *N4BiasFieldCorrection* [100] command from ANTs (see 3.4.3.5):

```
dwibiascorrect ants dwi_preproc.mif dwi_preproc_unbiased.mif
/
 -bias bias_field.mif
```

Non-brain Tissue Removal

DWI post-processing tasks such as tractographic reconstructions or estimations for advanced multicompartmental models like FWE-DTI [107] or NODDI [53], can be computationally demanding. To optimize efficiency, it's advised to confine intensive computations to just brain tissue regions, which can be achieved by defining a brain tissue mask to constrain the analysis volume. This mask can be estimated using the *dwi2mask* command:

```
dwi2mask dwi_preproc_unbiased.mif brain_mask.mif
```

Storing Your Data in NIFTI Format

The .mif format is generally not compatible with other neuroimaging software. Hence, saving your preprocessed data in NIFTI format is more convenient. This can be done using *mrinfo* to extract diffusion metadata (.bval and .bvec files):

```
mrinfo dwi_preproc_unbiased.mif -export_grad_fsl /
dwi_preproc_unbiased.bvec dwi_preproc_unbiased.bval
```

and DWI acquisition metadata (.json):

```
mrinfo dwi_preproc_unbiased.mif -json_all /
dwi_preproc_unbiased.json
```

On the other hand, *mrconvert* can be used to transform your main volume to NIFTI extension:

```
mrconvert dwi_preproc_unbiased.mif dwi_preproc_unbiased.nii.
gz
```

and also to isolate the first volume of the image:

```
mrconvert dwi_preproc_unbiased.mif -coord 3 0 -axes 0,1,2 /
b0_preproc.nii.gz
```

This isolated image is typically employed to register DWI scans to other imaging spaces.

3.3.7 Preprocessing for Functional Magnetic Resonance Imaging

fMRI preprocessing is complex, involves multiple steps, and is highly sensitive to errors. For those reasons, fmriprep [11] implementation is strongly recommended. This tool automates all steps applying a robust and consistent methodology, considerably streamlining fMRI preprocessing. In order to use fmriprep, some prerequisites are required. First, your data should be structured following BIDS standards. Run any BIDs validator (e.g., *https://github.com/bids-standard/bids-validator*) to verify that your data is correctly allocated before launching fmriprep. If you want to preprocess task-based fMRI data, keep in mind that your "func" folder should include a .tsv file with all the events. In addition, pre-installation of additional neuroimaging software, such as FSL, ANTs, AFNI, or Freesurfer, is also mandatory, since fmriprep exploits many functionalities provided by these packages. Finally, along with the fMRI data, a structural T1w is also requested to carry out the preprocessing. Once all prerequisites have been fulfilled, the entire preprocessing pipeline can be easily run with the following command (assuming that you are running fmriprep via docker container):

```
fmriprep-docker input_folder output_folder \
 participant --participant-label subject_id \
 --fs-license-file path_tofreesurfer/license.txt
 --stop-on-first-crash -w working_folder
```

where "*input_folder*" is the path to the folder where all your data in BIDS format is allocated, "*output_folder*" the directory where all your preprocessed data will be stored, "*subject_id*" is the subject to be preprocessed (e.g., "01," "02," ...), "*path_tofreesurfer/license.txt*" is the location where you have allocated your Freesurfer license and "*working_folder*" the working directory where intermediate results will be stored. This last folder can be automatically erased once the preprocessing is finished if you add the option "*--clean-workdir*," but it is not recommended. You may probably want to check intermediate files in order to verify that everything ran correctly. The preprocessing workflow implemented by fmriprep is outlined as follows:

1. Non-brain tissue removal on the structural T1w applying an atlas-based segmentation approach (*antsBrainExtraction.sh*).
2. Brain tissue segmentation applying FAST (FMRIB's Automated Segmentation Tool, *fast* from FSL).
3. T1w spatial normalization to MNI152 template space to facilitate the posterior normalization of the fMRI data (*antsRegistrationSyN.sh*).
4. Cortical surface reconstruction applying *recon-all* from Freesurfer.
5. Estimation of a reference fMRI volume to be used as a landmark in the following preprocessing step. This volume is contrast-enhanced and skull-stripped.
6. Head-motion correction across all fMRI volumes applying rigid-body transformations (*mcflirt* from FSL [108]).
7. Slice-timing correction to correct for differences in acquisition time between slices within each volume (*3dTshift* from AFNI). This correction is needed because each slice is not acquired simultaneously during fMRI volume acquisition. Instead, the slices are acquired sequentially, which can result in differences in the temporal sampling of the signal between slices.
8. Susceptibility-induced distortion correction. The applied routine varies depending on whether you have acquired auxiliary data with opposite phase encoding direction (blip-up/blip-down) or exported a supplementary fieldmap.
9. fMRI to T1w registration. This step is calculated using a rigid-body approach based on a boundary-based cost function (*bbregister* from FreeSurfer).
10. Spatial normalization of the fMRI data to MNI152 template space.

Some of these steps are computationally intensive. To speed up the preprocessing we suggest adding the "*--fs-no-reconall*" option in order to skip cortical surface reconstruction (**step 4**). In addition, fmriprep may monopolize all your workstation resources at certain preprocessing stages. To restrict resource usage, you can add the options "*--mem_mb dedicated_RAM*" and "*--nthreads dedicated_CPU*," where "*dedicated_RAM*" is the maximum RAM in MB that fmriprep will have access to and "*dedicated_CPU*" the maximum number of utilized CPU threads. All preprocessed data will be stored in "output_folder/fmriprep." Within this directory, a new folder labeled with the subject's ID will be created along with an .html file summarizing the performance of all preprocessing stages, providing thus a comprehensive overview that is very useful for quality control. This folder will include both preprocessed anatomical and functional data ("anat" and "func" directories), the preprocessing log ("log"), and the figures displayed by the .html report.

Although fmriprep runs a standard preprocessing pipeline for fMRI data, it does not assume how they will be analyzed during subsequent stages. Therefore, smoothing, a fairly standard step in fMRI preprocessing, is not carried out. Before going on, gather some information on whether this step should be applied to your dataset. In order to smooth your data just run:

```
fslmaths bold_preproc.nii.gz -kernel gauss sigma -fmean /
bold_smoothed.nii.gz
```

This command will smooth each 3D volume of the BOLD time series by applying a Gaussian kernel with a standard deviation of "*sigma*" (Fig. 16). This parameter is typically derived from the full width at half maximum (FWHM) of the Gaussian distribution, where FWHM $= 2.355 \times$ sigma.

3.3.8 Preprocessing of Other Imaging Modalities

So far, we have outlined the most standardized preprocessing pipelines within the neuroimaging community, enabling robust preprocessing for structural MRI, DWI, and fMRI data. However, there are other imaging modalities that are valuable for investigating brain pathologies, as we outlined previously. These modalities also often require a certain degree of preprocessing. Unfortunately, in the absence of well-defined preprocessing pipelines, there are no tools yet available to facilitate their user-friendly execution. This gap is gradually being filled with the implementation of new tools with similar philosophy to fmriprep, such as ASLPrep [109] for

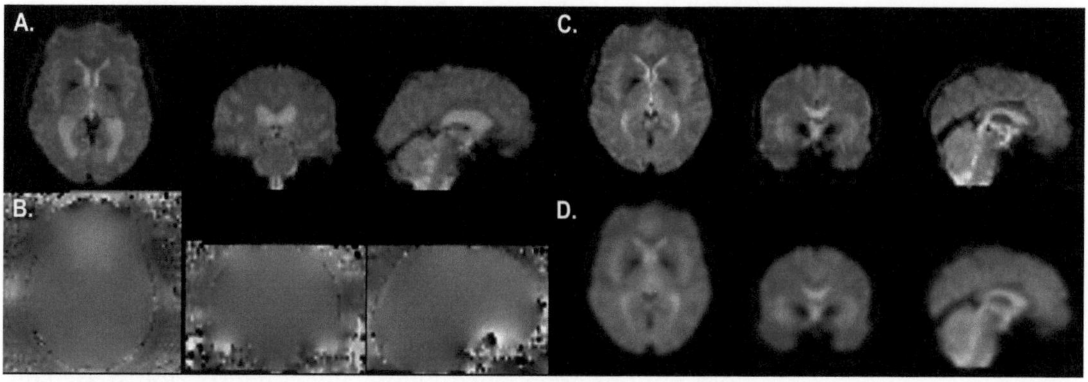

Fig. 16 Illustration of fMRI data before and after preprocessing. (**a**) Original fMRI image. (**b**) Fieldmap. (**c**) Preprocessed fMRI after fmriprep workflow. (**d**) Smoothed fMRI

ASL (*https://github.com/PennLINC/aslprep*), but they are still in early development stages or implement non-standardized preprocessing pipelines. Therefore, preprocessing for other imaging modalities tends to be more "handcrafted," often leveraging robust computation languages like MATLAB or Python, which have facilitated the development of multiple image processing tools that unfortunately we won't detail here due to the prerequisite proficiency in these programming languages. If you want to explore other imaging modalities and you have no programming expertise, check if your facility has access to any commercial processing software. Major imaging vendors such as Siemens, GE, Philips, and Toshiba may offer this kind of software with their scanners, as well as may provide technical support. In any case, it is always recommended to seek advice from a neuroimaging expert.

4 Notes

1. Maintain consistency with your acquisition and preprocessing protocols. Take your time to define the ideal settings for your experiment and maintain them throughout the whole study. While minor setup adjustments may seem harmless, they can result in significant fluctuations in your quantitative parameters, compromising the quality of your study by introducing noise and potentially degrading your ability to discern differences between cohorts.
2. Collect your data always using the same scanner if possible. Harmonizing data across different imaging devices remains challenging, and there is currently no consensus on how to optimally overcome this. Then, combining data from different devices is not indicated (even if they are identical scanners) as it may introduce unpredictable noise, thereby diminishing the sensitivity of your study.
3. If you have doubts about how to execute any command on the terminal or want to apply more advanced settings, just type the name of the command followed by --help or -h in the terminal (e.g., bet --help or bet -h). Most software will display a comprehensive description, as well as all configurable settings.
4. Although registration is typically conducted with skull-stripped images to speed up computations, providing the algorithm with whole-head images may increase registration performance as more information is available to the algorithm. Other options that could optimize coregistration performance are the initial moving transform and the histogram matching. Both options are applied before any coregistration step. This first option can be tuned by adding "*--initial-moving-transform [ref_image.nii.gz, moving_image.nii.gz, value]*" to the

main command and will align both images according to the geometric center (*value* = 0), their image intensities (*value* = 1) or the origin of coordinates (*value* = 2). The second option can be adjusted with "*--use-histogram-matching value*". Set "*value*" to 1 if you are registering images of the same modality, or to 0 if they are different. If registration keeps failing after a few tries, check *antsRegistration* documentation (---help) to explore more advanced features.

5. Diffeomorphic coregistration steps are computationally demanding. Nevertheless, computing efficiency can be boosted by activating parallelization with the "*-n #threads*" option, where "#threads" is the number of CPU threads to be enabled. Aside from parallelization, we recommend compute this step just once per subject, utilizing the reference structural image previously defined as the common link for all modalities or sessions.
6. Sometimes, either because of poor data quality, artifacts or image anomalies, automatic segmentation using "*run_first_-all*" may fail. If this happens, retry this process using a skull-stripped T1w instead of the original T1w. If this keeps failing, you can attempt segmenting the structures individually using the "*-s struct_name*" option as described in subheading "FIRST-Based Segmentation".

5 Concluding Remarks

Neuroimaging has become an integral part of our clinical daily routines. Over the years, these tools have evolved into effective in vivo approaches for examining multiple diseases, serving as valuable diagnostic aids and enabling characterization of pathological features. Throughout this chapter, we have provided an introduction to most neuroimaging techniques, outlining protocols that can serve as instruments in the diagnosis and/or pathological study of HE. In addition, we have described a set of basic preprocessing techniques intended to enhance image quality and facilitate image analysis. However, it is important to keep in mind that these are only basic concepts for data preprocessing that are intended to be applied to relatively good-quality data that do not require complex or unconventional preprocessing steps. Before proceeding, remember to carefully inspect your imaging for potential abnormalities. If they are noticed, know that many non-standard artifacts can be corrected, although they may require some time and expertise in advanced preprocessing techniques. Finally, seeking guidance from a neuroimaging expert is advisable in any scenario, as they can provide valuable assistance and insights.

References

1. Rose CF et al (2020) Hepatic encephalopathy: Novel insights into classification, pathophysiology and therapy. J Hepatol 73:1526–1547
2. Patidar KR, Bajaj JS (2015) Covert and overt hepatic encephalopathy: diagnosis and management. Clin Gastroenterol Hepatol 13: 2048–2061
3. Kumar S, Modi R, Bhandari BM, Sass DA, Rothstein KD (2018) A head CT is unnecessary in the initial evaluation of a cirrhotic patient with recurrent hepatic encephalopathy. Ann Hepatol 17:810–814
4. Rahimi RS, Rockey DC (2016) Overuse of head computed tomography in cirrhosis with altered mental status. Am J Med Sci 351:459–466
5. Romero-Gómez M, Montagnese S, Jalan R (2015) Hepatic encephalopathy in patients with acute decompensation of cirrhosis and acute-on-chronic liver failure. J Hepatol 62: 437–447
6. Li X, Morgan PS, Ashburner J, Smith J, Rorden C (2016) The first step for neuroimaging data analysis: DICOM to NIfTI conversion. J Neurosci Methods 264:47–56
7. Jenkinson M, Beckmann CF, Behrens TEJ, Woolrich MW, Smith SM (2012) FSL. NeuroImage 62:782–790
8. Avants B, Tustison NJ, Song G (2009) Advanced normalization tools: V1.0. Insight J. https://doi.org/10.54294/uvnhin
9. Fedorov A et al (2012) 3D Slicer as an image computing platform for the Quantitative Imaging Network. Magn Reson Imaging 30: 1323–1341
10. Tournier J-D et al (2019) MRtrix3: A fast, flexible and open software framework for medical image processing and visualisation. NeuroImage 202:116137
11. Esteban O et al (2019) fMRIPrep: a robust preprocessing pipeline for functional MRI. Nat Methods 16:111–116
12. Osborn AG, Hedlund GL, Salzman KL (2017) Osborn's brain e-Book. Elsevier Health Sciences
13. Reis E, Coolen T, Lolli V (2020) MRI findings in acute hyperammonemic encephalopathy: three cases of different etiologies. J Belgian Soc Radiology 104:9
14. Alonso J, Córdoba J, Rovira A (2014) Brain magnetic resonance in hepatic encephalopathy. Semin Ultrasound, CT MRI 35:136–152
15. Cordoba J, Blei A (1996) Brain edema and hepatic encephalopathy. Semin Liver Dis 16: 271–280
16. Butterworth RF, Spahr L, Fontaine S, Layrargues GP (1995) Manganese toxicity, dopaminergic dysfunction and hepatic encephalopathy. Metab Brain Dis 10:259–267
17. Watanabe A et al (1993) Reduction of increased signal intensity in the basal ganglia on Tl-weighted MR images during treatment of hepatic encephalopathy. Intern Med 32: 10–14
18. Lewis M, Howdle PD (2003) The neurology of liver failure. QJM 96:623–633
19. Rovira A et al (2002) Normalization of T2 signal abnormalities in hemispheric white matter with liver transplant. Neurology 59: 335–341
20. Pujol A et al (1993) Hyperintense globus pallidus on T_1-weighted MRI in cirrhotic patients is associated with severity of liver failure. Neurology 43:65–65
21. García Martínez R et al (2010) A long-term study of changes in the volume of brain ventricles and white matter lesions after successful liver transplantation. Transplantation 89: 589–594
22. Lee J, Lacomis D, Comu S, Jacobsohn J, Kanal E (1998) Acquired hepatocerebral degeneration: MR and pathologic findings. Am J Neuroradiol 19:485–487
23. Inoue E et al (1991) Portal-systemic encephalopathy: presence of basal ganglia lesions with high signal intensity on MR images. Radiology 179:551–555
24. Zeneroli ML, Cioni G, Crisi G, Vezzelli C, Ventura E (1991) Globus pallidus alterations and brain atrophy in liver cirrhosis patients with encephalopathy: an MR imaging study. Magn Reson Imaging 9:295–302
25. Rovira A, Alonso J, Córdoba J (2008) MR imaging findings in hepatic encephalopathy. AJNR Am J Neuroradiol 29:1612–1621
26. Aschner JL, Aschner M (2005) Nutritional aspects of manganese homeostasis. Mol Asp Med 26:353–362
27. McPhail MJW, Patel NR, Taylor-Robinson SD (2012) Brain imaging and hepatic encephalopathy. Clin Liver Dis 16:57–72
28. Binesh N et al (2006) Hepatic encephalopathy: a neurochemical, neuroanatomical, and neuropsychological study. J Appl Clin Med Phys 7:86–96

29. Hegde AN, Mohan S, Lath N, Lim CCT (2011) Differential diagnosis for bilateral abnormalities of the basal ganglia and thalamus. Radiographics 31:5–30
30. Ejima A et al (1992) Manganese intoxication during total parenteral nutrition. Lancet 339: 426
31. Fabiani G, Rogacheski E, Wiederkehr JC, Khouri J, Cianfarano A (2007) Liver transplantion in a patient with rapid onset parkinsonism-dementia complex induced by manganism secondary to liver failure. Arq Neuropsiquiatr 65:685–688
32. Cooper AJ, Plum F (1987) Biochemistry and physiology of brain ammonia. Physiol Rev 67: 440–519
33. Lockwood AH, Yap EWH, Wong W-H (1991) Cerebral ammonia metabolism in patients with severe liver disease and minimal hepatic encephalopathy. J Cereb Blood Flow Metab 11:337–341
34. Hazell AS, Normandin L, Norenberg MD, Kennedy G, Yi J-H (2006) Alzheimer type II astrocytic changes following sub-acute exposure to manganese and its prevention by antioxidant treatment. Neurosci Lett 396:167–171
35. Norenberg MD (1998) Astroglial dysfunction in hepatic encephalopathy. Metab Brain Dis 13:319–335
36. Andersen JV et al (2021) Glutamate metabolism and recycling at the excitatory synapse in health and neurodegeneration. Neuropharmacology 196:108719
37. Albrecht J, Jones EA (1999) Hepatic encephalopathy: molecular mechanisms underlying the clinical syndrome. J Neurol Sci 170:138–146
38. Grippon P et al (1986) Evidence for the role of ammonia in the intracerebral transfer and metabolism of tryptophan. Hepatology 6: 682–686
39. Häussinger D, Kircheis G, Fischer R, Schliess F, Dahl SV (2000) Hepatic encephalopathy in chronic liver disease: a clinical manifestation of astrocyte swelling and low-grade cerebral edema? J Hepatol 32:1035–1038
40. Kleinschmidt-DeMasters BK, Filley CM, Rojiani AM (2006) Overlapping features of extrapontine myelinolysis and acquired chronic (non-Wilsonian) hepatocerebral degeneration. Acta Neuropathol 112:605–616
41. Minguez B, Rovira A, Alonso J, Cordoba J (2007) Decrease in the volume of white matter lesions with improvement of hepatic encephalopathy. Am J Neuroradiol 28:1499–1500
42. Gawne-Cain ML, Silver NC, Moseley IF, Miller DH (1997) Fast FLAIR of the brain: the range of appearances in normal subjects and its application to quantification of white-matter disease. Neuroradiology 39:243–249
43. Rovira A et al (2007) Decreased white matter lesion volume and improved cognitive function after liver transplantation. Hepatology 46:1485–1490
44. Beaulieu C (2002) The basis of anisotropic water diffusion in the nervous system - a technical review. NMR Biomed 15:435–455
45. McKinney AM et al (2010) Acute hepatic encephalopathy: diffusion-weighted and fluid-attenuated inversion recovery findings, and correlation with plasma ammonia level and clinical outcome. AJNR Am J Neuroradiol 31:1471–1479
46. Lodi R et al (2004) Diffusion MRI shows increased water apparent diffusion coefficient in the brains of cirrhotics. Neurology 62:762–766
47. Basser PJ, Mattiello J, LeBihan D (1994) MR diffusion tensor spectroscopy and imaging. Biophys J 66:259–267
48. Goel A et al (2010) Cerebral oedema in minimal hepatic encephalopathy due to extrahepatic portal venous obstruction: Cerebral oedema in extrahepatic portal venous obstruction. Liver Int 30:1143–1151
49. Kumar R et al (2008) Voxel-based diffusion tensor magnetic resonance imaging evaluation of low-grade hepatic encephalopathy. Magn Reson Imaging 27:1061–1068
50. Chavarria L et al (2011) Biexponential analysis of diffusion-tensor imaging of the brain in patients with cirrhosis before and after liver transplantation. AJNR Am J Neuroradiol 32: 1510–1517
51. Qi R et al (2012) Structural and functional abnormalities of default mode network in minimal hepatic encephalopathy: a study combining DTI and fMRI. PLoS One 7: e41376
52. Jensen JH, Helpern JA (2010) MRI quantification of non-Gaussian water diffusion by kurtosis analysis. NMR Biomed 23:698–710
53. Zhang H, Schneider T, Wheeler-Kingshott CA, Alexander DC (2012) NODDI: practical in vivo neurite orientation dispersion and density imaging of the human brain. NeuroImage 61:1000–1016
54. Gupta P et al (2022) Whole brain atlas-based diffusion kurtosis imaging parameters for

evaluation of minimal hepatic encephalopathy. Neuroradiol J 35:67–76

55. Chen H-J, Liu P-F, Chen Q-F, Shi H-B (2017) Brain microstructural abnormalities in patients with cirrhosis without overt hepatic encephalopathy: a voxel-based diffusion kurtosis imaging study. Am J Roentgenol 209:1128–1135
56. Sun Q et al (2020) Characterization of brain microstructural abnormalities in cirrhotic patients without overt hepatic encephalopathy using diffusion kurtosis imaging. Brain Imaging Behav 14:627–638
57. Sato T et al (2019) Decreased mean kurtosis in the putamen is a diagnostic feature of minimal hepatic encephalopathy in patients with cirrhosis. Intern Med 58:1217–1224
58. Li J-L et al (2019) Microstructural brain abnormalities correlate with neurocognitive dysfunction in minimal hepatic encephalopathy: a diffusion kurtosis imaging study. Neuroradiology 61:685–694
59. Dong Q-Y et al (2023) White matter microstructural disruption in minimal hepatic encephalopathy: a neurite orientation dispersion and density imaging (NODDI) study. Neuroradiology 65:1589–1604
60. Benson JC et al (2016) Delineation of microhemorrhage in acute hepatic encephalopathy using susceptibility-weighted imaging. Eur J Radiol 85:629–634
61. Achiriloaie AF, Kido D, Wycliffe ND, Jacobson JP (2011) White matter microsusceptibility changes in patients with hepatic encephalopathy. Radiol Case 5:1–7
62. Ozawa E et al (2011) Ferritin/alanine aminotransferase ratio as a possible marker for predicting the prognosis of acute liver injury. J Gastro Hepatol 26:1326–1332
63. Taylor-Robinson SD et al (1996) MR imaging and spectroscopy of the basal ganglia in chronic liver disease: Correlation of T1-weighted contrast measurements with abnormalities in proton and phosphorus-31 MR spectra. Metab Brain Dis 11:249–268
64. Moffett J, Ross B, Arun P, Madhavarao C, Namboodiri A (2007) N-Acetylaspartate in the CNS: from neurodiagnostics to neurobiology. Prog Neurobiol 81:89–131
65. Van Der Graaf M (2010) In vivo magnetic resonance spectroscopy: basic methodology and clinical applications. Eur Biophys J 39: 527–540
66. Geissler A et al (1997) Cerebral abnormalities in patients with cirrhosis detected by proton magnetic resonance spectroscopy and magnetic resonance imaging: Cerebral abnormalities in patients with cirrhosis detected by proton magnetic resonance spectroscopy and magnetic resonance imaging. Hepatology 25:48–54
67. Pichili VBR, Rao KVR, Jayakumar AR, Norenberg MD (2007) Inhibition of glutamine transport into mitochondria protects astrocytes from ammonia toxicity. Glia 55:801–809
68. Desjardins P, Du T, Jiang W, Peng L, Butterworth RF (2012) Pathogenesis of hepatic encephalopathy and brain edema in acute liver failure: Role of glutamine redefined. Neurochem Int 60:690–696
69. Córdoba J, Gottstein J, Blei AT (1996) Glutamine, *myo* -inositol, and organic brain osmolytes after portocaval anastomosis in the rat: Implications for ammonia-induced brain edema: Glutamine, *myo* -inositol, and organic brain osmolytes after portocaval anastomosis in the rat: Implications for ammonia-induced brain edema. Hepatology 24:919–923
70. Spahr L et al (2000) Magnetic resonance imaging and proton spectroscopic alterations correlate with parkinsonian signs in patients with cirrhosis. Gastroenterology 119:774–781
71. Kreis R, Ross BD, Farrow NA, Ackerman Z (1992) Metabolic disorders of the brain in chronic hepatic encephalopathy detected with H-1 MR spectroscopy. Radiology 182: 19–27
72. Rovira A, Grivé E, Pedraza S, Rovira A, Alonso J (2001) Magnetization transfer ratio values and proton MR spectroscopy of normal-appearing cerebral white matter in patients with liver cirrhosis. Am J Neuroradiol 22:1137–1142
73. Laubenberger J et al (1997) Proton magnetic resonance spectroscopy of the brain in symptomatic and asymptomatic patients with liver cirrhosis. Gastroenterology 112:1610–1616
74. Naegele T et al (2000) MR imaging and ^{1}H spectroscopy of brain metabolites in hepatic encephalopathy: time-course of renormalization after liver transplantation. Radiology 216:683–691
75. Córdoba J et al (2001) The development of low-grade cerebral edema in cirrhosis is supported by the evolution of 1H-magnetic resonance abnormalities after liver transplantation. J Hepatol 35:598–604
76. Thomas MA et al (1998) Cerebral 1H MR spectroscopy and neuropsychologic status of patients with hepatic encephalopathy. Am J Roentgenol 171:1123–1130

77. Huda A et al (1998) Clinical correlation of neuropsychological tests with1 H magnetic resonance spectroscopy in hepatic encephalopathy. Psychosom Med 60:550–556
78. Bjerring PN, Gluud LL, Larsen FS (2018) Cerebral blood flow and metabolism in hepatic encephalopathy—a meta-analysis. J Clin Exp Hepatol 8:286–293
79. Li Y et al (2017) Combining arterial-spin labeling with functional magnetic resonance imaging measurement for characterizing patients with minimal hepatic encephalopathy. Hepatol Res 47:862–871
80. Zheng G et al (2013) Cerebral blood flow measured by arterial-spin labeling MRI: a useful biomarker for characterization of minimal hepatic encephalopathy in patients with cirrhosis. Eur J Radiol 82:1981–1988
81. Raichle ME (2015) The brain's default mode network. Annu Rev Neurosci 38:433–447
82. Zhang L et al (2012) Brain default-mode network abnormalities in hepatic encephalopathy: a resting-state functional MRI study. Hum Brain Mapp 33:1384–1392
83. Chen H-J et al (2013) Brain dysfunction primarily related to previous overt hepatic encephalopathy compared with minimal hepatic encephalopathy: resting-state functional MR imaging demonstration. Radiology 266:261–270
84. García-García R et al (2017) Reduced resting state connectivity and gray matter volume correlate with cognitive impairment in minimal hepatic encephalopathy. PLoS One 12: e0186463
85. Qi R et al (2014) Abnormal functional connectivity within the default mode network in patients with HBV-related cirrhosis without hepatic encephalopathy revealed by resting-state functional MRI. Brain Res 1576:73–80
86. Chen H-J et al (2015) Resting-state functional connectivity abnormalities correlate with psychometric hepatic encephalopathy score in cirrhosis. Eur J Radiol 84:2287–2295
87. Chen H-J, Chen Q-F, Liu J, Shi H-B (2016) Aberrant salience network and its functional coupling with default and executive networks in minimal hepatic encephalopathy: a resting-state fMRI study. Sci Rep 6:27092
88. Yang Z-T, Chen H-J, Chen Q-F, Lin H (2018) Disrupted brain intrinsic networks and executive dysfunction in cirrhotic patients without overt hepatic encephalopathy. Front Neurol 9:14
89. Qi R et al (2014) Default mode network functional connectivity: a promising biomarker for diagnosing minimal hepatic encephalopathy: CONSORT-compliant article. Medicine 93:e227
90. Wang Y et al (2023) Identifying minimal hepatic encephalopathy: a new perspective from Magnetic Resonance Imaging. Magnet Resonan Imaging jmri.29179. https://doi.org/10.1002/jmri.29179
91. Zhang LJ et al (2012) Altered brain functional connectivity in patients with cirrhosis and minimal hepatic encephalopathy: a functional MR imaging study. Radiology 265: 528–536
92. Qi R et al (2013) Disrupted thalamic resting-state functional connectivity in patients with minimal hepatic encephalopathy. Eur J Radiol 82:850–856
93. Zafiris O et al (2004) Neural mechanism underlying impaired visual judgement in the dysmetabolic brain: an fMRI study. NeuroImage 22:541–552
94. Liao L, Zhou L, Le H, Yin J, Ma S (2012) Spatial working memory dysfunction in minimal hepatic encephalopathy: an ethology and BOLD-fMRI study. Brain Res 1445:62–72
95. Lin W-C et al (2014) Reestablishing Brain Networks in Patients without Overt Hepatic Encephalopathy after Liver Transplantation. J Cereb Blood Flow Metab 34:1877–1886
96. Cheng Y et al (2018) Impact of previous episodes of hepatic encephalopathy on short-term brain function recovery after liver transplantation: a functional connectivity strength study. Metab Brain Dis 33:237–249
97. McPhail MJW et al (2013) Modulation of neural activation following treatment of hepatic encephalopathy. Neurology 80: 1041–1047
98. Ahluwalia V et al (2014) Enhancement of functional connectivity, working memory and inhibitory control on multi-modal brain MR imaging with Rifaximin in Cirrhosis: Implications for the gut-liver-brain axis. Metab Brain Dis 29:1017–1025
99. Manjón JV, Coupé P, Martí-Bonmatí L, Collins DL, Robles M (2010) Adaptive non-local means denoising of MR images with spatially varying noise levels. Magn Reson Imaging 31: 192–203
100. Tustison NJ et al (2010) N4ITK: improved N3 bias correction. IEEE Trans Med Imaging 29:1310–1320

101. Smith SM (2002) Fast robust automated brain extraction. Hum Brain Mapp 17:143–155
102. Avants BB et al (2011) A reproducible evaluation of ANTs similarity metric performance in brain image registration. NeuroImage 54: 2033–2044
103. Fonov V, Evans A, McKinstry R, Almli C, Collins D (2009) Unbiased nonlinear average age-appropriate brain templates from birth to adulthood. NeuroImage 47:S102
104. Avants B, Epstein C, Grossman M, Gee J (2008) Symmetric diffeomorphic image registration with cross-correlation: Evaluating automated labeling of elderly and neurodegenerative brain. Med Image Anal 12:26–41
105. Horn A, Kühn AA (2015) Lead-DBS: a toolbox for deep brain stimulation electrode localizations and visualizations. NeuroImage 107:127–135
106. Manjón JV et al (2013) Diffusion weighted image denoising using overcomplete local PCA. PLoS One 8:e73021
107. Pasternak O, Sochen N, Gur Y, Intrator N, Assaf Y (2009) Free water elimination and mapping from diffusion MRI. Magn Reson Med 62:717–730
108. Jenkinson M, Bannister P, Brady M, Smith S (2002) Improved Optimization for the Robust and Accurate Linear Registration and Motion Correction of Brain Images. NeuroImage 17:825–841
109. Adebimpe A et al (2022) ASLPrep: a platform for processing of arterial spin labeled MRI and quantification of regional brain perfusion. Nat Methods 19:683–686

Chapter 5

Immunoenzymatic Assays to Target Central and Peripheral Inflammatory Markers in Preclinical and Clinical Studies of Hepatic Encephalopathy

Brener Cunha Carvalho, Heliana de Barros Fernandes, Antonio Lucio Teixeira, Aline Silva de Miranda, and Eliana Cristina de Brito Toscano

Abstract

Inflammation has emerged as an important mechanism in the pathophysiology of cognitive and motor impairment related to hepatic encephalopathy (HE). Systemic inflammatory response syndrome (SIRS) associated with liver disease is common among patients with acute liver failure (ALF), leading to high mortality risk and the development of HE. Peripheral proinflammatory mediators cross the blood-brain barrier, leading to the activation of astrocytes and microglia, oxidative stress, and changes in neurotransmitter levels. Translational investigation of the inflammatory mechanisms underlying experimental HE represents a promising way for the elucidation of the pathophysiology of human HE, as well as the development of potential biomarkers and novel therapeutic targets. HE-related cognitive dysfunction is associated with worse employment, income, and caregiver burden, impacting family functioning and medical adherence. The development of potential biomarkers for HE screening, diagnosis, prognosis, and prediction/monitoring clinical response to an intervention can bring benefits to patient care, clinical research, and therapeutic development. Immunoenzymatic assays, especially ELISA, but also immunoblotting and immunohistochemistry are among the most employed methods to detect or measure these molecular biomarkers. In the current chapter, we will provide an overview of these immunoenzymatic techniques covering aspects ranging from sample collection in experimental models and humans to methodological procedures and potential pitfalls.

Key words Hepatic encephalopathy, Neuroinflammation, Immunoenzymatic assays, Biomarkers

1 Introduction

Hepatic encephalopathy (HE) is a potentially reversible neurological syndrome marked by altered arousal levels and cognitive and motor symptoms. HE is a consequence of liver insufficiency or shunting between the portal and systemic vasculature, leading to a failure of blood detoxification in the portal circulation

Tiziano Balzano (ed.), *Experimental and Clinical Methods in Hepatic Encephalopathy Research*, Neuromethods, vol. 219,
https://doi.org/10.1007/978-1-0716-4354-9_5,

[1]. Approximately 30–70% of patients with cirrhosis are affected by overt or subclinical HE [2]. In addition to the impact on the health and quality of life of patients, HE represents an increasing burden on the health care system [1]. The direct costs related to HE ranged between US5,370 and US50,120 per patient per year [2]. The severity of HE is one of the most important predictors of cost, charge, and length of hospitalization. Patients with a history of HE have a higher risk for relapse, and HE is the main cause of cirrhosis-related rehospitalization [3–5].

The severity of HE often influences treatment and is a strong predictor of mortality among cirrhotic patients [6, 7]. Cirrhosis-related HE impairs the ability to perform daily activities requiring more caregiver support [8]. Moreover, HE and the consequent cognitive dysfunction are associated with worse employment, income, and caregiver burden, impacting family functioning and medical adherence [9]. In addition, patients with severe HE waiting for liver transplantation had a 66% higher risk of death than patients without HE. Despite these facts, the presence of HE is not considered a major criteria for liver transplantation [10].

1.1 Inflammation in the Pathophysiology of Hepatic Encephalopathy

Liver failure leads to an accumulation of ammonia and other toxic substances in the blood, which in turn reach the central nervous system (CNS), affecting its functioning and promoting neurological symptoms. The mechanisms underlying the HE-related neurotoxicity are not fully understood. Although hyperammonemia is always present in HE patients, it does not correlate with the clinical severity of this neurological syndrome [11, 12].

A growing body of evidence has indicated that hyperammonemia and inflammation act synergically in the pathophysiology of HE [13–16]. Ammonia, a toxic molecule, is converted into urea by the liver, and urea is excreted by the kidneys [16]. During acute or chronic liver failure, the urea cycle is impaired, leading to hyperammonemia which is metabolized mainly in the CNS [17]. Ammonia neutralization occurs in the astrocytes through the conjugation of this molecule with glutamate by the glutamine synthase enzyme (GS) [16]. Increased brain levels of ammonia induce glutamine accumulation and enhancement of osmotic pressure, prompting swelling of astrocytes and brain edema [12, 15]. In addition to the imbalance in osmotic pressure, hyperammonemia may lead to chronic neuroinflammation [15, 17, 18].

Systemic inflammatory response syndrome (SIRS) associated with liver disease is also common among patients with acute liver failure (ALF), leading to high mortality risk and the development of HE [19]. Increased expression of proinflammatory cytokines ("cytokine storm"), mainly tumor necrosis factor (TNF), interleukin (IL)-1β, and IL-6, is observed in human and experimental ALF [15, 18, 20]. Peripheral proinflammatory mediators cross the blood-brain barrier, leading to the activation of astrocytes and microglia and oxidative stress [20–22]. Gliosis contributes to the maintenance of the increased expression of proinflammatory cytokines and oxidative stress in the CNS [20–22]. It is worth mentioning that activated microglia play an important role in inflammatory responses and the restoration of homeostasis. However, when chronically activated, microglia assume a non-resolving phenotype, not contributing to inflammatory resolution [20]. HE-related neuroinflammation induces cell depolarization and ion imbalance in astrocytes and microglia, disturbing glial reuptake of glutamate and γ-aminobutyric acid (GABA) and promoting, as a consequence, impairment in cognitive and motor functions [23, 24]. In line with this, anti-inflammatory strategies have demonstrated neuroprotective effects, reducing motor impairments and behavioral changes in experimental HE [25].

Minimal HE is related to a shift in the immune profile, with the expansion and activation of T helper (TH) 17, TH22, and T CD4+ lymphocytes as well as increased serum levels of pro-inflammatory markers, such as IL-6, IL-21, IL-17, IL-18, TNF, IL-1β, IL-15, IL-22, CCL20, CXCL13, and CX3CL1 [26]. In addition, the severity of HE shows a positive correlation with levels of serum TNF, IL-1β, IL-2R, IL-6, IL-8, and glial fibrillary acid protein (GFAP)—a marker of astrocyte activation [27, 28]. Microglial activation has also been reported in post-mortem brains of patients with HE. Conversely, this increased microglial activation was not linked to an increase in inflammatory cytokines. Such discrepancy may be explained by different stages of HE progression and also the evaluation of different brain areas [26]. In this scenario, translational investigation of the inflammatory mechanisms underlying experimental HE represents a promising way for the elucidation of the pathophysiology of human HE, as well as the development of potential biomarkers and novel therapeutic targets [22].

1.2 Rodent Models of HE

Experimental models can reproduce HE-induced psychomotor and cognitive symptoms, as well as inflammation in the pre-frontal cortex, hippocampus, and cerebellum observed in patients [29]. The administration of hepatotoxic drugs is considered a valuable model of acute liver failure (ALF)-related HE (type A), while surgical models (such as bile duct ligation) are widely used for the reproduction of HE due to portosystemic shunting (type B). To date, there has been no satisfactory model mimicking HE related to chronic liver disease (type C) [18, 30]. Type A HE has been developed to induce ALF and subsequent overt progressive neurological decline. Models of Type B and C HE mimic features of covert/minimal HE in which overt symptoms do not develop and are not observed [31].

The administration of thioacetamide (TAA), acetaminophen, and azoxymethane can cause acute liver injury and HE through the induction of oxidative stress and systemic inflammation, similar to that observed in acute HE patients. Therefore, these models are suitable for studying the crosstalk between peripheral and neuroinflammation in HE in rodents [18, 32]. Both plasma and brain levels of IL-1β, IL-6, and TNF were higher in TAA-treated rats than in control rats [33–35]. Acetaminophen intoxication was also associated with activation of brain transcription factors (such as NF-kB). TAA, acetaminophen, and azoxymethane intoxication lead to an increase in ammonia levels in the brain and consequent morphological changes in astrocytes and microglia [31].

Hepatic devascularization in rats is another established model of Type A HE. Anastomosis of the portal vein to the vena cava and ligating the hepatic artery induces ALF and reproduces the progression of Type A HE [30]. A consistent and progressive increase in blood and brain ammonia as well as astrocyte swelling, and brain edema have been characterized in this model. Altered gene expression and CNS inflammation have also been identified [31].

1.3 Immunoenzymatic Assays to Detect Possible Predictors and Prognosis Biomarkers in HE

A biomarker has been defined as "a characteristic that is objectively measured and evaluated as an indicator of a normal biological process, a pathological process, or a biological response to a therapeutic intervention" [36]. Biomarkers can be used for disease screening, diagnosis, prognosis, and prediction/monitoring clinical response to an intervention. Based on the quantification methods and sample source, they can be classified as molecular/biological, physiologic, histologic, radiographic (imaging), and digital biomarkers. Surrogate biomarkers can also be easier and less expensive to measure than final clinical endpoints [37, 38]. The appropriate application of biomarkers can bring benefits to patient care, clinical research, and therapeutic development. Over the past years, substantial effort was put into the search for potential biomarkers for neurological and neuropsychiatric conditions, especially Alzheimer's disease and related dementias. The complexity and heterogeneity of the pathophysiology of these disorders pose a great challenge in discovering and validating biomarkers [39–41]. One of these challenges focuses on the reliability, reproducibility, specificity, and sensitivity of the methods/techniques employed to measure a potential biomarker [40, 42]. Recent systematic reviews revealed that molecular biomarkers are the most frequently assayed in neuropsychiatric studies including cytokines, growth factors, measures of oxidative stress, and hormones [42, 43]. Immunoenzymatic assays, especially ELISA, but also immunoblotting and immunohistochemistry are among the most employed methods to detect or measure these molecular biomarkers [42]. In the current chapter, we will provide an overview of these immunoenzymatic techniques covering aspects ranging from sample collection in experimental models and humans to methodological procedures and potential pitfalls.

2 Materials

2.1 Immunohistochemistry

2.1.1 Animal Samples: Brain Collection, Histological Treatment, and Storage

- Dissection forceps: curved iris forceps with delicate teeth, 10 cm long (Moria Surgical, France, code MC31).
- Iris surgical scissor 115 mm—straight (Altomed, code A1120).
- Decapitation scissor 21.5 cm (AgnThos, code 14200-21).
- Saline solution: sodium chloride 0.9% (Laborclin, code 510183).
- Syringe 1 mL with 26G1/2 hypodermic needles (Descarpack, code C2161).
- Polypropylene tubes 1.5 mL (Olen, code K30-1015).
- Formalin solution 10% (Dinâmica, CAS 50-00-0).
- Mouse brain slicer matrix (Insight LTDA, Ribeirão Preto, SP, Brazil).
- Histological cassettes (MyLabor, code 2921).
- Ethanol solution 70%, 80%, 90%, and 100% (Dinâmica, CAS 64-17-5).
- Xylene (Êxodo, CAS 1330-20-7).
- Paraffin for histology (MerckMillipore, CAS 8042-47-05).
- Histology stainless steel embedding molds 24 × 24 × 5 mm (Easypayh, code EP-51-20,053).
- Microtome blades 0.1 × 0.03 × 0.06 cm (Easypath, code HK-NH50).
- Microscope slides 26 × 76 mm (Starfrost Advanced Adhesive, code 112721).
- Automated vacuum tissue processor (Leica BioSystems, code TP1020).
- Tissue embedding station (Leica BioSystems, code EG 1150).
- Tissue water bath (LupeTech, code BH2015).
- Rotary microtome (Leica BioSystems, code RM2255).
- Laboratory oven (SolidSteel, code SKU SSD64L).
- Paraffin block storage cabinets.
- Histological slide storage cabinets.

2.1.2 Immunohistochemistry Assay

(a) Hydration and clearing of the samples

- Xylene (Êxodo, CAS 1330-20-7).
- Ethanol solution 100%, 95%, and 80% (Dinamica, CAS: 64-17-5).
- Phosphate Buffered Saline (PBS) 10x (Sigma-Aldrich, code P7059), Tween® 20 (Sigma-Aldrich, code P1379).

- Tris Buffered Saline (TBS) 10x: $C_4H_{11}NO_3$ 0.05 M (Synth, CAS 77-86-1), NaCl 0.3 M (Synth, CAS 7647-14-5), add HCl for pH 7.6 (Synth, CAS 7647-01-0), Tween® 20 (Sigma-Aldrich, code P1379).
- Manual staining set with 12 reservoirs (Easypath, code EP-31-05101).
- Staining dishes (Global Trade Technology, code GT209-0530).
- Racks for staining dishes (Global Trade Technology, code GT209-1630).
- Analytical balance (Marte, code AD 330).
- Digital pHmeter (Kasvi, code MINTK39-1410A).
- Refrigerator (Consul, model practice 230).

(b) Antigen retriever and blockages

- Electric pressure cookie 900 W (Mondial, code PE38).
- Citrate buffer: $C_6H_8O_7$ 2 Mm (Synth, CAS 77-92-9), $Na_3C_6H_5O_7.2H_2O$ 8 mM (Synth, CAS 6132-04-3) pH 6.0.
- Tris/EDTA buffer: $C_4H_{11}NO_3$ 10 Mm (Synth, CAS 77-86-1), $C_{10}H_{14}N_2O_8Na_2.2H_2O$ 1 mM (Synth, CAS 6381-92-6) pH 9.0.
- Hydrogen Peroxide block solution: 0.3% hydrogen peroxide (Abcam, code ab236466).
- Protein block solution: 0.5% bovine serum albumin and 0.5%.
- casein (Abcam, code ab236466).
- Wash buffer: Phosphate Buffered Saline (PBS) 10x (Sigma-Aldrich, code P7059), Tween® 20 (Sigma-Aldrich, code P1379).
- Wash buffer: Tris Buffered Saline (TBS) 10x: $C_4H_{11}NO_3$ 0.05 M (Synth, CAS 77-86-1), NaCl 0.3 M (Synth, CAS 7647-14-5), add HCl for pH 7.6 (Synth, CAS 7647-01-0), Tween® 20 (Sigma-Aldrich, code P1379).
- Staining dishes (Global Trade Technology, code GT209-0530).
- Racks for staining dishes (Global Trade Technology, code GT209-1630).
- Hydrophobic Barrier Pap Pen (Easypath, code 18194).
- Slide staining tray (Easypath, code EP-51-05022).

(c) Protein detection and revelation

- Staining dishes (Global Trade Technology, code GT209-0530).

- Racks for staining dishes (Global Trade Technology, code GT209-1630).
- Slide staining tray (Easypath, code EP-51-05022).
- Tips 0.5–10 μL (Global Trade Technology, code 18291), 0–100 μL (Global Trade Technology, code 18170) e 100–1000 μL (Global Trade Technology, code 18191).
- Adjustable micropipettes (HTL, code 7903).
- Primary antibody specific for the target protein.
- Mouse Specifying Reagent—Complement (Abcam, code ab236466).
- Goat anti-rabbit HRP Conjugate (Abcam, code ab236466).
- 50x Diaminobenzine (DAB) chromogen (Abcam, code ab236466).
- DAB substrate (Abcam, code ab236466).
- Harris' Hematoxylin (MerckMillipore, code 1092531000).
- Bright field microscope (Leica MicroSystems, LM500).

(d) Sample dehydration and slide mounting.

- Ethanol solution 70%, 90%, and 100% (Dinamica, CAS: 64-17-5).
- Xylene (Êxodo, CAS 1330-20-7).
- Coated glass slides 24 × 50 mm (Perfecta, code 303.100.506).
- Permanent mounting medium Entellan® (MerckMillipore, code 107961).
- Manual staining set with 12 reservoirs (Easypath, code EP-31-05101).
- Fume hood (Nalgon, model 3720).

(e) Data processing and statistical analysis

- Digital slide scanner (3D Histech, model Pannoramic Scan II).
- Case viewer software (3D Histech, Budapest, Hungary).
- Image J 1.53e software (National Institute of Health, USA).
- GraphPad Prism 8.4.2 software (Dotimatics, USA).

- Decapitation scissor 21,5 cm (AgnThos, code 14200-21).
- Dissection forceps: curved iris forceps with delicate teeth, 10 cm long (Moria Surgical, code MC31).

2.2 Enzyme-Linked Immunosorbent Assay (ELISA)

2.2.1 Animal Samples: Brain Collection, Protein Extraction, and Storage

- Ethanol solution 75% (Dinâmica, CAS: 64-17-5).
- Filter 0.2 μm (Sarstedt, code 83.1826001).
- Iris surgical scissor 115 mm—straight (Altomed, code A1120).
- Tips 0.5–10 μL (Global Trade Technology, code 18291), 100–1000 μL (Global Trade Technology, code 18191).
- Adjustable micropipettes (HTL, code 7903).
- Petri plate (Jprolab, code 0303-8).
- Polypropylene tubes 1.5 mL (Olen, code K30-1015).
- Polypropylene tubes 15 mL (Olen, code K30-0015S).
- Protein extraction solution: Tris–HCl 20 mM (Tris-base, Merk, CAS 77-86-1, HCl, Synth, CAS 7647-01-0), NaCl 137 mM (Merk, CAS 7647-14-5), igepal 1% (Sigma-Aldrich, CAS 9002-93-1), glycerol 10% (Sigma-Aldrich, CAS 56-81-5), EDTA 10 mM (Sigma-Aldrich, CAS 6381-92-6), protease inhibitors: E-64 1 μM (Sigma-Aldrich, CAS 66701-25-5), PMSF 1 mM (Sigma-Aldrich, CAS 329-98-6), pepstatin A 1 μM (Sigma-Aldrich, CAS 26305-03-3), sodium orthovanadate 0.5 mM (Sigma-Aldrich, CAS 13721-39-6).
- Saline solution: sodium chloride 0.9% (Laborclin, code 510183).
- Syringe 1 mL with 26G1/2 hypodermic needles (Descarpack, code C2161).
- Syringe 10 mL (Descarpack, code C2161).
- Ultrasonic probe cleaning solution: HCl 0.04% (Synth, HCl P.A. 37%, CAS 7647-01-0).
- Ultrasonic processor (Cole-Parmer, model GE 130 PB).

2.2.2 Animal Samples: Blood Collection, Serum Treatment, and Storage

- Anatomical dissection forceps straight serration (Mekne, code 1646477).
- Syringe 1 mL with 26G1/2 hypodermic needles (Descarpack, code C2161).
- Hypodermic needles 21Gx1 (Labor Import, code 80082).
- Polypropylene tubes 1.5 mL (Olen, code K30-1015).
- Ketamine 0,116 g/mL (Ceva).
- Xylazine 0,1 g/mL (JA Saúde Animal).
- Micropipette 20–200 μL (HTL, code 11511).
- Tips 0.5–10 μL (Global Trade Technology, code 18291).
- Support for mouse immobilization.
- Iris surgical scissor 115 mm—straight (Altomed, code A1120).
- Ethanol 70% (Dinâmica, CAS: 64-17-5).

- Ultra Low-Temperature Freezer 700 Series (Thermo Scientific, Model 705).
- Refrigerated microcentrifuge (Eppendorf, model 5415C).

2.2.3 Human Samples: Plasma and Serum Collection, Treatment, and Storage

- Syringe 20 mL (Descarpack, code 0324701).
- Hypodermic needles 25 × 0.7 mm 22G (Descarpack, code 0353201).
- Silica tube 10 mL 16 × 100 mm (BD Vacutainer, code 367820).
- EDTA K2 tube 4 mL 13 × 75 mm (BD Vacutainer, code 367861).
- Ethanol 70% (Dinâmica, code P0100240060684).
- Microcentrifuge tubes (Axygen, code MCT-150-C-S).
- Tips 100–1000 μL (Global Trade Technology, code 18191).
- Adjustable micropipettes (HTL, code 7903).
- Refrigerated microcentrifuge (Eppendorf, model 5415C).
- Ultra Low-Temperature Freezer 700 Series (Thermo Scientific, Model 705).

2.2.4 ELISA Assay

(a) Plate coating with the capture antibody

- Capture antibody (DuoSet® ELISA Development System kit, R&D Systems).
- Phosphate Buffered Saline (PBS) 10x (Sigma-Aldrich, code P7059) 0.2 μm filtered.
- Polypropylene tubes 15 mL (Olen, code K30-0015S).
- Proline Plus Mechanical 8-channel Pipette 10–100 μL (Sartorius, code 43587382).
- Polystyrene Clear Flat-Bottom Immuno Nonsterile 96-Well Plates (Thermo Scientific, code 439454).
- Reagent reservoir for multichannel pipette, V-shaped bottom (StarTub, code S4026-5806).
- SealPlate® film (Sigma-Aldrich, code Z369659).
- Wet chamber: plastic box with lid 18x13x7 cm with wet paper towel (Garden, code 6610).
- Tips 0.5–10 μL (Global Trade Technology, code 18291), 0–100 μL (Global Trade Technology, code 18170) e 100–1000 μL (Global Trade Technology, code 18191).
- Micropipette 10–100 μL (HTL, code 5664).
- Refrigerator (Consul, model practice 230).

(b) Blocking and standard and sample incubation

- Blocking solution: Bovine Serum Albumin 1% (BSA -Sigma-Aldrich, CAS 9048-46-8) in PBS 1×, 0.2 μm filtered.
- Diluent solution: BSA 0.1% (Sigma-Aldrich A2153) in PBS, 0.2 μm filtered.
- Mass-calibrated standard (DuoSet® ELISA Development System kit, R&D Systems).
- Wash buffer: Tween® 20 0.05% (Sigma-Aldrich, CAS 9005-64-5) (Synth, CAS 9005-64-5) in PBS 1 × .
- Wet chamber: plastic box with lid 18x13x7 cm with wet paper towel (Garden, code 6610).
- Polypropylene tubes 1.5 mL (Olen, code K30-1015).
- Polypropylene tubes 15 mL (Olen, code K30-0015S).
- Reagent reservoir for multichannel pipette, V-shaped bottom (StarTub, code S4026-5806).
- SealPlate® film (Sigma-Aldrich, code Z369659).
- Tips 0.5–10 μL (Global Trade Technology, code 18291), 0–100 μL (Global Trade Technology, code 18170) e 100–1000 μL (Global Trade Technology, code 18191).
- Adjustable micropipettes (HTL, code 7903).
- Proline Plus Mechanical 8-channel Pipette 10–100 μL (Sartorius, code 43587382).
- AgileWasher™ Microplate washer (Avans Biotechnology).
- Refrigerated microcentrifuge (Eppendorf, model 5415C).

(c) Target detection and colorimetric measurement

- Alumni paper.
- Biotinylated detection antibody (DuoSet® ELISA Development System kit, R&D Systems).
- Diluent solution: BSA 0.1% (Sigma-Aldrich, CAS 9048-46-8) in PBS, 0.2 μm filtered.
- Reagent reservoir for multichannel pipette, V-shaped bottom (StarTub, code S4026-5806).
- Polypropylene tubes 15 mL (Olen, code K30-0015S).
- SealPlate® film (Sigma-Aldrich, code Z369659).
- Stop solution: H_2SO_4 1 M (Merk, CAS 7664-93-9).
- Streptavidin-HRP (DuoSet® ELISA Development System kit, R&D Systems).
- Substrate solution: OPD (*o*-phenylenediamine dihydrochloride) 0.4 mg/mL (Sigma-Aldrich, CAS 615-28-1) and H_2O_2 0.02% in citrate buffer ($C_6H_8O_7$ 27.014 mM,

Synth, CAS 77-92-9; NaH_2PO_4 111 mM, Merk, CAS 10049-21-5) pH 5.0.
- Tips 0–100 μL (Global Trade Technology, code 18170).
- Proline Plus Mechanical 8-channel Pipette 10–100 μL (Sartorius, code 43587382).
- Microplate washer (AgileWasher™, Avans Biotechnology).
- Microplate Photometer (Biosan, model HiPo MPP-96).

(d) Data processing and statistical analysis
- Microsoft Office Excel (Microsoft Windows, United States).
- GraphPad Prism 8.4.2 (Dotmatics, United States).

2.3 Western Blot

2.3.1 Protein Extraction

- Lysis solution: Tris–HCl 50 mM pH 7.4 (Sigma-Aldrich, Tris-base CAS 77-86-1, HCl CAS 7647-01-0), NaCl 150 mM (Sigma-Aldrich, CAS 7647-14-5), NaF 50 mM (Sigma-Aldrich, CAS 7681-49-4), β-glycerophosphate 10 mM (Sigma-Aldrich, CAS 154804-51-0), EDTA 0.1 mM, pH 8.0 (Sigma-Aldrich, CAS 60-00-4); glycerol 10% (Sigma-Aldrich, CAS 56-81-5), Triton X-100 1% (Sigma-Aldrich, CAS 9036-19-5).
- Pierce Protease Inhibitor Mini Tablets (Thermo Fisher Scientific, code A32953).
- Sodium Orthovanadate (Sigma-Aldrich, CAS 13721-39-6).
- 1 mL needle syringe (Descarpack, code C2161).
- Adjustable micropipettes (HTL, code 7903).
- Tissue homogenizer (Omni International, model Tissue Master 125).
- Ultrasonic homogenizer (sonicator) with probe (Cole-Palmer, model GE 130 PB).
- Microcentrifuge tubes (Axygen, code MCT-150-C-S).
- Refrigerated microcentrifuge (Eppendorf, model 5415C).
- Ultra Low-Temperature Freezer 700 Series (Thermo Scientific, Model 705).

2.3.2 Protein Quantification

- Quick start Bradford dye reagent (Bio-Rad, code 5000205).
- BSA (Sigma-Aldrich, CAS 9048-46-8) pH = 7.0.
- Sodium Orthovanadate (Sigma-Aldrich, CAS 13721-39-6).
- Mili-Q® water.
- 96 wells polystyrene microplates (Kasvi, code K12-096).
- Microcentrifuge tubes (Axygen, code MCT-150-C-S).
- Adjustable micropipettes (HTL, code 7903).
- Spectrophotometer for microplate reading (Thermo Scientific, model Multiskan FC).

2.3.3 Gel Electrophoresis and Transferring

- 6× sample buffer: Tris–HCl 375 mM (Sigma-Aldrich, Tris-base CAS 77-86-1, HCl CAS 7647-01-0), SDS 6% (Sigma-Aldrich, CAS 151-21-3), glycerol 48% (Sigma-Aldrich, CAS 56-81-5), 2-mercaptoethanol 9% (Sigma-Aldrich, CAS 60-24-2), bromophenol blue 0.03% (Sigma-Aldrich CAS 115-39-9) pH 6.8.
- Mini-PROTEAN® TGX™ Precast Protein Gels 4–15% (Bio-Rad, code 4561086).
- Running buffer 10×: Tris-base 0.25 M (Sigma-Aldrich, CAS 77-86-1), Glycine 1.92 M (Sigma-Aldrich, CAS 56-40-6), SDS 1% w/v (Sigma-Aldrich, CAS 151-21-3).
- Precision Plus Protein Kaleidoscope™ pre-stained protein standards (Bio-Rad, code 1610375).
- Transfer buffer 10×: Tris base 0.25 M (Sigma-Aldrich, CAS 77-86-1), Glycine 1.92 M (Sigma-Aldrich, CAS 56-40-6), SDS 0.35% w/v (Sigma-Aldrich, CAS 151-21-3). Methanol should be added in the proportion of 20% v/v at the moment of dilution.
- Methanol (Sigma-Aldrich, CAS 67-56-1).
- Ponceau S solution (Sigma-Aldrich, CAS 6226-79-5).
- PVDF membrane 0.45 μm (Millipore, code IPVH00010).
- Filter paper (Bio-Rad, code 1703932).
- Adjustable micropipettes (HTL, code 7903).
- Plastic container for western blot membrane (Sigma-Aldrich, code Z742091).
- Filter forceps with blunt end (Merck Millipore, code XX6200006P).
- Western Blot Roller (Thermo Fisher Scientific, code 84747).
- Blot assembly tray (Bio-Rad, code 1704089).
- Heating thermoblock (Kasvi, model K80-0102).
- Mini-PROTEAN® tetra vertical electrophoresis cell with mini Trans-Blot® module and PowerPac™ basic power supply (Bio-Rad, model 1658034).

2.3.4 Antibody Incubating

- Non-fat powdered milk.
- Tris Buffered Saline (TBS) 10x: Tris–HCl 0.2 M pH 7.6 (Sigma-Aldrich, Tris-base CAS 77-86-1), HCl (CAS 7647-01-0), NaCl 1.37 M (Sigma-Aldrich, CAS 7647-14-5), Tween® 20 (Sigma-Aldrich, CAS 9005-64-5).
- Primary antibody specific for the target protein.
- Horseradish peroxidase (HRP) conjugated secondary antibody.

- Clarity™ western ECL substrate (Bio-Rad, code 1705061).
- Mild stripping buffer: Glycine 200 mM (Sigma-Aldrich, CAS 56-40-6), SDS 0.1% (Sigma-Aldrich, CAS 151-21-3), Tween® 20 1.0% (Sigma-Aldrich, CAS 9005-64-5).
- PBS 10× (Sigma-Aldrich, code P7059).
- Tips 0.5–10 μL (Global Trade Technology, code 18291), 0–100 μL (Global Trade Technology, code 18170) e 100–1000 μL (Global Trade Technology, code 18191).
- Adjustable micropipettes (HTL, code 7903).
- Orbital shaker (Biosan, model PSU-10i).
- Amersham™ Imager (GE, model 600 QC).

2.3.5 Data Processing and Statistical Analysis

- Image J 1.53e software (National Institute of Health, USA).
- GraphPad Prism 8.4.2 software (Dotimatics, USA).

3 Methods

3.1 Immunohistochemistry

3.1.1 Animal Samples: Brain Collection, Histological Treatment, and Storage

In deeply anesthetized mice, use the decapitation scissor to remove the head cutting the neck behind the ears. Make a midline incision in the free skin and separate it from the skull with the help of forceps (or flip the skin over the eyes). Make an incision with the iris surgical scissor in the parietal region of the skull and remove it carefully. Gently remove the meninges, cut the optic chiasma, and place the brain in a Petri plate containing cold PBS to remove excess blood from the tissue. Place the brain over a filter paper placed in a Petri plate over the ice to preserve tissue consistency.

The process of tissue autolysis begins immediately after death. Therefore, rapid and adequate fixation after sampling is essential. Under the fume hood, procedure the immersion of the tissue sample in an adequate volume of formalin 10%. It is recommended that a volume ratio of tissue to fixative of at least 1:10 for optimal fixation. Stir gently the fixative container for a few seconds to make sure that the tissue sample does not stick to the container surface. Add an identification label on each container. Keep the tissue in the fixator for 72 h at room temperature. After, position the entire brain coronal brain matrix system, trimming it at 2 cm intervals and placing them into cassettes under the fume hood (*see* **Note 1**). Routinely, three transverse sections are obtained:

1. The first section at the level of optic chiasm including the basal ganglia, septum, cortex, and anterior hypothalamus.

2. The second section at the level of hippocampus containing the cortex and brain stem at the transition of diencephalon to mesencephalon.
3. The third section contains the cerebellum and brain stem (medulla oblongata) [44].

Location 4 is recommended for special investigations of the pons. The readers are highly recommended to use the comprehensive material of the Registry of Industrial Toxicology Animal-data (RITA) group [45].

Submitted the sliced samples to the pre-inclusion, a sequential process that consists of the dehydration of tissues in increased concentrations of alcohol solutions and a paraffin solvent, using the automated vacuum tissue processor. Dehydrate the tissue in a graded series of ethanol: 70% ethanol for 30 min, 80% ethanol for 30 min, and absolute ethanol for 30 min (two times). After, clear tissues using xylene (a paraffin solvent) for 30 min (two times) and infiltrate with paraffin for 30 min (two times) (*see* **Note 2**).

Once tissue samples are infiltrated by paraffin, carefully position them inside a base mold filled with melted paraffin (at 56–58 °C) on the warm work surface (*see* **Note 3**). Using heated forceps, orientate the tissue inside the mold to obtain the desired position to the cutting axis, the specimen surface in contact with the base of the mold being the one that will be on the slide after sectioning. Center the specimen in the mold ensuring that paraffin surrounds the edge of the tissue and gently transfer the mold onto the cool plate (*see* **Note 4**). Allow a few seconds for paraffin solidification. If the samples move during this phase, put the mold back onto the warm work surface until the whole paraffin liquefies, reorganize the tissues, and re-solidify the paraffin. Immediately place the base of the original cassette on top of the mold, allowing the specimen identification and use of the cassette as a holder during sectioning. Snap off the mold, bring the paraffin blocks together, and store them at room temperature until sectioning.

For the section of the blocks, carefully clean the water bath and heat it to 40 °C. Use ice to harden the cut surface of the block for a short time, avoiding cracking on the block surface. Install a disposable blade in the microtome. Adjust the angle between the blade edge and the block to approximately 2–5°, locking the blade in this position. Pruning the edges of the block parallel to the edges of the microtome blade, keeping at least 2 mm of paraffin wax around the tissue. Fit the paraffin block attached to the cassette onto the cassette holder of the microtome. Orientate the block on the perpendicular axis to the edge of the blade, positioning the smallest edge to be cut first. Cut the block at 15 μm until the whole surface of the embedded tissue can be sectioned. Put an ice rock on the cut surface of the block for a short time again. Set the microtome section thickness to 5 μm. Install a new disposable blade in the microtome and cut a series of paraffin sections.

Separate the ribbon from the blade edge using a paintbrush and transfer the ribbon to the surface of the water bath heated to 40 °C. When the ribbons are stretched, gently separate the floating sections on the water bath with the tips of forceps and collect them on clean advanced adhesive slides. Label slides with a pencil and keep them vertically in a laboratory oven overnight at 37 °C to dry and remove paraffin excess. Store the slides at 4 °C until the immunohistochemistry assay.

3.1.2 Immunohistochemistry Assay

(a) Hydration and clearing of the samples

Using the manual staining set, immerse the slides in xylene, in the following order: 1. heated xylene at 80 °C for 30 min, 2. xylene at room temperature for 15 min (two times). Submerge the slides in an ethanol series to hydrate them: 1. absolute ethanol for 7 min (two times), 95% ethanol for 7 min, 3. 80% ethanol for 7 min. Wash the slides twice in the wash buffer for 5 min (two times). The wash buffer is a mix of tween 20 and PBS 1× or TBS 1× at the ratio of 1/2000 (*see* **Note 5**).

(b) Antigen retriever and blockages

Introduce the slides into a heat and pressure-resistant staining dish containing citrate or Tris/EDTA buffer (retriever buffer) according to the recommendation of the primary antibody manufacturer. Put the dish in the electric pressure cooker and set it to normal pressure for 8 min. After the antigen retriever, keep the slides in the buffer at room temperature for approximately 30 min for progressive cooling. Wash the slides with the wash buffer for 5 min (two times).

Before performing the endogenous peroxidase blocking, dry around the tissue sections and delineate them with the hydrophobic pen. Be cautious about maintaining tissue moisture during the assay. Incubate the slides with the hydrogen peroxidase block solution (read-to-use) for 15 min. Wash the slides for 5 min (two times). Apply the protein block solution (read-to-use) for 10 min. Do not wash the slides, just remove excessive reagent.

(c) Protein detection and revelation

Immediately after the protein blocking, incubate the slides with the primary antibody appropriate for the target protein at 4 °C overnight. For that, dilute the antibody in a cold-blocking PBS or TBS buffer using the proportion recommended by the manufacturer or according to preliminary tests.

After 16–18 h, wash the slides using the wash buffer for 5 min (three times) and incubate them with 100 μL of the mouse specifying reagent-complement solution (read-to-use) for 10 min at room temperature. Do not wash, just remove the excess reagent, and incubate with 100 μL the goat anti-rabbit HRP conjugate solution (read-to-use) for 15 min at room temperature. Wash the slides using the wash buffer for 5 min (two times) (*see* **Note 6**).

Prepare the DAB solution using a mix of the pure chromogen and the substrate buffer at 1/20 (v/v). DAB is photosensitive and this solution should be prepared at the time of use, avoiding direct light exposure. Pipette around 100 μL of chromogen solution for slide. The pipetted volume of the solutions can vary according to the section's size. It is essential to cover all the tissue. Based on preliminary tests, incubate the slides with the chromogen for 1–10 min. Dispose of the DAB solution in a designated area and wash the slides with distilled water for 5 min. It is recommended that the immunostaining be checked using a bright field microscope during the tissue exposition to DAB.

Procedure the counter-staining with hematoxylin. Filter the commercial solution through filter paper to remove the metallic precipitant that forms in the solution upon standing. Stain the tissue sections for 1–3 min and wash in running tap water for 5 min.

(d) Sample dehydration and slide mounting

Using the manual staining set, immerse the slides in an ethanol series to dehydrate them: 70% ethanol for 5 min, 90% ethanol for 5 min, and absolute ethanol for 5 min (two times). Clear the slides by immersing them in xylene for 5 min (two times).

After, a coverslip should be placed over the tissue section to protect it and improve the optical evaluation of the tissue. Under the fume hood, keep the tissue covered by xylene and apply two or three drops of Entellan®. Place a coverslip on the slide and press gently with forceps to remove any bubbles. Clean the surface of the slides with xylene and dry them overnight at room temperature within the fume hood. Store stained slides in appropriate boxes, avoiding prolonged exposure to light until morphometric analysis.

(e) Data processing and statistical analysis

Histopathological analysis of the brain is essential to establish the crosstalking between central and peripheral inflammation, as well as to investigate potential biomarkers in HE. Although manual examination of histological slides on the microscope is a classically used method, it is considered time-consuming, limited, and associated with intra- and inter-observer variability. Emerging technologies such as digital microscopy have been used to improve the assessment of histopathological findings with applications in both clinical and research laboratories. Images obtained by digital

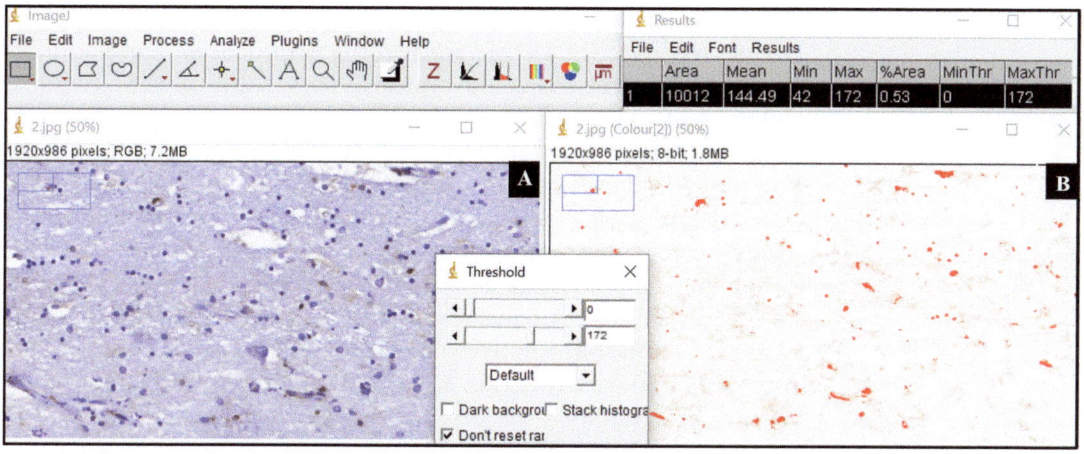

Fig. 1 Automatic morphometric analysis of immunostaining area and gray value. Diaminobenzidine (DAB)-stained cell areas of the tissue subjected to immunohistochemistry (**a**) are selected, digitally processed, and measured using the plugin color deconvolution and the vector Hematoxylin/DAB available in the software Image J (**b**). The red regions represent immunopositive cells (**b**)

microscopy have high resolution, can be easily shared, and enable quantitative and automatic analysis.

For high-quality image acquisition, digitalize the slides using a digital scanner and objectives of 20 or 40×. Proceed with the histopathological analysis of the whole image slide using virtual software and acquire the images of the brain area or subarea of interest using the appropriate magnificence. Quantify the immunostaining area and or immunostaining optical density, using the plugin color deconvolution from Image J software (Fig. 1), following the steps:

1. Select the vector hematoxylin and DAB (H DAB) (Fig. 1a).
2. Set the threshold including positive staining and exclude background and false positive staining (Fig. 1b).
3. Perform **step 2** for three slides per group to standardize the best threshold for the evaluated protein.
4. Run the morphometric analysis applying the same selected threshold for each histologic subfield. The area can be directly measured, while the optical density is calculated using the mean of the gray value [46].

To counter the positive cells, use the plugin counter cells and select the positive cells manually. Finally, the quantified immunostaining can be compared between the groups and represented in a graph using GraphPad Prism software [47].

3.2 ELISA

3.2.1 Animal Samples: Brain Collection, Protein Extraction, and Storage

Proceeded with the brain collection as described in Subheading 3.1, **step 3**. Use the curved iris forceps with delicate teeth and Iris scissors to dissect the brain regions. To identify specific regions, follow the coordinates established by Paxinos and Franklin [44]. Separate the interest regions (for HE, the key regions are cortex and hippocampus, and cerebellum) and put them in a 1.5 mL polypropylene tube. Immediately freeze the tissue to avoid protein degradation [48]. Use the fresh tissue right after the extraction or storage at −70 °C.

To obtain the proteins from the intended brain regions, pipette cold extraction solution over the tissue (*see* **Note 8**). Use a proportion of 100 μL of extraction solution for 100 mg of tissue (for the cortex, hippocampus, or cerebellum use around 175 μL for the entire hemisphere). Macerate the tissue with the tip of the sonicator probe. To avoid foam, gently press the tissue on the bottom of the tube and macerate using pulse mode, 60 V. During this procedure, keep the bottom of the tube inside the ice. Clean up the probe between each sample by washing it in three-step processes: water, ultrasonic probe cleaning solution, and water again. Before macerating the next sample, dry the probe with a paper towel. Centrifuge the samples for 20 min, 2500× *g*, 4 °C, collect the supernatant (yields around 200 μL), and dispense in a new tube. Storage the sample at −70 °C until use.

3.2.2 Animal Samples: Blood Collection, Serum Treatment, and Storage

For analysis of blood components such as the serum, the mice cannot be infused with PBS, PFA, or any other solutions (*see* **Note 9**). At the end of the experimental protocol, induce terminal anesthesia by injecting intraperitoneally a ketamine (16 mg/Kg)/xylazine (60 mg/Kg) solution, diluted in PBS. Only initiate the blood collection procedure after checking the anesthetic plane of the animal (based on the loss of the paw withdrawn reflex).

Turn the animal in the prone position, securing the paws on the immobilization support, allowing the exposure of the abdominal region. Clean up the abdominal region with ethanol 70% and make a V-shaped incision, exposing the peritoneal cavity. Gently remove the intestine using the dissecting forceps, push forward the liver, and expose the posterior vena cava (located between the kidneys). Gently insert the 21G needle with the bevel facing up and collect the blood by pushing slowly, to avoid vase collapse. Dismiss the collected sample in a 1.5 mL polypropylene tube. Repeat this procedure three or four times to obtain a suitable amount of blood.

Incubate the blood at room temperature for 30 min to clot and centrifuge the tube for 20 min, 2350× *g*, 4 °C. Gently collect the supernatant (serum) using a 200 μL micropipette, avoiding disturbing the clot on the bottom, and store at −70 °C until use.

3.2.3 Human Samples: Plasma and Serum Collection, Treatment, and Storage

Identify the individual and position him/her with the arm extended to form a straight line from shoulder to wrist. Select the appropriate vein, clean the area using 70% ethanol, and adjust the tourniquet above this area. Place a sheathed needle on the syringe and remove the cap of protection. Holding the needle in line with the vein, penetrate the skin and vein in one motion and draw the desired amount of blood by pulling back slowly on the syringe plunger. Release the tourniquet. Remove the needle and apply pressure on the area. Transfer blood drawn into the appropriate tubes immediately after the collection. Gently invert tubes containing EDTA 5–8 times. Dispose of the syringe and needle into an appropriate sharp container. Incubate tubes with blood at room temperature for 30 min and centrifuge them at 1008× *g* and 4 °C for 10 min. Gently collect the supernatant (serum in the case of silica tube and plasma, in EDTA tubes) and make samples aliquots of 1 mL to avoid freeze-thawing cycles. Store aliquots at −70 °C until use.

3.2.4 ELISA Assay

(a) Plate coating with the capture antibody

Prepare 5.5 mL of the capture antibody diluted in PBS 1× (*see* **Note 10**) in a 15 mL polypropylene tube following the DuoSet® ELISA kit (R&D Systems) dilution. Dispense the capture antibody solution on the reservoir for a multichannel pipette and pipette 50 μL of the solution in all wells of an ELISA 96-well plate. Seal the plate with a new adhesive seal plate and incubate it overnight at 4 °C inside a wet chamber to prevent the evaporation of the solution. In this incubation step, the capture antibody is immobilized in the bottom of the plate well, which has a treated surface (MaxiSorp, Thermo Scientific) that allows hydrophilic binding of the Fc portion of the antibody, leaving the antigen-binding portion free.

(b) Blocking and sample incubation

On the next day, wash the plate in the plate washer with 300 μL of the wash buffer to remove any excess capture antibody (three times). Remove any remaining liquid by tapping the plate over a fluffy surface covered with a paper towel (repeat this procedure in all washing steps). With the multichannel pipette, add 200 μL/well of blocking solution and incubate the plate at room temperature for at least 1 h. During this time, organize the samples and defrost them over ice. Design the plate, identifying the wells of standard, blank, and samples (Fig. 2). For details of sample preparation, *see* **Note 11**.

Repeat the washing step described above. In a 1.5 mL polypropylene tube, prepare 1 mL of the first point of the standard curve,

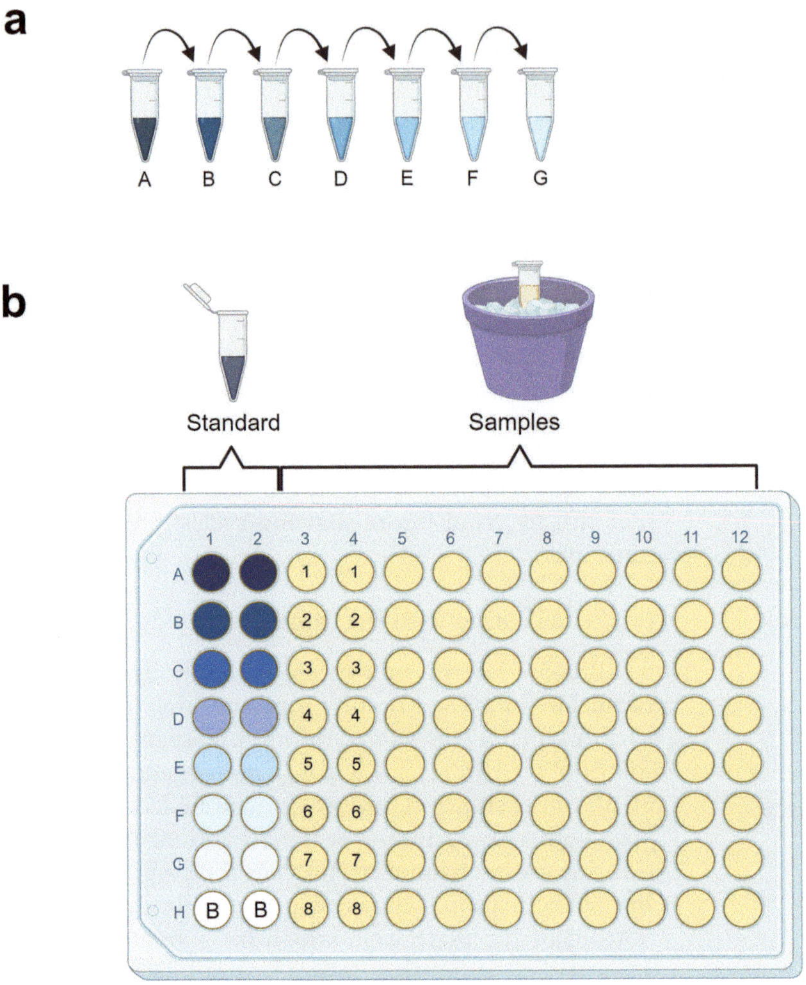

Fig. 2 Schematic illustration of standard dilution and plate design. a) Serial 1:2 dilution of standard solution. In the first tube (**a**), dilute 1 mL of the higher concentration of the standard solution with the diluent solution (PBS/BSA 0.1%). In the next six tubes (**b**–**g**), pipette 500 μL of the diluent solution. Pipette 500 μL from tube **a** to tube **b** and homogenize, then pipette 500 μL from tube **b** to tube **c** and repeat the same procedure until the last tube (**g**). (**b**) Suggestion of plate design. Pipette the standard solutions following the sequence of the tubes (from row **a** to **g**), in duplicate (columns 1 and 2). In the last row, pipette the diluent solution as the blank (**b**). Use the rest of the wells to pipette the samples, in duplicates, as exemplified

using the diluent solution (the information about the concentration of the first point is available in the certificate of analysis of the ELISA kit). To obtain the other six points of the standard curve, make a 1:2 serial dilution using equal volumes from the diluent solution and the standard solution with higher concentration (Fig. 2a). Pipette 50 μL/well of the standard in duplicate using the 100 μL micropipette. Pipette diluent solution as the negative control. In the remaining wells, pipette 50 μL of the samples, preferentially in duplicates. Seal the plate and incubate inside the wet chamber overnight at 4 °C.

(c) Target detection and colorimetric measurement

On the next day, repeat the washing step to remove the excess solutions. Prepare 5.5 mL of biotinylated detection antibody solution, diluted in diluent solution, following the manufacturer's suggestion. Pipette 50 μL/well of this solution, seal the plate, and incubate for 2 h, at room temperature. If there are target proteins immobilized in the plate, the biotinylated antibody will bind in the free epitope of the target protein during this incubation step.

Repeat the washing step. Prepare 5.5 mL of streptavidin-HRP in dilution solution (usually the stock solution requires a 1:200 or 1:40 dilution) and add 50 μL/well using the multichannel micropipette. Seal the plate, cover it with alumni paper, and incubate at room temperature for 30 min. The alumni paper will protect the streptavidin solution from light exposure, considering that HRP (horseradish peroxidase) is photosensitive. During the streptavidin incubation, prepare a fresh substrate solution (*see* **Note 12**). Repeat the washing step, add 50 μL/well of substrate solution, seal the plate, incubate at room temperature for 20–30 min, under agitation, and protected from light. At this moment, it should be possible to visualize a yellow-colored solution, mainly in standard wells. At the end of incubation time, stop the reaction with 50 μL/well of the stop solution, read the plate content in O.D. 490 nm on the Microplate Photometer, and export the absorbance values to an Excel sheet.

(d) Data processing and statistical analysis

The DuoSet ELISA kit protocol from the R&D Systems suggests using the four parameter-logistic non-linear regression curve fit (*see* other calculation forms in www.rndsystems.com/resources/how-to-analyze-elisa-data) which can be calculated using the software GraphPad Prism, for quantify the protein levels. The first step to calculate the concentrations of target cytokines in the samples is to do the mean value of standard and sample duplicates. Using the GraphPad Prism software or create an XY table with single values of X and Y. On the opened table, plot known standard concentrations in the X-column cells and the mean absorbance values of standard in the matched Y-column cell. In this last column, plot the absorbance of the samples that will be calculated. In the software Analysis tab, select "Fit a curve with nonlinear regression" and then select "Sigmoidal, 4PL, X is concentration" curve fit, picking the box "Interpolate unknown values." After confirming this setup, the resulting table will show the interpolated X values corresponding to the respective absorbance. If the samples were diluted, multiply for dilution factor before plotting the final results. For serum samples, the data is expressed as pg/mL. For brain samples, express data as pg/100 mg of tissue (*see* **Note 13**).

3.3 Western Blot

3.3.1 Animal Samples: Brain Collection, Protein Extraction and Storage

Proceeded with the brain collection and dissection as described in Subheadings 3.1, **step 3** and 3.2.1, respectively. For protein extraction, dilute a Pierce Protease Inhibitor Mini Tablet according to manufacturer instructions and sodium orthovanadate so the final concentration will be 2 mM in the lysis solution. Thaw samples on ice and then add 100 μL of ice-cold lysis solution per 100 mg of brain tissue (*see* **Note 14**). Macerate the tissue with the tip of the sonicator probe. To avoid foam, gently press the tissue on the bottom of the tube and macerate using pulse mode, 60 V. During this procedure, keep the bottom of the tube inside the ice. Clean up the probe between each sample by washing it in three-step processes: water, ultrasonic probe cleaning solution, and then water again. Before macerating the next sample, dry the probe with a paper towel. Usually, the use of a sonicator is sufficient, but if the tissue dissolution is more difficult, *see* **Note 15** for alternatives. Once the tissues are lysed, place the samples in an orbital shaker for 20 min keeping the lysates cold. After centrifuging the samples at 15,000× *g* and 4 °C for 20 min, extract the supernatants and make samples aliquots to avoid freeze-thawing cycles. Store the lysates at −70 °C or proceed to the next steps.

3.3.2 Protein Quantification

Load 1 μL of each sample in a microplate well and dilute to a 1/10 ratio by adding 9 μL of Mili-Q® water. Optionally, a serial dilution can be made to guarantee that the absorbance values will be included in the standard curve in case the samples are too concentrated. The standard curve can be done using seven different concentrations (2, 1.5, 1, 0.75, 0.5, 0.25, and 0.125 mg/mL) of BSA. Pipette 10 μL of each standard dilution and 10 μL of Mili-Q® water (blank control) into different wells. Load 190 μL of Bradford reagent to all sample dilutions, standards, and blank controls. Homogenize well and leave for at least 5 min at room temperature (it should not exceed 1 h). Measure the absorbance in a microplate spectrophotometer at 595 nm. It is recommended that the assay be carried out in triplicate.

3.3.3 Western Blot Assay

(a) Gel Electrophoresis

Prepare the proteins by transferring the proper volume of each sample according to the desired protein amount to a new microcentrifuge tube. Usually, 20–30 μg of protein is a suitable amount, but it can be adjusted according to the abundance of the specific protein being investigated. The samples in the same gel must be loaded with equal amounts of protein. Also, reach the same final volume for all samples by adding the proper amount of lysis solution. Then, add the 6× sample buffer so the concentration of it reaches 1×. Homogenize the prepared samples and place them in a heating thermoblock or water bath at 95 °C for 5 min. Take the 4–15% precast gel with the number of wells sufficient for the number of samples and rinse it with distilled water. Remove the comb from the top of the gel and the green tape stuck to the bottom. Insert the gel cassette in the electrode chamber (running gel holder) and then into the buffer tank. Be aware that the wells from the gel must be positioned facing the center of the tank. Dilute the running buffer to 1× with the mili-Q water (preferentially cold) and fill all sides of the tank with it. Load the protein standard and samples into the wells (*see* **Note 16**). Use the volume of the standard indicated by the manufacturer. For the samples, the maximum volume of the wells is about 30–35 μL. Place the tank lid fitting the correct cathode and anode ports of the electrode chamber. Connect the lid cable to the power supply and adjust to 100 V until the samples completely enter the gel, then adjust to 130–150 V. Stop the electrophoresis right before the samples reach the bottom of the gel so they will not come out. The buffer in the tank must not overheat during electrophoresis.

(b) Transfer from gel to membrane

Dilute 100 mL of the 10× transfer buffer by adding 700 mL of mili-Q® water and then 200 mL of methanol. The diluted 1× transfer buffer must be prepared on the same day of use and must be ice-cold at the moment of transfer. Cut the PVDF membrane in a shape and size proportional to the gel (a rectangle with about 9 × 7 cm). Using a small plastic boat, activate the membrane by soaking it in pure methanol for about 0.5–1 min (*see* **Note 17**). Once activated, equilibrate the membrane by immediately placing it in another container filled with 1x transfer buffer for 10–20 min. Carefully take out the gel from the electrophoretic apparatus, remove the well guides, and trim the gel excess. After that, place the gel for 5–10 min in a 1× transfer buffer. Using an assembly tray, soak the transfer cassette, foam pads, and filter papers into the transfer buffer. With the transfer cassette opened, assembly the transfer sandwich inside the tray filled with transfer buffer by placing the items one on top of the other in the following order: cathode side of the transfer cassette (usually identified in black color); foam pad; filter paper; gel; activated and equilibrated PVDF membrane; filter paper; foam pad; anode side of the transfer cassette. Always be careful to not let bubbles form between one layer and another, especially between the gel and the membrane. A western blot roller or a 15 mL centrifuge tube can be rolled over the sandwich when necessary (after adding one of the layers) and when the sandwich is ready, so any residual bubble can be eliminated. Carefully close tight the transfer cassette, insert it in the electrode chamber (transfer cassette holder), and place it in the electrophoresis tank filled with cold transfer buffer. Put the ice-cool pack in the appropriate place inside the tank and close it by connecting the electrodes to the respective inputs in the lid. The transference must be done at 4 °C, and since the current applied can warm the buffer, put the tank inside the ice letting just the top of it out. Connect the lid cables to the power supply and set it to 100 V for 1 h.

(c) Antibody Incubating

Dilute the TBS 10× to 1× using mili-Q® water, add 1 mL of Tween® 20 per liter of TBS1×, and homogenize to create a TBS-T 1× solution. It is recommended to prepare TBS-T 1× at the moment of use. After finishing the transfer electrophoresis, carefully take the membrane out of the cassette and place it in a small plastic container with TBS-T 1× (*see* **Note 18**). The ladder should be visible, and the membrane must always be incubated with the front (the face that was in contact with the gel) up. Dilute 5% of skim powdered milk in TBS-T1× to produce a blocking solution (*see* **Note 19**). Discard the TBS-T1× from the container (being careful to not lose the membrane) and pour the blocking solution. Place the container in an orbital shaker with gentle agitation for at least 1 h so the membrane will be blocked for non-specific binding.

Once blocking is finished, do a cycle of washing: replace the blocking solution for TBS-T1× and place the container in an orbital shaker with gentle agitation for 5 min, then repeat this procedure twice replacing for fresh TBS-T1×. After the membrane was washed, incubate it with the primary antibody appropriate for the target protein. For that, dilute the antibody in a cold-blocking solution using the proportion recommended by the manufacturer, then incubate the membrane with the diluted antibody at 4 °C overnight under gentle agitation. Run another wash cycle in TBS-T1× and dilute the appropriate HRP secondary antibody in the blocking solution. Incubate the membrane in secondary antibody solution at room temperature for 1 h under gentle agitation. After another membrane wash cycle, prepare the ECL mixture by pipetting a 1/1 solution of peroxide and luminol reagents. Place a thin, transparent plastic sheet over the appropriate tray of the imaging system and put the membrane over the plastic sheet. Cover completely the membrane with a layer of the ECL solution (about 1.5–2.0 mL is sufficient) by dropping the mixture with the help of a micropipette. Place a thin, transparent plastic sheet over the membrane to prevent it from drying and eliminate possible bubbles between the membrane and the plastic sheet. Leave the membrane in the ECL mixture for 3–5 min and bring to the imaging system. Take the first photo with automatic exposure and if necessary, take some photos with different exposure times using the automatic time exposure as a reference. The advantage of taking more photos is that the one with the best quality can be chosen later to be analyzed.

Finally, label a constitutive protein (like β- actin or GAPDH) in the same membrane to use as a normalizer. Usually, before carrying out another antibody incubation process to a new protein (including the normalizer), the membrane may be "stripped" of the antibodies previously used (*see* **Note 20**). To that, wash the membrane twice for 10 min each with the mild stripping buffer (it is very important to use it fresh and that the pH of this buffer is adjusted to 2.2 at the moment of use), two more times in PBS 1× for 10 min each and finally wash twice in TBS-T1× for 5 min each. Repeat the antibody incubation procedure described above since membrane blocking but now with the antibodies for the new target protein.

(d) Data processing and analysis

The results from western blots are usually represented in the literature by the membrane photograph in grayscale and cropped to show the most relevant region where the bands corresponding to the target of interest are seen (some figure examples can be viewed in Carvalho et al. [49]. However, it is highly recommended to include an evident area above and below the bands, since tight cropping can be seen as an attempt to hide information. Also, brightness and contrast adjustments may be acceptable under some conditions. They should be as minimal as possible, applied uniformly to the entire image, the background from the membrane image must be seen and the adjustments should not faint or hide any band [50, 51]. Finally, blots from different membranes must be represented in different panels, while the normalizer protein must be labeled from the same membrane and its image is usually presented cropped in the same size and just below the protein being investigated.

Despite representing a more qualitative or semi-quantitative analysis, western blot images can be used to indirectly quantify the expression of the proteins by in silico analysis. ImageJ software can be used to give values based on a density histogram of a selected area in the image. After obtaining the densitometry for each band, the corresponding values can be divided by the ones of the bands from the normalizer. Finally, a fold change relative to a control sample can be represented in a graph with statistical analysis using GraphPad Prism software.

3.4 General Statistical Analysis

Results obtained are presented as mean ± standard deviation of the mean (SD). All variables are tested for Gaussian distribution by the Kolmogorov-Smirnov normality test. For variables normally distributed and with compliance of homogeneity of variances, differences will be compared by Student's t-tests or by analysis of variance (ANOVA). In case of variables not normally distributed, differences will be analyzed by Mann–Whitney or by Kruskal-Wallis non-parametric tests. Bonferroni post-test will be used as needed for multiple comparisons. All statistical tests will be two-tailed and performed using a significance level of $\alpha = 0.05$. Statistical analyses, including association and correlation, will be performed using GraphPad Prism 8.4.2 (Dotmatics, United States).

4 Notes, Major Source of Problems

4.1 Immunohistochemistry

1. Optimal fixation can be obtained by intracardiac perfusion of the mice with formaldehyde, paraformaldehyde, or glutaraldehyde. However, this method requires specialized training and time.
2. Histology-grade paraffin wax, with a melting point around 56 or 58 °C, maintains tissue structures and essential morphological characteristics, ensuring thorough microscopic examination by pathologists.
3. In case of electrical failure of the processor, the pre-embedding process could not be finished properly. The tissue can be air-dried or returned to the 70% ethanol accidentally. To overcome these situations, it is recommended to rehydrate the tissue immerging it for 48 h in a solution of absolute ethanol (30 mL), formaldehyde 37% (0.5 mL), sodium carbonate (0.2 g) water up to 100 mL. After this, proceed with standard dehydration and pre-embedding. If tissues returned to 70% ethanol after the wax infiltration, discard contaminated fluids and rinse them in 95% ethanol, absolute ethanol (two times), and xylene. Finally, perform paraffin infiltration two to three times for 30–60 min each.
4. The precise orientation of the tissues is fundamental for accurate microscopic analysis. The mold must be filled with melted paraffin and transferred to a cooling surface. To maintain the traceability of each tissue specimen, the cassette identification must be meticulously positioned atop the base mold and incorporated into the paraffin block before the solidification.
5. For the immunostaining of phosphorylated proteins, the TBS wash buffer should be used, as the presence of phosphorus from the PBS buffer interferes with the interaction between the proteins and their respective phospho-specific antibodies. Conversely, the use of TBS should be avoided when the evaluated protein has a binding target on an amine group, given that TBS is a polyamine. The use of a surfactant in the washing steps is recommended to potentialize the removal of non-specific binding, reducing background. Although Tween 20 is a strong surfactant due to the presence of lauric acid, brain slices typically resist it well. However, if tissue degradation occurs, Tween 20 can be replaced with a milder surfactant, such as Tween 80, which contains oleic acid instead of laurate.

6. The micro-polymer system detection is biotin-free (ideal for studying tissue rich in endogenous biotin such as the brain), which represents an important advantage in comparison to ABC-based detection. In addition, the use of a micro-polymer system has better tissue penetration compared with the traditional polymer detection system.
7. Table 1 comprises the major source of problems in the immunohistochemistry assay and how to identify them.

4.2 ELISA

8. The protein extraction solution can be stored for 3 months, but the efficiency of the protease inhibitors decays. So, try to use a fresh-prepared solution.
9. The analysis of soluble components in the blood obtained from animals can be made along with the experimental protocol execution. In this approach, it is possible to evaluate the kinetics of mediators production, but a disadvantage is that the amount of sample obtained is restricted (the volume varies between <15% in 28 days of interval until less than 1% of total blood volume of the animal). Here we described the procedure for collection of blood in a terminal technique, when it is possible to obtain a great amount of samples and make it possible to extract other tissues of the same animal.
10. Prepare ELISA solutions right before the procedure. During the assay, bring them to room temperature (except for the protein extraction solution which needs to be cold). The described protocol follows the recommendations of the manufacturer (R&D systems), with some modifications. For other ELISA kits, see the manufacturer's specifications for antibody dilution, incubation time, and temperature.
11. Unfreeze the samples over the ice to reduce protein degradation. Depending on the concentration of the target in the sample, dilute the sample using the diluent solution (PBS/BSA 0.1%) to improve the detection. Homogenize the samples before pipetting in the plate. Avoid many defrosting cycles of the samples to ensure sample quality.
12. The DuoSet ELISA kit suggests the use of a TMB substrate solution. Alternatively, an OPD solution is also used, generating a different cleavage product that is read in a different wavelength (490 nm).
13. A limitation of the ELISA assay is the sensitivity of each kit, which is shown by the range of detection of the mass-calibrated standard. Due to this, if the calculated target concentration was null or negative, the result should be represented as "not detected."

Table 1
Major source of problems during the immunohistochemistry and possible solutions

Problem	Major source	Possible solution
Absent or weak staining	Fixation of the tissue may be modifying the epitope recognized by the antibody	Reduce the time of fixation Use a longer antigen retrieval or increase the temperature/pressure
	Deparaffinization may be insufficient	Deparaffinize sections longer, and change the melted xylene
	The target protein is nuclear and the antibody cannot penetrate the nucleus	Add a surfactant agent to the blocking buffer and antibody dilution solution
	The bond between the primary antibody and the protein is weak	Use a more concentrated antibody solution
	The protein is not expressed in the tissue Primary or secondary antibody/amplification kit may have lost their activity due to improper storage or extensive freezing/thawing	Run a positive control recommended by the supplier of the primary antibody for comparison
	The protein of interest is not abundantly expressed in the tissue	Perform the incubation with secondary antibody and micro-polymer in an oven at 37 °C
Intense background	Fixation procedures are too strong and modify the epitope recognized by the antibody	Decrease the time and temperature/pressure of the antigen retrieval step
	Endogenous peroxidases are active	Increase the blocking incubation period and consider adding an incubation step with H_2O_2 (0.3% v/v) for 30 min after read-to-use the peroxidase block solution
	Blocking of non-specific binding may be insufficient	Increase the blocking incubation period and consider adding an incubation step with 1% BSA for 30 min after the read-to-use protein block solution
	The Fixative is present in the tissue	Wash extensively in the buffer between the steps
	Permeabilization has damaged the membrane protein	Remove surfactant agents from the wash buffers
	The primary antibody is very concentrated	Incubate the tissues for longer in a more dilute antibody
	The amplification is excessive	Reduce amplification incubation time and dilute the amplification solution
Non-specific staining	Primary/secondary antibody concentration may be too high	Decrease the antibody concentration and/or the incubation period. Use negative control to compare
	Endogenous peroxidases are active	Increase the blocking incubation period and consider adding an incubation step with H_2O_2 (0.3% v/v) for 30 min after read-to-use the peroxidase block solution
	The primary antibody was produced against the same species as the tissue stained	Use a primary antibody raised against a different species than your tissue
	The sections dried out during the assay	Keep sections at high humidity using a wet incubation chamber

4.3 Western Blott

14. To avoid protein degradation, it is very important that all steps of protein extraction are carried out with the samples on ice and that protease inhibitors are added to the lysis buffer. Sodium orthovanadate (a phosphatase inhibitor) and the protease inhibitor cocktail should be added at the moment of use.
15. Before sonicating, to help dissolve the tissue in the lysis solution, a tissue homogenizer with a motorized pestle can be used and/or some suction cycles with a 1 mL syringe. If homogenizing the tissues in the lysis buffer still proves difficult, more lysis buffer can be added gradually, taking care so the samples do not become too diluted. It is highly recommended to take breaks of ~10 s between every 4–6 s of homogenizer or sonicator use to avoid the samples from heating. The homogenizer pestle and sonicator probe must be well-sanitized between samples.
16. If empty wells remain at the sides of the gel, fill these wells with a disposable sample to avoid dispersion of the samples of interest.
17. After the PVDF membrane is activated do not let it dry and handle it with the help of a filter forceps to avoid damage. A membrane of 0.45 μm pore size is suitable for most procedures, but when working with protein targets below 20 kDa, a PVDF membrane of 0.2 μm can be used to better retain these small proteins.
18. Optionally, after the transference and before blocking, the membrane can be incubated with Ponceau S 0.1% w/v to confirm transfer quality by total protein staining. Simply soak the membrane for about 1 min. in the Ponceau S solution under gentle shaking, remove ponceau excess and photo document if this control is needed. Wash the Ponceau S from the membrane with TBS-T1×. After this washing, some Ponceau may remain on the membrane, but this will not affect subsequent procedures.
19. Alternatively, the membrane can be incubated in 5% BSA in TBS-T1× instead of 5% skim powdered milk. Some proteins are better labeled using BSA and others skim milk. Also, when incubating with the antibodies dissolved in a blocking solution, the dilution proportion of the antibodies can be adjusted so the best signal is achieved.
20. Sometimes stripping the membrane is not necessary to label additional proteins, so it is sufficient to do another washing cycle, block the membrane again, and repeat the washes and incubation procedure using the antibody to the new target. However, some aspects of the western blot must be

considered so you can do the blotting process for the same membrane twice or more without stripping it, like the molecular weight of your proteins, the strength of the secondary antibody signal, and the host used to produce your primary antibodies. For example, if your target protein has a similar molecular weight to the protein to be used in normalization, the signal of the first will likely overlap with the second, so normalization will not be possible. Or, if both primary antibodies have the same host—which means that probably the same secondary antibody will be used for both—and one of them has a strong signal, there will be an interference in their chemiluminescence, making it not possible to analyze the second target. So, frequently, the membrane needs to be "stripped" before incubating for a new protein.

References

1. Yanny B, Winters A, Boutros S et al (2019) Hepatic encephalopathy challenges, burden, and diagnostic and therapeutic approach. Clin Liver Dis 23(4):607–623. https://doi.org/10.1016/j.cld.2019.07.001
2. Neff G, Zachry W III (2018) Systematic review of the economic burden of overt hepatic encephalopathy and Pharmacoeconomic impact of Rifaximin. PharmacoEconomics 36(7):809–822. https://doi.org/10.1007/s40273-018-0641-6
3. Chirapongsathorn S, Krittanawong C, Enders FT et al (2018) Incidence and cost analysis of hospital admission and 30-day readmission among patients with cirrhosis. Hepatol Commun 2(2):188–198. https://doi.org/10.1002/hep4.1137
4. Patidar KR, Thacker LR, Wade JB et al (2017) Symptom domain groups of the patient-reported outcomes measurement information system tools independently predict hospitalizations and re-hospitalizations in cirrhosis. Dig Dis Sci 62(5):1173–1179. https://doi.org/10.1007/s10620-017-4509-y
5. Tapper EB, Halbert B, Mellinger J (2016) Rates of and reasons for hospital readmissions in patients with cirrhosis: a multistate population-based cohort study. Clin Gastroenterol Hepatol 14(8):1181–1188. e1182. https://doi.org/10.1016/j.cgh.2016.04.009
6. Stepanova M, Mishra A, Venkatesan C et al (2012) In-hospital mortality and economic burden associated with hepatic encephalopathy in the United States from 2005 to 2009. Clin Gastroenterol Hepatol 10(9):1034–1041. e1031. https://doi.org/10.1016/j.cgh.2012.05.016
7. Bustamante J, Rimola A, Ventura PJ et al (1999) Prognostic significance of hepatic encephalopathy in patients with cirrhosis. J Hepatol 30(5):890–895. https://doi.org/10.1016/s0168-8278(99)80144-5
8. Amodio P (2018) Hepatic encephalopathy: diagnosis and management. Liver Int 38(6):966–975. https://doi.org/10.1111/liv.13752
9. Bajaj JS, Wade JB, Gibson DP et al (2011) The multi-dimensional burden of cirrhosis and hepatic encephalopathy on patients and caregivers. Am J Gastroenterol 106(9):1646–1653. https://doi.org/10.1038/ajg.2011.157
10. Wong RJ, Gish RG, Ahmed A (2014) Hepatic encephalopathy is associated with significantly increased mortality among patients awaiting liver transplantation. Liver Transpl 20(12):1454–1461. https://doi.org/10.1002/lt.23981
11. Jiang W, Desjardins P, Butterworth RF (2009) Direct evidence for central proinflammatory mechanisms in rats with experimental acute liver failure: protective effect of hypothermia. J Cereb Blood Flow Metab 29(5):944–952
12. Jayakumar A, Norenberg MD (2018) Hyperammonemia in hepatic encephalopathy. J Clin Exp Hepatol 8(3):272–280
13. Shawcross D, Jalan R (2005) The pathophysiologic basis of hepatic encephalopathy: central role for ammonia and inflammation. Cell Mol

Life Sci 62(19–20):2295–2304. https://doi.org/10.1007/s00018-005-5089-0

14. Shawcross DL, Wright G, Olde Damink SW et al (2007) Role of ammonia and inflammation in minimal hepatic encephalopathy. Metab Brain Dis 22(1):125–138. https://doi.org/10.1007/s11011-006-9042-1
15. Tranah TH, Vijay GK, Ryan JM et al (2013) Systemic inflammation and ammonia in hepatic encephalopathy. Metab Brain Dis 28(1):1–5. https://doi.org/10.1007/s11011-012-9370-2
16. Walker V (2014) Ammonia metabolism and hyperammonemic disorders. Adv Clin Chem 67:73–150. https://doi.org/10.1016/bs.acc.2014.09.002
17. Parekh PJ, Balart LA (2015) Ammonia and its role in the pathogenesis of hepatic encephalopathy. Clin Liver Dis 19(3):529–537. https://doi.org/10.1016/j.cld.2015.05.002
18. Lima LCD, Miranda AS, Ferreira RN et al (2019) Hepatic encephalopathy: lessons from preclinical studies. World J Hepatol 11(2):173–185. https://doi.org/10.4254/wjh.v11.i2.173
19. Kerbert AJC, Jalan R (2020) Recent advances in understanding and managing hepatic encephalopathy in chronic liver disease. F1000Res 9. https://doi.org/10.12688/f1000research.22183.1
20. Butterworth RF (2016) The concept of "the inflamed brain" in acute liver failure: mechanisms and new therapeutic opportunities. Metab Brain Dis 31(6):1283–1287. https://doi.org/10.1007/s11011-015-9747-0
21. Azhari H, Swain MG (2018) Role of peripheral inflammation in hepatic encephalopathy. J Clin Exp Hepatol 8(3):281–285. https://doi.org/10.1016/j.jceh.2018.06.008
22. Reus GZ, Fries GR, Stertz L et al (2015) The role of inflammation and microglial activation in the pathophysiology of psychiatric disorders. Neuroscience 300:141–154. https://doi.org/10.1016/j.neuroscience.2015.05.018
23. Limon ID, Angulo-Cruz I, Sanchez-Abdon L et al (2021) Disturbance of the glutamate-glutamine cycle, secondary to hepatic damage, compromises memory function. Front Neurosci 15:578922. https://doi.org/10.3389/fnins.2021.578922
24. Ochoa-Sanchez R, Rose CF (2018) Pathogenesis of hepatic encephalopathy in chronic liver disease. J Clin Exp Hepatol 8(3):262–271. https://doi.org/10.1016/j.jceh.2018.08.001
25. Santos RPC, Toscano ECB, Rachid MA (2023) Anti-inflammatory strategies for hepatic encephalopathy: preclinical studies. Arq Neuropsiquiatr 81:656–669
26. Häussinger D, Dhiman RK, Felipo V et al (2022) Hepatic encephalopathy. Nat Rev Dis Primers 8(1):43. https://doi.org/10.1038/s41572-022-00366-6
27. França MER, Peixoto CA (2020) cGMP signaling pathway in hepatic encephalopathy neuroinflammation and cognition. Int Immunopharmacol 79:106082. https://doi.org/10.1016/j.intimp.2019.106082
28. Gairing SJ, Danneberg S, Kaps L et al (2023) Elevated serum levels of glial fibrillary acidic protein are associated with covert hepatic encephalopathy in patients with cirrhosis. JHEP Rep Innov Hepatol 5(4):100671. https://doi.org/10.1016/j.jhepr.2023.100671
29. Wahab DYAA, Gau CH, Zakaria R et al (2019) Review on cross talk between neurotransmitters and Neuroinflammation in striatum and cerebellum in the mediation of motor behaviour. Biomed Res Int 2019:1767203. https://doi.org/10.1155/2019/1767203
30. Butterworth RF, Norenberg MD, Felipo V et al (2009) Experimental models of hepatic encephalopathy: ISHEN guidelines. Liver Int 29(6):783–788
31. DeMorrow S, Cudalbu C (2021) 2021 ISHEN guidelines on animal models of hepatic encephalopathy. Liver Int 41(7):1474–1488. https://doi.org/10.1111/liv.14911
32. Sepehrinezhad A, Shahbazi A, Negah SS et al (2021) Drug-induced-acute liver failure: a critical appraisal of the thioacetamide model for the study of hepatic encephalopathy. Toxicol Rep 8:962–970
33. El-Marasy SA, El Awdan SA, Abd-Elsalam RM (2019) Protective role of chrysin on thioacetamide-induced hepatic encephalopathy in rats. Chem Biol Interact 299:111–119. https://doi.org/10.1016/j.cbi.2018.11.021
34. Wang L-Q, Zhou H-J, Pan C-F et al (2011) Expression of IL-1β, IL-6 and TNF-α in rats with thioacetamide-induced acute liver failure and encephalopathy: correlation with brain edema. Asian Biomed 5(2):205–215
35. Zhang L, Tan J, Jiang X et al (2017) Neuron-derived CCL2 contributes to microglia activation and neurological decline in hepatic encephalopathy. Biol Res 50(1):1–11
36. Biomarkers Definitions Working Group (2001) Biomarkers and surrogate endpoints: preferred definitions and conceptual framework. Clin Pharmacol Ther 69(3):89–95. https://doi.org/10.1067/mcp.2001.113989

37. Aronson JK, Ferner RE (2017) Biomarkers-a general review. Curr Protoc Pharmacol 76: 9.23.21–29.23.17. https://doi.org/10.1002/cpph.19
38. Califf RM (2018) Biomarker definitions and their applications. Exp Biol Med (Maywood) 243(3):213–221. https://doi.org/10.1177/1535370217750088
39. Abi-Dargham A, Moeller SJ, Ali F et al (2023) Candidate biomarkers in psychiatric disorders: state of the field. World Psychiatry 22(2): 236–262. https://doi.org/10.1002/wps.21078
40. Berk M (2023) Biomarkers in psychiatric disorders: status quo, impediments and facilitators. World Psychiatry 22(2):174–176. https://doi.org/10.1002/wps.21071
41. Trares K, Wiesenfarth M, Stocker H et al (2024) Addition of inflammation-related biomarkers to the CAIDE model for risk prediction of all-cause dementia, Alzheimer's disease and vascular dementia in a prospective study. Immun Ageing 21(1):23. https://doi.org/10.1186/s12979-024-00427-2
42. Edalatfar M, Piri SM, Mehrabinejad MM et al (2021) Biofluid biomarkers in traumatic brain injury: a systematic scoping review. Neurocrit Care 35(2):559–572. https://doi.org/10.1007/s12028-020-01173-1
43. Parellada M, Andreu-Bernabeu Á, Burdeus M et al (2023) In search of biomarkers to guide interventions in autism Spectrum disorder: a systematic review. Am J Psychiatry 180(1): 23–40. https://doi.org/10.1176/appi.ajp.21100992
44. Paxinos G, Franklin KB (2019) Paxinos and Franklin's the mouse brain in stereotaxic coordinates. Academic
45. Ruehl-fehlet C (2000) Revised guides for organ sampling and trimming in rats and mice. Pathol India: Tixol
46. Seyed Jafari SM, Hunger RE (2017) IHC optical density score: a new practical method for quantitative immunohistochemistry image analysis. Appl Immunohistochem Mol Morphol 25(1):e12–e13. https://doi.org/10.1097/pai.0000000000000370
47. Toscano ECB, Vieira ÉLM, Grinberg LT et al (2023) Hyperphosphorylated tau in mesial temporal lobe epilepsy: a neuropathological and cognitive study. Mol Neurobiol 60(4): 2174–2185. https://doi.org/10.1007/s12035-022-03190-x
48. Spijker S (2011) Dissection of rodent brain regions. Neuroproteomics:13–26
49. Carvalho BC, Oliveira LC, Rocha CD et al (2019) Both knock-down and overexpression of Rap2a small GTPase in macrophages result in impairment of NF-κB activity and inflammatory gene expression. Mol Immunol 109:27–37. https://doi.org/10.1016/j.molimm.2019.02.015
50. Rossner M, Yamada KM (2004) What's in a picture? The temptation of image manipulation. J Cell Biol 166(1):11–15. https://doi.org/10.1083/jcb.200406019
51. Kroon C, Breuer L, Jones L et al (2022) Blind spots on western blots: Assessment of common problems in western blot figures and methods reporting with recommendations to improve them. PLoS Biol 20(9):e3001783. https://doi.org/10.1371/journal.pbio.3001783

Part II

Experimental Models and Methods

Chapter 6

Animal Models of Hepatic Encephalopathy

Mar Martínez-García, Ana Agustí, Yaiza M. Arenas, Paula Izquierdo-Altarejos, Gergana Mincheva, and Marta Llansola

Abstract

Hepatic encephalopathy (HE) is a complex neuropsychiatric syndrome induced by liver disease. HE is characterized by cognitive and motor disturbances, along with sleep disturbances, that progress and expands and finally altered mental status is obvious, leading to coma and death. Three types of HE have been defined: (1) Type A HE, associated with acute liver failure; (2) Type B HE, associated with portal-systemic shunting in the absence of liver disease or failure and (3) Type C HE, associated with chronic liver disease from different etiologies as alcohol consume, chronic hepatitis, or obesity. Different animal models have been developed to study the pathological mechanisms of HE. These models should reproduce etiopathology and HE symptoms observed in humans, but it results difficult due in part to the complex and variable etiopathology and multiple factors modifying disease progression, including presence of different complications after liver disease affecting different organs. In addition, it is yet difficult to reproduce in animals some features of mental alterations present in humans. In spite of this, the studies in animal models have allowed to known basic and common mechanisms of HE, mainly related to hyperammonemia and inflammation, and to propose several therapeutic targets. The efficacy of pharmacological modulation of these mechanism-based therapeutic targets has also been evaluated in these different preclinical animal models of HE. In this chapter, we describe the most used animal models of Type A and Type C HE.

Key words Hepatic encephalopathy, Acute liver failure, Chronic liver disease, Hyperammonemia, Anastomosis, Hepatotoxin, Bile duct ligation, Surgery

1 Introduction

Hepatic encephalopathy (HE) is a complex neuropsychiatric syndrome that appears as a common complication of liver disease/failure. HE is characterized by psychiatric, cognitive and motor disturbances. Pathological mechanisms are now best known due in great part to the use of different preclinical animal models to study them. Three types of HE have been defined based on degree and severity of liver failure or disease: (1) Type A HE, which is associated with acute liver failure (ALF) resulting from the rapid

Tiziano Balzano (ed.), *Experimental and Clinical Methods in Hepatic Encephalopathy Research*, Neuromethods, vol. 219, https://doi.org/10.1007/978-1-0716-4354-9_6,

onset of hepatocyte necrosis and severe inflammation without pre-existing liver impairment; (2) Type B HE, associated with portal-systemic shunting in the absence of liver disease or failure and (3) Type C HE, associated with chronic liver disease (CLD), including the recently named steatotic liver disease (SLD), with origin in steatosis in the liver, and chronic hepatitis on the other side. CLD usually starts presenting less severe chronic liver damage (steatosis, inflammation and fibrosis) and progressively reaches severe fibrosis and cirrhosis [1].

Although the three types imply different etiopathologies and underlying mechanisms, they share two important characteristics: the increase in ammonium in the blood, hyperammonemia, and the prominent role of peripheral inflammation that, when translated to the brain, induces neuroinflammation which leads to alterations of neurotransmission and cellular damage and death. This is responsible for many of the neurological alterations, such as cognitive and motor impairment of different degrees. Hyperammonemia is caused by the decreased capacity of the damaged liver to detoxify ammonia, which is a product of protein degradation, with toxic effects in the brain when levels are increased [2–8].

In 2017, globally 1.5 billion people suffered a CLD [9]. Some common causes of CLD are chronic hepatitis C, alcohol abuse and obesity (associated with SLD), either as a unique cause or in combination [10]. Several million patients with liver cirrhosis suffer covert or minimal HE (MHE), with mild cognitive and motor impairments that reduce their quality of life and life span and increase de probability of finally suffer overt or clinical HE [7].

No sufficiently efficient treatments has been developed at this moment to prevent or notably improve HE symptoms. However, the different animal models have allowed preclinical studies testing new therapeutic strategies, some of which are now being tested in clinical trials and this improves therapeutic perspectives for patients with HE [7, 8].

In this chapter, the different models of both, HE due to CLD and HE caused by ALF, will be described in detail. Animal models of chronic HE are (1) chronic hyperammonemia, without liver failure, in rats (2) porta-cava end-to-side shunt or porta-cava anastomosis, in rats (3) bile duct ligation, (4) chronic administration of the hepatotoxin carbon tetrachloride, and (5) administration of thioacetamide. The animal models described here for ALF-induced HE are all based in administration of high dose of different hepatotoxins (1) D − +-Galactosamine (2) Acetaminophen (3) Thioacetamide, and (4) Carbon tetrachloride.

2 Materials

2.1 Chronic Hyperammonemia Without Liver Failure

1. Commercial flour (SAFE: Scientific Animal Food & Engineering, Scientific diet brand-A04).
2. Ammonium acetate (PanReac AppliChem 131114.1211).
3. Trays.
4. Aluminum foil.
5. Balance.
6. Water.
7. Universal Oven UF with Forced-Air Convection UF750 (Memmert™, Fisher), that has a minimum temperature of 42 °C and a maximum temperature of 60 °C.

2.2 Porta-Cava End-to-Side Shunt

All surfaces should be disinfected and the material must be previously autoclaved in order to maintain sterile conditions.

General equipment:

1. Analgesia: meloxicam 2 mg/Kg and 0.1 mg/Kg buprenorphine.
2. Appropriately sized needles and syringes for subcutaneous (s.c.) injections.
3. Inhalation anesthesia: isoflurane 1000 mg/g liquid (IsoVet).
4. Vaporizer with Flowmeter (Medical Supplies & Services) and O_2 supply and a Fluosorber (Harvard Apparatus) for absorption of isoflurane gas.
5. Hot bead sterilizer.
6. Electric hair clipper.
7. Surgical magnifying glass.
8. Corneal protectant (lacrilube, isoptotears, TearGel, etc).
9. Warmed subcutaneous fluids (sterile saline solution (0.9% NaCl)).
10. Paper towels and gauze.
11. Cotton swab sticks.
12. Drape with appropriately sized fenestration (hole) for surgical incision.
13. Scalpel (blade size 10).
14. Surgical scissors—curved, sharp-blunt.
15. Muscle forceps.
16. Skin forceps.
17. Suture: silk 4/0, 6/0, 7/0, absorbable suture 4/0.

Table 1
Specific material to perform the end-to side- portacaval anastomosis

Tool	Image
a. Retractors	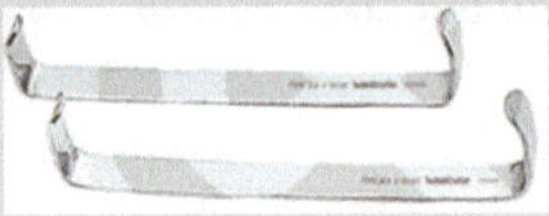
b. Moria spring scissors—5 mm blades	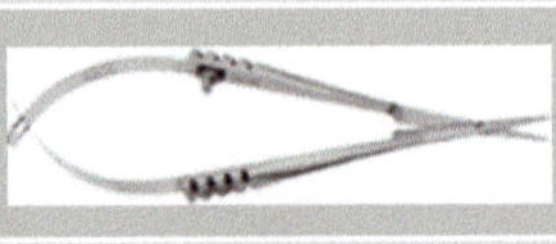
c. Moria MC31 iris forceps serrated, 0.5 mm tip	
d. Moria MC44 bulldog clamp curved 53 mm	
e. Castaneda neonatal clamp, very Delicated curved, 12 cm	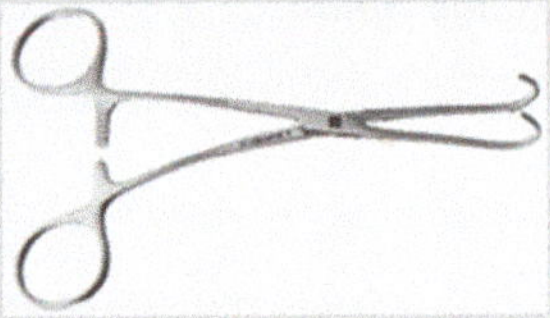
f. Dumont #5 mirror finish forceps	
g. Mosquito forceps curved, 14 cm	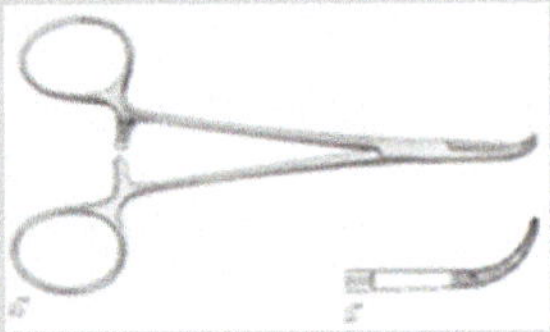
h. Castroviejo needle holder, with lock tungsten carbide 14 cm	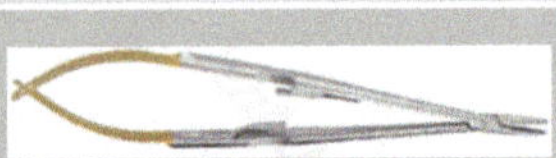

Specific material to perform the end-to side- portacaval anastomosis is shown in Table 1.

2.3 Bile Duct-Ligation

1. 9 mm staples.
2. Staple applicator.
3. Curved scissors.
4. Spring-loaded scissor pin.
5. Retractor.
6. Gauze.
7. Shaving machine.
8. Tissue clamps.
9. Very fine round tip curved tweezers.

10. 6/0 braided silk suture to ligate the duct and 4/0 to sew the muscle.
11. Anesthesia chamber.
12. Inhalation anesthesia: isoflurane 1000 mg/g liquid (IsoVet).
13. Vaporizer with Flowmeter (Medical Supplies & Services) and O_2 supply and a Fluosorber (Harvard Apparatus) for absorption of isoflurane gas.
14. Buprenorphine.
15. Meloxicam.
16. 15 mL tube.

Reptile heating plate to keep warm.

2.4 Intraperitoneal Injections of Hepatotoxins

1. Hepatotoxin solution (range of doses and solvent is specified in each method).
2. Syringes sterile and disposables.
3. Needles: 25-gauge needle for rats or 25–27 for mice.
4. Infra-red lamp.
5. Olive or Corn oil, as solvent and vehicle for control rats.
6. Saline (0.9% NaCl), as solvent and vehicle for control rats.

2.5 Thioacetamide

Supportive therapy: to avoid hypoglycemia, hypovolemia, hypokalemia, or electrolyte imbalance, s.c. injections of a solution containing a) 10% glucose water mixed with lactate ringer (25 mL/kg, 100 mM NaCl, 3.7 mM KCl, 1.3 mM $CaCl_2$, 1.2 mM $MgSO_4$, 28 mM $NaHCO_3$, and 20 mM sodium lactate) or b) 10% glucose or 5% dextrose/0.45% saline, containing 20 mEq/L (0.2%) of KCl.

3 Methods

Models of Chronic HE

3.1 Chronic Hyperammonemia Without Liver Failure

One of the most important factors that trigger the motor and cognitive alterations in MHE is hyperammonemia [7]. In order to differentiate which alterations are due to hyperammonemia and which are due to other alterations associated with liver failure, at the end of the 1980s our research group developed a model of "pure" chronic hyperammonemia without liver failure. This model is simple, no invasive and allows hyperammonemia to be maintained for a long time without altering liver function [11, 12]. The hyperammonemia model does not require surgery but it lacks liver failure and it does not have portal-systemic shunting. It is a useful animal model to study the effect of ammonia as a neurotoxin.

The model consists on a daily diet enriched in ammonia content.

Ammonium-enriched diet is made with commercial flour and ammonium acetate. First, the flour and ammonium are weighed based on the desired percentage of ammonium, then water is added and kneaded until a homogeneous dough remains. The prepared dough is then placed in several trays with the base covered with aluminum foil. The dough should be approximately 1 cm thick and cut into square "cookies." Next, they are placed in an oven put at 42 °C. They are left in the oven for two days and then they are turned over so that they can also be baked in the other part of the dough and left again for two more days. Finally, the oven is turned off and they are allowed to cool so that they can be collected and stored at −4 °C.

The rats are first fed with a diet with a 20% ammonium acetate (2 kg of flour +533 grams of ammonium acetate) so that it is less unpleasant and they get used to eating it. After 5–7 days, the rats are changed to 30% ammonium diet (2 kg of flour +1 kg of ammonium acetate). The ammonium cookies in the rat cages are replaced every two or three days so that they are in good condition and do not have humidity, which softens them and makes them more unpleasant to gnaw. Hyperammonemia is induced quick after the first three days, and studies have shown that rats can tolerate this ammonia-supplemented diet for up to 100 days. After three-four weeks, the rats are considered to have chronic hyperammonemia and then experiments with these rats can start. Controls for these hyperammonemic rats are fed with a standard commercial chow, in pellets. In diet-induced models, pair-fed animals are also recommended as controls. Ammonium can also be administered parenterally [13] or induced by urease treatment [14], but they are time-consuming and not suitable for long-term studies.

Hyperammonemic rats have a 35%, 41%, and 70% increase in ammonium levels in liver, brain, and muscle, respectively, compared to control rats, while in blood the concentration doubles. In addition, there is an increase in ammonium excretion. The levels of urea in the blood and its elimination are around double the values corresponding to control rats [12]. These values are in the range of the values found in cirrhotic patients.

The motor and cognitive alterations in this model are similar to those presented by MHE surgical models (PCS and BDL rats) and patients with MHE [15]. Cognitive impairment and microglia activation can be yet observed after only seven days of ammonia diet. Recently, it has reported that this model also show peripheral inflammation, altered plasma cytokine content, at only two days after starting ammonium-diet [16, 17].

3.2 Porta-Cava End-to-Side Shunt

The portal-systemic shunt occurs when the usual flow of blood from the portal vein is redirected, either partially or entirely, into the systemic circulation, bypassing the liver. This results in a reduction of the liver's ability to remove ammonia, and other toxic compounds, from the blood, leading to hyperammonemia, but without underlying liver disease. Congenital shunt is found in both human [18] and dogs [19]. While in mice there is a specific strain of congenital shunt, C57BL/6 J mice [18] in rats, to obtain a shunt model, it is necessary to perform a surgical procedure called end-to side- portacaval anastomosis. This model reproduces neurological alterations observed in patients with MHE. The end-to side- portacaval anastomosis (PCS) leads to elevated blood ammonia levels, neurological impairment, and eventually induces liver atrophy [20–22]. The neurological impairment in PCS rats includes cognitive impairment, motor incoordination, altered motor activity and alterations in circadian rhythms and sleep fragmentation [23–28].

Although this surgical technique can also be performed in larger animals, like pigs [29], we are focusing in describe the end-to side- portacaval anastomosis rat surgery.

3.2.1 Preparation and Stabilization of Animal for Surgery

For this surgery to be successful, it should be performed on rats weighing between 200 and 250 grams. Thirty minutes before the surgery commenced, the analgesia protocol was initiated through the s.c. injection of 2 mg/kg of meloxicam and 0.1 mg/Kg of buprenorphine by tenting the skin of the back and injecting into the space between the skin and underlying muscle. Thirty minutes after, the rat is placed in a small chamber to induce the anesthesia by inhalation of isoflurane (5% isoflurane in oxygen) until the animal achieves a suitable level of nociception, as indicated by the absence of reaction in the pinch-reflex test applied to the foot, with ongoing reflex monitoring at subsequent stages of the surgery. Then the rat is removed from the anesthesia chamber and its abdomen is shaved as much as possible with an electric hair clipper and quickly. Shave at least 2 cm perimeter around planned surgical site. Remove all loose hair using gauze. The rat is placed on the sterile drape on the surgery table positioned lying down and face upwards. The rat is covered with a drape with appropriately sized hole for surgical incision and the anesthesia mask is placed on the muzzle, reducing the isoflurane to a maintenance dose of 2% in oxygen. Then, corneal protectant in both eyes is applied. In terms of body temperature, although for surgeries it is generally recommended to use a safe heat source (i.e., homeothermic feedback blanket, warm-water circulating pad, infrared heater) to avoid hypothermia and maintain the body temperature, we observed that this surgery in particular is more successful if the body temperature is the same as the room temperature, avoiding the use of these devices. As the surgery lasts 20–25 min, the risk of hypothermia is relative. Then, clean the skin at the planned surgical site with 2% chlorhexidine solution (or equivalent) using gauze. Commence cleaning directly over where the planned incision will be and extend outwards without touching the center again. Dry the surface of the animal with gauzes using the same technique and discard used gauzes.

3.2.2 Surgical Procedure

Surgical Procedure to Access and Clamp the Portal and Cava Veins

Once the surface of the abdomen rats is cleaned we start the laparotomy. Using a scalpel, an incision is made from top to bottom of the abdomen (approximately 4 cm), taking care not to cut the muscle. The skin is separated from the muscular tissue with the help of scissors. The abdominal muscle is stretched with forceps and a small cut is made vertically with the scissors in order to make a hole from which the abdomen is cut from the bottom to the top. The cuts should be long, taking care not to tear the muscle. The viscera are pulled apart with the retractors[a], exposing the portal vein and the vena cava. Clean the vena cava with cotton swab sticks. If necessary, cut any adhesions between the vena cava and the liver with small scissors[b]. The vena cava should be cleaned carefully, taking care to remove as much fat as possible. A 50 mL falcon tube was positioned on the rat's back, causing the body to arch, thereby exposing the portal vein for enhanced visibility. Locate the hepatic artery which is located just behind the highest branch of the portal artery and to the right. The hepatic artery has a "Y" shape. It is very important not to rupture the hepatic artery as the animal will bleed to death. Clean the portal vein thoroughly and separate the right branch of the vein from the hepatic artery. To do this, use fine forceps[c] with a curved, blunt tip to not perforate the artery. Tie the right branch of the portal vein. Firstly, we will tie two simple and strong knots on the right end of the branch, ensuring they are as distant as possible from the opposite end. Utilize 6/0 black silk suture and when tightening the strands to form the simple knots, ensure do it in the intended direction. In this case extend the suture ends to the right. Secondly, we tie two simple and strong knots at the left end, close to the portal vein, proceeding in the same way as before. Once we have the two knots in the branch and separated by a sufficient distance, we cut with very small scissors[b] between the two knots (Fig. 1b, c). We tie the portal vein at its highest end with two simple and strong knots, using 6/0 black silk suture and stretching the suture ends upwards. Clamp the portal vein from left to right with the clamp[d] downwards just below the knot that remains in the portal vein after cutting its right branch. Cut the portal vein as high up as possible, so as close as possible to the knot we have just made, taking care not to cut it (Fig. 1c). Use the small scissors[b]. Immediately lower the anesthesia from 2% to 0.5% of isoflurane in oxygen. From this point on, we have a maximum of *15 minutes* to perform the end-to side- portacaval anastomosis.

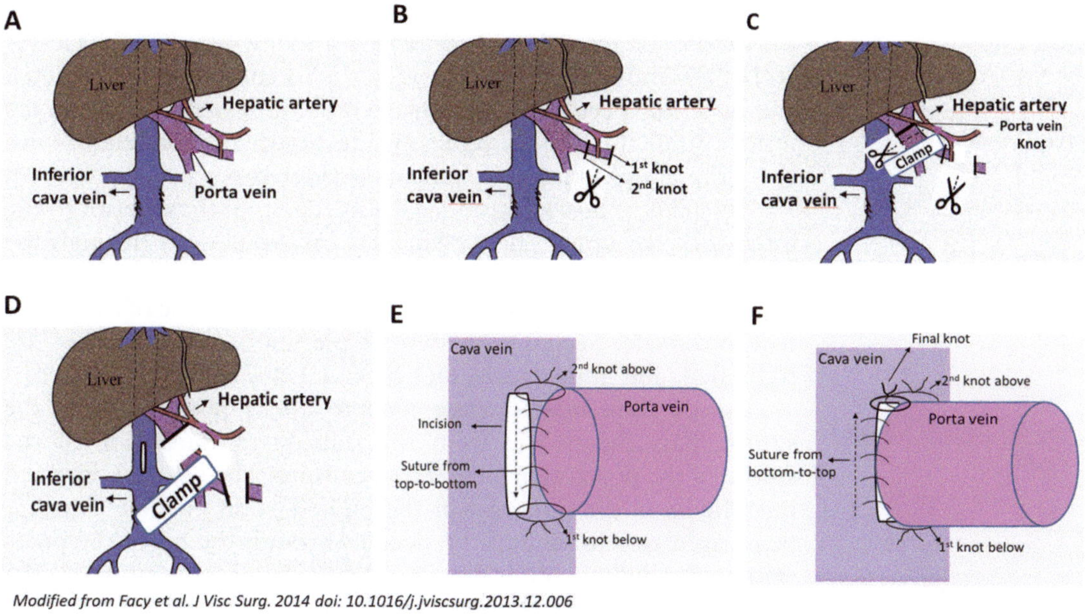

Fig. 1 Representative scheme of the end-to-side portacaval shunt. Scheme of the normal cava and porta veins (**a**). Diagram showing how to tie two knots in the right branch of the portal vein in order to cut this branch (**b**). Diagram of how to knot the portal vein at its highest point and clamp it to cut in between (**c**). Diagram of the approach to the portal vein and the vena cava and the incision in the vena cava for suturing (**d**). Scheme of the porta and cava vein suture: first know below, the second knot above and the suture from top-to-bottom (bottom lips or floor (**e**). Scheme of the porta and cava vein suture: suture from bottom-to-top (upper lips) (**f**)

Surgical Procedure to Approach the Portal and Cava Veins

To suturing the portal and the cava veins, we must first position the animal and both veins to facilitate our vision and work as much as possible. If we are right-handed, we will turn the rat around (180°), leaving the vena cava to our right. If we are left-handed we will leave it in the same position. With fine forceps we stretch the vena cava and with a scissor-shaped clamp[e] with a very curved end we take a pinch of the vena cava being careful not to pinch the two walls of the vena, we only have to pinch the upper wall. We position the end of the cut portal vein in proximity to the vena cava, securing it in that position by holding the clamp with an extralong mosquito forceps curved[g] (Fig. 1d). Using a magnifying glass fur surgery, we verify precise focus and ensure alignment with our intended view. From this moment on, we will work with the magnifying glass until we finish suturing both veins. Subsequently, we measure the diameter of the cut portal vein's end by delicately stretching it with two fine-tipped forceps. Once measured, we proceed to make a small incision in the vena cava matching the diameter with the cut end of the portal vein. Employing forceps, we gently stretch the pinch of the vena cava executing a small vertical incision. Subsequently, we cut upwards and downwards through this small aperture. A reassessment of the diameter ensures the cut is of the correct size. This is a crucial step. We start the suture of the portal vein with the vena cava using 7/0 black silk suture.

Suture of the Portal and Cava Veins

First knot below: using a small needle holder or castroviejo [h], grasp the needle and insert it from the exterior to the interior of the lower end of the stump, pulling it through the recently made cut. Stretch and insert the needle into the lower end of the portal vein, from the inside (inside the stump) to the outside. Secure the connection with two simple, tight knots. Cut the ends (Fig. 1e).

Second knot above: introduce the needle from the exterior to the interior of the upper end of the vena cava, pulling it through the incision in the vena cava. Extend the thread and once more insert the needle into the portal vein, moving from the inside of the stump to the outside. Secure with two simple knots, ensuring they are tightened, and importantly, DO NOT CUT THE TIES (Fig. 1e).

Suture from top-to-bottom (bottom lips or floor): to start the suture, pass the needle from the outside to the inside of the stump of the portal vein and start the suture. The needle is inserted into the lip of the vena cava at the highest point possible. Moving from right to left we pull the needle through the lip of the porta vein at the highest point possible. We continue the suture all the way down. We will make approximately five stitches (Fig. 1e).

Suture from bottom-to-top (upper lips): we pull the needle from the inside to the outside of the portal stump and initiate the suture from bottom to top. We insert the needle into the vena cava from right to left, stretching. We the insert the needle from the outside to the inside of the porta vein and from the inside to the outside and we stretch. We proceed with the next stitch and we performed approximately five stitches (Fig. 1f).

Final knot: Once we have reached the top, we will tie a knot with the remaining end and with the end we left when making the first top knot. We tie with two simple and tightly secured knots and cut off the excess. End of the stitch (Fig. 1f).

Completion of Surgery and Post-surgery Analgesia Protocol

Then, we carefully unclamp the vena cava. There might be a slight amount of blood, which we'll clean with a cotton swab. Next, we unclamp the portal vein with great care. Once again, a small amount of blood may be released. If the anastomosis is well-executed, the portal vein should transition from a dark color to a reddish color. We readjust the viscera, adding two mL of saline solution to prevent dehydration. The abdominal closure is done using absorbable 4/0 suture, employing individual stitches with 3 knots (double, single, single). Then, the skin is sutured with 4/0 silk suture. In this case, a continuous or intradermal suture can be performed, using 6/0 black silk for the latter. Following the surgery, the animal is housed in an individual cage until it awakens. Subsequently, they will be reintegrated with their cage mate's post-operation. An analgesia protocol will be implemented, involving the s.c. administration of 0.1 mg/Kg buprenorphine every 8–12 h, beginning 8 h after the initial meloxicam 2 mg/Kg and 0.1 mg/Kg buprenorphine dose, during three days.

With the described protocol a favorable outcome of 60–80% of the rats surviving the surgery is achieve.

3.3 Bile-Duct Ligation

This is a model of cholestatic-induced liver disease. Experimental bile duct ligation has been performed in rats and mice and also in dogs and pigs [30–33]. The surgery procedure consists in perform a double ligation in the common bile duct and cut between these two ligatures.

Cholestatic liver disease progresses from mild to more severe disease by progressive increase of hepatobiliary injury, inducing cholangitis (primary biliary cholangitis (PBC)) and ultimately liver fibrosis and secondary biliary cirrhosis [34]. This model is possibly the most widely used to induce CLD in rodents, mainly to study the process of hepatic fibrosis, but also for studies of HE induced by CLD [1]. After biliary ligation, the animals suffer cholestasis and rapidly liver inflammation appears leading to hepatic fibrosis and finally cirrhosis, around four weeks after surgery. The timeframe in which these processes occur is greatly dependent on the mouse strains [34]. Rats with bile-duct ligation survive between five-six weeks, if biliary reconduction has not occurred, but this is a possibility and in a small percentage of bile-duct ligated rats biliary reconduction occurs with subsequent physiological improvement and survival. A visual typical characteristic of these rats is to show jaundice due to increased bilirubin levels with yellow urine and plasma. Usually ascites is also developed after long time (four weeks at least). To prevent the dilation of the ligated bile ducts, formalin (1 μL/mg) can be injected intracholedochally before the surgery is performed [35, 36].

Rats with bile-duct ligation (BDL) are not hyperammonemic two to three weeks after ligation. Therefore, a hyperammonemic diet was introduced to create a BDL + hyperammonemic model, which replicates the brain alterations observed in cirrhotic patients [37–39].

First analgesia is applied by an intradermal injection of a mixture of Buprenorphine (0.1 mg/kg) and Meloxicam (2 mg/kg) 15–20 min before surgery. Then, anesthesia is induced in the anesthetic chamber through an isoflurane gas dispenser with the pressure gauge at 5% (max). When the rat has fallen asleep, pressure is lowered to 2.5%. Next, place the rat with the mask connected to de isoflurane dispenser, on the operating table face up, on top of the heating plate. During operation isoflurane is maintained at 2–2.5%. Now, shave the abdominal area and clean it with Chlorexidine. Apply lidocaine cream to have additional local analgesia and wait 5–10 min. Now surgery procedure start: open the abdomen along the Alba line about 2–3 cm, first the skin is cut and then the muscle, taking care of the internal organs; put the 15 mL tube under the rat to bend it and then open with the retractor, separating the liver and leaving the ducts visible; separate the common bile duct with small curved forceps and tie a strong knot in the upper part of the bile duct and another in the lower part with the 6/0 suture; now the duct is cut between the nodes (Fig. 2); remove the 15 mL tube

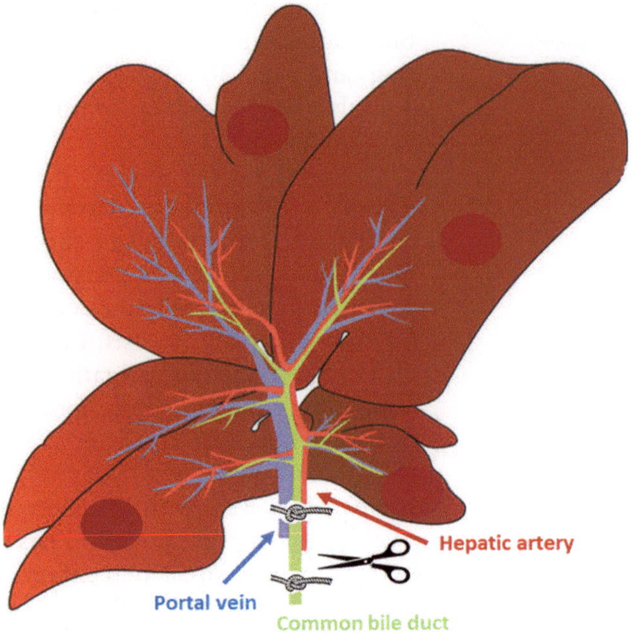

Modified from Tag et al. J Vis Exp. 2015 doi: 10.3791/52438

Fig. 2 Scheme of bile duct ligation

from under the rat and the retractor and reposition the organs and infuse saline solution (approximately 3 mL) to prevent dehydration; sew the muscle with 4/0 braided silk suture loose stitches, staple the skin and finally clean and let the animal wake up. The surgery itself should not last more than 10 min. For 3 days, the mixture of analgesics for pain will be injected intradermally. The staples will be removed after 10 days if the wound is dry. It will be necessary to monitor daily so that the animal itself or its cage mate does not remove the staples and open the wound.

Controls of BDL animals are shams. In these animals, a midline incision is performed and bile duct localized, but ligation is not performed.

3.4 Chronic and Progressive Liver Damage by Administration of Hepatotoxins

These HE animal models are based on the induction of liver injury by chronic administration of different hepatic toxins.

3.4.1 Carbon Tetrachloride

The administration of the hepatotoxin carbon tetrachloride (CCl_4) for the induction of cirrhosis in the liver was reported in 1936 by Cameron and Karunaratne [40]. Since then, several techniques of administration and animal species have been used for experimental production of cirrhosis [41]. Chronic administration of CCl_4 in rats reproduces the progression of liver disease with similar characteristics to patients [42]. Centrilobular liver damage is caused by the toxic metabolite trichloromethyl (CCl_3-) produced after CCl_4 metabolization via cytochrome p450. Continued exposure to CCl_4 in rats or mice induces liver damage, ductular reaction, myofibroblast and hepatic stellate cell activation, and an imbalance in extracellular matrix proteins, leading to progressive liver fibrosis [1]. This is a model frequently used to study hepatic fibrosis. Yamamoto and Sugihara [43] showed that administration of CCl_4 for over 10 weeks produced HE in 80% of the treated rats. Long-term administration of CCl_4 induces in rats motor incoordination and impaired memory, hyperammonemia after six weeks of toxin administration, increased peripheral inflammation, disruption of the blood-brain barrier (BBB), neuroinflammation with glial cell activation, altered glutamatergic and GABAergic neurotransmission, and a decrease in hippocampal neurogenesis. Most of these physiologic and biochemical alterations are also present in mice after prolonged administration of CCl_4 [1, 44].

This model has the advantage to reproduce the progression of liver damage, and thus, can be used also as a model of SLD and of MHE. Administration of CCl_4 for 4 weeks, which leads to mild liver damage, with low level of hepatic fibrosis and without reach hepatic cirrhosis, already induces motor incoordination and altered locomotor gait and hippocampus-dependent neurological impairment. This is associated to altered GABAergic neurotransmission in the cerebellum and alteration of glutamatergic neurotransmission in the hippocampus, neuroinflammation, and increased systemic inflammation and BBB disruption in rats, although they did not show yet hyperammonemia [44–46].

This model is usually chosen due to the easy application and its practicality, and also due to its low mortality [1]. However, the response to the toxin of individual animals is variable.

For the administration of CCl_4 in rodents, different routes of administration, dosages, frequency of dosing and duration can be used. Routes of administration can be intraperitoneal (i.p.) or subcutaneous (s.c.) injections or oral gavage. In rats, CCl_4 is commonly administered by i.p. injections at 1–3 mL/kg (of body weight) diluted in olive or corn oil, two to three times per week. Rats injected with 1 mg/kg three times per week show cirrhosis at 12 weeks [42]. In mice, doses used range from 0.5 mL/kg to 1 mL/kg, twice per week for up to 16 weeks [1]. Different strains show different susceptibility to CCl_4-induced liver fibrosis [47].

Intraperitoneal Injection

(This Procedure Is Only Explained Here and Is the Same for All i.p. Injections Included in the Different Animal Models Described Below in the Text)

Fill the syringe with the corresponding hepatotoxin solution and attach the needle to the syringe, making certain that there are no bubbles in the line. Infuse hepatotoxin solution with gentle hand pressure on the syringe. Syringes should be prepared previously to animal restriction. Hepatotoxin dilution should be at room or body temperature. The i.p. injection site should be in the lower left or right quadrant of the abdomen of rodents restrained with the head lowered to avoid vital organs [48]. The lower right quadrant is preferred to avoid damaging the urinary bladder, cecum and other abdominal organs. The needle should be inserted with the bevel facing up, towards the head of the animal and at a 30–40° angle to horizontal. Although the lower right quadrant is preferred, due to frequent injections in this model, sides may be switched between injections. Small bleeding in the injection site may occur, in that case pressure should be applied and the site should be cleaned.

Control rats always receive same number of i.p. injections with the same volume and concentration of the corresponding solvent.

3.4.2 Thiocetamide

Thioacetamide (TAA) is an organosulfur compound that was first identified as hepatotoxic to rats in 1948 by Fitzhugh and Nelson [49]. Later on, TAA was established as a reliable and reproducible experimental model of fibrosis and cirrhosis by oral or i.p. routes [50–52]. It can be used to study the biology of fibrosis and cirrhosis development and the functional complications of this liver damage, such as portal hypertension or HE.

The administration of TAA generates a dose- and time-dependent pattern of liver damage that reproduces human fibrosis and cirrhosis [53]: at 6 weeks of TAA oral administration, rats show histopathological features of hepatitis, which progresses to advanced fibrosis at 12 weeks and to cirrhosis at 18 weeks. Histological alterations typical of cirrhosis include parenchymal cell damage and death, nodular parenchymal regeneration and complete fibrous vascularized septa. Dysplasia of cholangiocytes and hepatocytes can also be observed, a premalignant change frequently observed in human chronic liver disease. From 12 weeks of treatment, portal hypertension and systemic pressure also start to be altered [54]. The TAA model has been described to increase ammonia in blood and brain, as well as ALT and AST levels in serum [55, 56]. Compared to BDL and PCS models, which induce hyperammonemia passively, TAA-induced liver cirrhosis is likely to represent more closely the actual condition found in patients with liver diseases [57].

A daily TAA i.p. administration during 10 days can be considered as mild-grade type C HE or rats with MHE, and it has been observed to impair performance in the Morris Water Maze [56]. The TAA model impairs cognitive functions to a greater extent than other HE models: prolonged TAA administration in drinking water (more than 12 weeks) severely affected learning and both spatial reference and working memory in the Morris Water Maze [57]. Other symptoms like motor deficits and anxiety are not so clearly affected in this model and contradictory results can be found.

In this model, is common to observe mild body weight loss. Mortality of up to 15% can be expected when TAA is administered i. p., while oral administration leads to less extrahepatic toxicity and higher survival rates, achieving a more continuous dosing [58].

According to the 2021 ISHEN guidelines [1] on animal models of HE, this TAA administration in rats is recommended as Type C HE model. Its main advantage would be an excellent practicality, while the main disadvantage is a not so good reproducibility [1].

Two main routes of administration are used: oral and i.p. administration.

Intraperitoneal Injections

Dissolve TAA in saline solution to a total volume of 1–1.5 mL per rat 1 h before injection and keep the solution at room temperature. Following the above described procedure (*see* Subheading 2.4.1), with the dominant hand, inject the TAA at a dose of 100–150 mg/kg.

Different injection frequencies and total time of the study have been used, ranging from daily injections during shorter times (10 days to 3 weeks) to 2 or 3 injections per weeks during longer times (10–18 weeks), depending on the stage of liver damage that the authors aimed to study. It can be considered that, after 10–12 weeks with 3 injections per week, virtually all animals have already developed all cirrhotic features [58].

Oral Administration

In this case, TAA is dissolved in drinking water at a general dose of 300 mg/L. It is recommended to filter and acidify the drinking water (pH around 4.0) to reduce the risk of infections. Treated animals are given free access to drinking water. The water volume is determined daily and control animals are given the same volume consumed by the animals in the TAA group. To better monitor water intake, animals should be placed in individual cages. However, after the model is already stablished (around 12 weeks), it is recommended to place more than 1 animal per cage before performing behavioral studies to minimize the effects of isolation. A usual period to stablish the model is 12 weeks, with studies ranging from 2 to 4 months.

Models of Acute Liver Failure-Induced HE

Different animal models of acute liver failure (ALF) has been developed and used to investigate causes and mechanisms of HE. HE, including brain alterations as edema and increased intracranial pressure are the main causes of death after ALF. Existence of numerous animal models of ALF is due to the fact than any of them reproduces optimally the clinic characteristics of ALF in humans. Animal models of ALF primary are based in to induce an extensive liver injury in a short time leading to liver failure. Two principal ways to induce this liver failure are through surgery and by administration of high doses of hepatotoxic compounds or drugs [60].

Two different surgery-based models of ALF exist: (1) resection of hepatic artery or devascularization and (2) resection of a great proportion of liver tissue. Surgery models reproduce clinical features of ALF patients better than hepatotoxicity models, present HE and are reproducible. The main inconvenient of surgery models is that this is not reversible with adequate treatment and survival time is very short, if liver transplantation is not performed. These models has been used mainly to evaluate artificial liver support [61, 62], but not for mechanistic and therapeutic studies. We don't describe these models in this chapter.

Thus, the main difference between the different models described in this chapter is the hepatotoxin used. Most used hepatotoxins to investigate pathogenesis of acute HE in animal models of ALF are D-galactosamine, Acetaminophen, Carbon tetrachloride and Thioacetamide.

Chemical hepatotoxin-based animal models are mostly reversible (some animals would survive the process if a suitable treatment is administered), non-reproducible, and show a variable interval between damage and death, and also species and age variability. Thus, no optimal model is yet developed. These models reproduce the main clinical characteristics of ALF in humans (hypoglycemia, increased blood levels of hepatic enzymes, hyperammonemia and HE), but intracranial hypertension and cerebral edema is not observed in some chemical models [62, 63].

Rats and mice have been widely used and also rabbits, dogs, pigs, and non-human primates have been occasionally used. Large animal models are more useful, even necessary, to research and development of artificial and bioartificial liver support [61, 62, 64–67].

To better study progression of brain pathology until coma and death occurs after hepatotoxin administration, lower sublethal doses has been used in several studies using these models of ALF, which delay progression of symptoms and death. The survival at the maximal dose is usually around 48 h. This implies intense monitoring during that time to evaluate the mechanisms that trigger HE after liver failure and also the possible therapeutic treatments that can delay or prevent death. Thus, a common and important feature of all these models is the evaluation of the progression of clinical symptoms of HE. Behavioral and functional evaluation is performed by defined established criteria in rats and mice. Animals are monitored clinically and neurologically by testing 14 different reflexes including pain, grasping, position and corneal reflexes. For each reflex present one point is given; thus, a normal rat has a score of 14 points. The proposed protocol for behavioral and functional analysis of mouse phenotype is SHIRPA, that allows to obtain a comprehensive phenotype assessment. Animals should be closely observed until death [66, 68, 69].

Progression of clinical symptoms of HE are classified in four grades. These defined stages are (1) loss of scatter instinct, decreased motor activity, unbalanced posture, mild ataxia, piloerection and hypothermia (stage I), (2) severe ataxia (back pins), very calm, animal is able to turn to the 4 pins position from the back (stage II), (3) animal is not able to turn to the 4 pins position from the back, it does not have reaction to a pinch, it does not have corneal reflex, loss of righting reflex (stage III, pre-coma), and (4) does not react to tail pinching, coma (stage IV) [68]. These grades can be preceded by a prodromal phase characterized by decreased spontaneous activity and decreased food and water intake [70].

In this chapter, we describe the characteristics and doses of the different most used hepatotoxins in rats and mice.

3.5 Galactosamine

D-Galactosamine (D-Gal) is a highly hepatospecific compound that does not directly affect other organs. After in vivo administration of D-Gal, it is eliminated by the pathway responsible for galactose metabolism in hepatocytes. During the first phase, D-Gal is phosphorylated to galactosamine-1-phosphate (GalN-1-P), then converted to UDP-D-Gal, which is an inhibitor of UDP-glucose pyrophosphorylase-catalysed reaction. This leads to depletion of uridine phosphate pool and uridine deficiency inhibits mRNA synthesis and then protein synthesis [71, 72]. However, reactive oxygen species also contribute to D-Gal-induced hepatotoxicity [73].

The doses usually used in rats ranges from 1 to 3 g/kg, injected i.p. At a dose of 1.5 g/kg, histochemical and biochemical liver alterations in rats, mice, pigs and rabbits, and males and females were compared. No sex differences in liver injury induced by D-Gal were found. Mice, and also rabbits and pigs, resulted less affected than rats in these experimental conditions [64]. This observation was also later reported by Czekaj et al. [72], showing that rats are more susceptible to D-Gal intoxication than mice. To develop full ALF dose of 80–140 mg/100 g of body weight in rats and 150–270 mg/100 g in mice are necessary. Furthermore, the effect of D-Gal intoxication is highly dependent on the animal strain.

A dose of 1 g/kg allows to perform evaluation of the progression of symptoms and brain alterations as cerebral edema or electroencephalogram alterations for seven days [74, 75], whereas with a dose of 2.5 g/kg most of the rats death in 48 h. Even a lower dose of 250 g/kg injected twice a week for 30 days has been used as sublethal dose model of acute HE [76].

D-(+)-Galactosamine hydrochloride is injected i.p. in a dose of 2.5 g/kg. The galactosamine can be dissolved in sterile water (for example as a 0.28 M solution) and adjusted to pH 7.4 with 5 M sodium hydroxide immediately before injection or in saline. Control rats receive an equivalent volume of saline. Twenty-four hours after galactosamine injection the drinking water is replaced by 5% glucose to prevent hypoglycemia [77, 78]. Mortality in these conditions was of 69% and occurred between 43 and 60 h after galactosamine injection [77].

Various dose combinations of LPS and D-Gal produce sublethal liver failure which is relevant to clinical situations in viral, drug or alcohol-induced, immune-induced or ischemia reperfusion hepatitis. The combination of 10 μg/kg LPS with 400 mg/kg D-Gal produce a nonlethal hepatitis as proved histologically and biochemically evidenced. Therefore, D-Gal/LPS combinations with different amounts of D-Gal/LPS appear very well reproducible model for in vivo experimental hepatitis studies [79]. Some studies with D-Gal model of ALF add a LPS injection (or a main proinflammatory cytokine, tumor necrosis factor-alpha (TNF-α)) to induce a septic-shock, as precipitating factor for HE, than often occurs in patients, in addition to hepatic failure. In these studies, D-Gal dose used was between 1 and 1000 mg/kg as an i.p. injection. LPS is diluted in saline solution [67, 75, 80, 81]. LPS can be used at a dose of 20–50 μg/kg (LPS, Klebisella) [67, 80] and TNF-α at 0.10 μg/g [81].

3.6 Acetaminophen

Acetaminophen (paracetamol) is a commonly used analgesic drug with hepatotoxic potential. Acetaminophen hepatic metabolism is produced by glucuronidation and sulfation reactions, with formation of metabolites which are later excreted through the kidney. When an excess of the drug is present, normal detoxifying pathways are saturated and the drug is metabolized through cytochrome P-450 to N-acetyl-p-benzoquinoneimine which, unless conjugated with glutathione, is thought to interrupt mitochondrial calcium flux and to induce cell damage by the formation of hydroxyl radicals, nitrites and nitrates, leading to apoptosis and cell necrosis [82]. Its toxicity is also mediated by c-jun kinases (JNK). The toxic effect of acetaminophen is dose-dependent and can be exacerbated by fasting, cytochrome P-450 inducer drugs and especially by alcohol [62, 83].

Rats are injected i.p. with 10–25 mg/kg of 3-methylcholanthrene as a 10 ml solution in corn oil 16 h–24 h prior to use and then fasted overnight. This treatment makes rats more susceptible to acetaminophen by the induction of cytochrome p-450. Then rats are injected i.p. with 750–1000 mg/kg of acetaminophen in dimethyl sulphoxide [84, 85]. A single dose of acetaminophen (1 g/kg) induced mild degree hepatotoxicity. This model of mild hepatic impairment more resembles the human acetaminophen-induced liver injury [79, 86].

3.7 Thioacetamide

Thioacetamide causes hepatocellular necrosis following biotransformation by mono oxygenases.

This model need repeated administration and a support therapy but has the advantage of a large time window before death and higher reproducibility.

ALF is induced by i.p. injection of TAA (350–600 mg/kg/day in 5 mL/kg saline) daily for three consecutive days. Control animals received an equal volume of the vehicle (5 mL/kg).

Supportive therapy: to avoid hypoglycemia and electrolyte imbalance, s.c. injections of a solution containing 10% glucose water mixed with lactate ringer (25 mL/kg) or 5% dextrose/ 0.45% saline containing 20 mEq/L of potassium chloride, are performed every 12 h after the first injection of TAA [87, 88]. With the high dose of 600 mg/kg/day and applying the second supportive therapy, the mean interval between the last dose of TAA and the onset of overt HE was 40 ± 10 (S.D.) min for the unsupplemented model and 120 ± 30 min for the supplemented model. The duration of encephalopathy from onset to death was 60 ± 20 min in the unsupplemented model and 7 ± 3 h in the supplemented model [87]. With a dose of 350 mg/kg/day, and glucose plus lactate ringer as supportive therapy about 50% of animals survive [88].

ALF can be also induced by administration of 300 mg/kg of TAA by gavage on two consecutive days. Control rats received 2 mL normal saline by gavage. The first dose was given to overnight fasted rats. Thereafter, rats had free access to food and water. To prevent hypoglycemia, rats receive a mixture of 5 mL saline and 5 mL 10% glucose with 20 mEq/L KCl by s.c. injection within the first 2 days every 8 h, thereafter every 4 h. Body temperature can be maintained at 37 °C by the intermittent use of an infra-red lamp. Controls should be always treated identically. Monitoring of behavior and neurological symptoms for clinical classification of HE grades start after receive all dose of TAA [68].

In this model rats became lethargic and displayed a delayed response to noise or passive movement 64.7 ± 2.1 h (range: 49 to 85 h) after the first dose of TAA, but not present yet neurological alterations (HE grade I). This stage lasted 2.1 ± 0.2 h (range: 1 to 5 h). Then the animals became ataxic, but the reflex score was normal (HE grade II). After 2.9 ± 0.3 h (range: 1 to 6 h), the condition of the animals deteriorated progressively and the reflexes gradually disappeared, starting with the disappearance of position reflexes, then of the grasping and pain reflexes (HE grade III). Thus, the reflex score decreased. Grade III HE lasted 1.7 ± 0.2 h (range: 0.5 to 4.5 h). Finally, the corneal reflex was absent, and the rats were in deep coma (HE grade IV). The duration of Stage IV HE is not predictable and vary from 5 min

Table 2
Staging of hepatic encephalopathy in rats with thioacetamide-induced liver failure

Stage of HE	Prominent neurohehavioral change	Reflex score
I	Lethargy	14
II	Ataxia	13–14
III	Gradual loss of reflexes	2–12
IV	Coma	0–2

to 2 h. Some animals with Stage IV HE can present tonic clonic convulsions prior to death, whereas a proportion of animals survive without signs of HE [68] (Table 2).

This rat model has been also adapted to mice. TAA dissolved in sterile saline solution is injected i.p. as a single dose of 200 mg/kg. Twenty-four hours after injection of TAA all animals (including controls) were injected s.c. with 0.5 mL of a solution containing 0.45% NaCl, 5% dextrose and 0.2% KCl in order to prevent hypovolemia, hypokalemia and hypoglycemia. The mice were intermittently exposed to infrared light in order to prevent hypothermia [89]. In these conditions mice survive until 12 days, at least. In this case neurological function was assessed by a 10-point scale based on reflexes and task performance: exit from a 1 m in diameter circle in less than 1 min, seeking, walking a straight line, startle reflex, grasping reflex, righting reflex, placing reflex, corneal reflex, maintaining balance on a beam 3, 2, and 1 cm in width, climbing onto a square and a round pole. For each task failed or abnormal reflex reaction, a score of 1 was assigned. Thus, a higher score indicates poorer neurological function. The neurological score is assessed 1 day after induction of ALF by TAA (day 2) to start treatment so that all groups with different treatments and controls groups had similar baseline neurological scores after TAA induction. The post-treatment neurological score is assessed 1 day after administration of treatment or vehicle (day 3) [89].

3.8 Carbon Tetrachloride (CCl_4)

The main mechanism of liver injury by CCl_4 involves its biotransformation leading to generation of a free radical, CCl_3, that induces cell membrane oxidation and DNA damage. Free radicals are generated by cytochrome P450 (CYP) enzymes, in particular the CYP2E1 isozyme, which reduce CCl_4 to CCl_3 and then covert it to superoxides [90]. Histological evidence of CCl_4-induced liver damage can be observed mainly in zone 3 of the hepatic acinus. The CCl_4 model of liver injury is one of the most widely used, but unfortunately, in terms of its mechanism of action, it has no direct equivalent in human liver injury. However, histopathological changes in acute CCl_4 intoxication are very similar to acetaminophen overdose, which is one of the causes of ALF in humans.

To induce ALF in mice and rats, the CCl_4 dose should be 40–750 μL/100 g body weight [72]. To be a useful model for evaluation of therapeutic agents is necessary to use lower dose with the purpose of decrease mortality and to have wide therapeutic window. For this purpose, Czekaj et al. [72] used dose of CCl_4 (dissolved 1:1 in olive oil) of 200 μL/100 g in rats and 50 μL/100 g in mice, in a single i.p. injection.

This hepatotoxin has extrahepatic toxicity, mainly in lung and kidney and a small time window before death. Depending on the dose administered, exposure time, the presence of exacerbating agents, or the age of the organism affected, regeneration can occur and lead to the total recovery of the damaged liver.

A lethal dose is 2.5 mL/kg of CCl_4 injected i.p. At this dose most of the rats death 36 h post-injection and all are death in a maximum of 96 h [65]. Other authors use this same i.p. dosage, but of a dilution of CCl_4 1:4 v/v in mineral oil, for induction of ALF. This dose provoked death at 4–5 days post-injection. A 3.0 mL/kg induced death at 2–3 days and lower dose of 1–2 mL/kg lead to death only of 30–50% of rats and death start at 7–5 days post-injection [91].

3.9 Azoxymethane (AOM)

Azoxymethane (AOM) has been used only in mice and induces alterations similar to those encountered in humans with ALF [70, 81, 92, 93].

Mice receive a single i.p. injection of AOM (100 μg/g) dissolved in 100 μL saline. Control animals receive an equivalent volume of saline. Mice showed decreased locomotor activity followed by loss of righting and corneal reflexes, hyperammonemia and spontaneous hypothermia after AOM administration, and then can be a reproducible model suitable for the study of the cerebral complications of ALF [92]. Changes in expression of zonula occludens-2 and increased BBB permeability has been reported in this model, suggesting a role of this process in the induction of brain edema in ALF [94].

3.10 Acute-on-Chronic Liver Failure-HE

Acute-on-chronic liver failure (ACLF) is a catastrophic syndrome that occurs in patients with cirrhosis who decompensate acutely, usually due to a precipitating event, such as infection and inflammation, and rapidly progress to multiorgan failure [80]. One example model of ACLF consists on mice gavaged with carbon tetrachloride (CCl_4 0.5 mg/mL dissolved in olive oil—dose 0.5 mL/kg) twice weekly for a total of 6 weeks to induce chronic liver injury. LPS derived from Klebsiella pneumonia (4 mg/kg) is injected i.p. to mimic ACLF. Animals are sacrificed 24 h after LPS injections [80].

3.11 Animal Models of HE Induced by Alcohol Liver Disease (ALD) and High Fat Diet-Induced Liver Damage

Some studies have been carried out with animal models that specifically include some factors that cause liver damage. For example, it is necessary to include alcohol consumption as a factor responsible for chronic liver damage [1], but taking into account that alcohol also causes brain damage. Thus, some alcoholic-liver disease models have been used, but HE has rarely been investigated in them [95–97]. The existing mouse and rat models of ALD are based on chronically administering in the intestinal tract a percentage of around 30% alcohol, either through a specific diet or through the implantation of an intragastric tube, which is more invasive. Controls for these models must take a supplement that has the same caloric value as the ethanol added to the diet [97–99].

As a model of steatotic liver disease (SLD), high fat diet-induced liver disease has been often used in rodents. However, only few articles afford brain alterations [100].

3.12 Conclusions

We describe here all the most used animal models for both HE type C and A. They are numerous but each of them has its advantages and are based on different etiologies and reproduce the most outstanding characteristics of the different types and etiologies of HE. This variety is due to the numerous factors that trigger HE and that contributes to the complex symptoms, despite the fact that the origin is always liver disease or failure, which are also due to very different causes.

All models have been and are useful to unravel mechanisms and triggering factors of HE and also to carry out preclinical studies of therapeutic strategies. These studies are very necessary given the current lack of truly effective treatments to prevent or improve HE, which is a cause of significant deterioration in the quality of life, hospitalization, and death of patients with liver diseases or failure.

4 Notes

4.1 Chronic Hyperammonemia Without Liver Failure

It must be taken into account that the oven must be a space thermocycler and that it does not exceed 45 °C; otherwise the vitamins present in the flour will be spoiled. In addition, the trays where the ammonium cookies are placed must have an order so that the heat eddies are distributed evenly. It should be noted that ammonium cookies must be stored for use in closed buckets at 4 °C; otherwise the ammonium evaporates.

4.2 Porta-Cava End-to-Side Shunt

- The weight of the rats should be at least 200 grams.
- No use of heat blanket.
- It is very important not to rupture the hepatic artery; otherwise the animal will die.

- Success in this surgery hinges on both speed and precision; completing the suturing of the portal and vena cava veins within a brief timeframe, ideally no more than 15 min, is crucial.
- Achieving an effective suture of both the portal vein and the vena cava requires precise measurements of the portal vein diameter and a carefully proportioned incision in the vena cava. An overly large incision in the cava vein may result in stenosis of the porta, while an incision that is too small may lead to inadequate blood flow. In either scenario, the rat is at risk of mortality within 48 h.

4.3 Thioacetamide

- The handling of TAA involves some safety concerns. Due to its high toxicity, this hepatotoxin should be manipulated wearing eyeglasses, protective gloves, appropriate clothing, and respiratory mask.
- TAA should be stored in a cool, dry, and safe place to avoid chemical decomposition.
- To reduce the possible side-effects of body weight loss, hypoglycemia, and renal failure induced by TAA administration, supportive therapy consisting in 5% dextrose containing 0.45% NaCl and 20 meq l^{-1} KCl can be co-administered in the drinking water [59].

References

1. DeMorrow S, Cudalbu C, Davies N, Jayakumar AR, Rose CF (2021) 2021 ISHEN guidelines on animal models of hepatic encephalopathy. Liver Int 41(7):1474–1488. https://doi.org/10.1111/liv.14911
2. Butterworth RF (2011) Hepatic encephalopathy: a central neuroinflammatory disorder? Hepatology 53(4):1372–1376. https://doi.org/10.1002/hep.24228
3. Butterworth RF (2015) Pathogenesis of hepatic encephalopathy and brain edema in acute liver failure. J Clin Exp Hepatol 5 (Suppl 1):S96–S103. https://doi.org/10.1016/j.jceh.2014.02.004
4. Butterworth RF (2016) The concept of "the inflamed brain" in acute liver failure: mechanisms and new therapeutic opportunities. Metab Brain Dis 31(6):1283–1287. https://doi.org/10.1007/s11011-015-9747-0
5. Jayakumar AR, Rama Rao KV, Norenberg MD (2015) Neuroinflammation in hepatic encephalopathy: mechanistic aspects. J Clin Exp Hepatol 5(Suppl 1):S21–S28. https://doi.org/10.1016/j.jceh.2014.07.006
6. Donnelly MC, Hayes PC, Simpson KJ (2016) Role of inflammation and infection in the pathogenesis of human acute liver failure: clinical implications for monitoring and therapy. World J Gastroenterol 22(26):5958–5970. https://doi.org/10.3748/wjg.v22.i26.5958
7. Cabrera-Pastor A, Llansola M, Montoliu C, Malaguarnera M, Balzano T, Taoro-Gonzalez L, García-García R, Mangas-Losada A, Izquierdo-Altarejos P, Arenas YM, Leone P, Felipo V (2019) Peripheral inflammation induces neuroinflammation that alters neurotransmission and cognitive and motor function in hepatic encephalopathy: underlying mechanisms and therapeutic implications. Acta Physiol (Oxf) 226(2):e13270. https://doi.org/10.1111/apha.13270
8. Lima LCD, Miranda AS, Ferreira RN, Rachid MA, Silva AC SE (2019) Hepatic encephalopathy: lessons from preclinical studies. World. J Hepatol 11(2):173–185. https://doi.org/10.4254/wjh.v11.i2.173
9. Moon AM, Singal AG, Tapper EB (2020) Contemporary epidemiology of chronic liver

disease and cirrhosis. Clin Gastroenterol Hepatol 18(12):2650–2666. https://doi.org/10.1016/j.cgh.2019.07.060
10. Pinzani M, Rosselli M, Zuckermann M (2011) Liver cirrhosis. Best Pract Res Clin Gastroenterol 25(2):281–290. https://doi.org/10.1016/j.bpg.2011.02.009
11. Felipo V, Miñana MD, Grisolía S (1988) Long-term ingestion of ammonium increases acetylglutamate and urea levels without affecting the amount of carbamoyl-phosphate synthase. Eur J Biochem 176(3):567–557
12. Azorín I, Miñana MD, Felipo V, Grisolía S (1989) A simple animal model of hyperammonemia. Hepatology 10(3):311–314
13. Rodrigo R, Cauli O, Gomez-Pinedo U, Agusti A, Hernandez-Rabaza V, Garcia-Verdugo JM, Felipo V (2010) Hyperammonemia induces neuroinflammation that contributes to cognitive impairment in rats with hepatic encephalopathy. Gastroenterology 139(2):675–684
14. Llansola M, Montoliu C, Cauli O, Hernández-Rabaza V, Agustí A, Cabrera-Pastor A, Giménez-Garzó C, González-Usano A, Felipo V (2013) Chronic hyperammonemia, glutamatergic neurotransmission and neurological alterations. Metab Brain Dis J28(2): 151–154. https://doi.org/10.1007/s11011-012-9337-3
15. Balzano T, Dadsetan S, Forteza J, Cabrera-Pastor A, Taoro-Gonzalez L, Malaguarnera M, Gil-Perotin S, Cubas-Nuñez L, Casanova B, Castro-Quintas A, Ponce-Mora A, Arenas YM, Leone P, Erceg S, Llansola M, Felipo V (2020) Chronic hyperammonemia induces peripheral inflammation that leads to cognitive impairment in rats: reversed by anti-TNF-α treatment. J Hepatol 73(3):582–592. https://doi.org/10.1016/j.jhep.2019.01.008
16. Ferenci P, Pappas SC, Jones EA (1982) L-leucine prevents ammonia-induced changes in glutamate receptors in the brain and in visual evoked potentials (abstract). Hepatology 2:708
17. Prior RL, Visek WJ (1972) Effects of urea hydrolysis on tissue metabolite concentrations in rats. Am J Phys 223(5):1143–1149
18. Cudalbu C, McLin VA, Lei H, Duarte JMN, Rougemont A-L, Oldani G, Terraz S, Toso C, Gruetter R (2013) The C57BL/6J mouse exihibits sporadic congenital portosystemic shunts. PLoS One 8(7):e69782
19. Or M, Devriendt N, Kitshoff AM, Peremans K, Vandermeulen E, Paepe D, Polis I, Martlé V, de Rooster H (2017) Ammonia concentrations in arterial blood, venous blood, and cerebrospinal fluid of dogs with and without congenital extrahepatic portosystemic shunts. Am J Vet Res 78(11):1313–1318
20. Gandhi CR, Murase N, Subbotin VM, Uemura T, Nalesnik M, Demetris AJ, Fung JJ, Starzl TE (2002) Portacaval shunt causes apoptosis and liver atrophy in rats despite increases in endogenous levels of major hepatic growth factors. J Hepatol 37(3): 340–348
21. Apelqvist G, Hindfelt B, Andersson G, Bengtsson F (1999) Altered adaptive behaviour expressed in an open-field paradigm in experimental hepatic encephalopathy. Behav Brain Res 106(1–2):165–173
22. Coy DL, Mehta R, Zee P, Salchli F, Turek FW, Blei AT (1992) Portal-systemic shunting and the disruption of circadian locomotor activity in the rat. Gastroenterology 103(1):222–228
23. Erceg S, Monfort P, Hernández-Viadel M, Rodrigo R, Montoliu C, Felipo V (2005) Oral administration of sildenafil restores learning ability in rats with hyperammonemia and with portacaval shunts. Hepatology 41(2):299–306
24. Agusti A, Cauli O, Rodrigo R, Llansola M, Hernández-Rabaza V, Felipo V (2011) p38 MAP kinase is a therapeutic target for hepatic encephalopathy in rats with portacaval shunts. Gut 60(11):1572–1579. https://doi.org/10.1136/gut.2010.236083
25. Hernandez-Rabaza V, Agusti A, Cabrera-Pastor A, Fustero S, Delgado O, Taoro-Gonzalez L, Montoliu C, Llansola M, Felipo V (2015) Sildenafil reduces neuroinflammation and restores spatial learning in rats with hepatic encephalopathy: underlying mechanisms. J Neuroinflammation 12:195. https://doi.org/10.1186/s12974-015-0420-7
26. Agusti A, Hernández-Rabaza V, Balzano T, Taoro-Gonzalez L, Ibañez-Grau A, Cabrera-Pastor A, Fustero S, Llansola M, Montoliu C, Felipo V (2017) Sildenafil reduces neuroinflammation in cerebellum, restores GABAergic tone, and improves motor in-coordination in rats with hepatic encephalopathy. CNS Neurosci Ther 23(5):386–394. https://doi.org/10.1111/cns.12688
27. Llansola M, Cantero JL, Hita-Yañez E, Mirones-Maldonado MJ, Piedrafita B, Ahabrach H, Errami M, Agusti A, Felipo V (2012) Progressive reduction of sleep time and quality in rats with hepatic encephalopathy caused by portacaval shunts. Neuroscience 201:199–208. https://doi.org/10.1016/j.neuroscience.2011.11.009

28. Felipo V, Piedrafita B, Barios JA, Agustí A, Ahabrach H, Romero-Vives M, Barrio LC, Rey B, Gaztelu JM, Llansola M (2015) Rats with minimal hepatic encephalopathy show reduced cGMP-dependent protein kinase activity in hypothalamus correlating with circadian rhythms alterations. Chronobiol Int 32(7):966–979. https://doi.org/10.3109/07420528.2015.1057640
29. Cuschieri A, Baker PR, Holley MP, Hanson C (1974) Portacaval shunt in the pig 1. Effect on survival, behavior, nutrition, and liver function. J Surg Res 17(6):387–396
30. Layden TJ, Schwarz, Boyer JL (1975) Scanning electron microscopy of the rat liver. Studies of the effect of taurolithocholate and other models of cholestasis. Gastroenterology 69(3):724–738
31. Melman A (1978) Effect of angiotensin II inhibition in the bile-duct-ligated dog. J Surg Res 24(4):277–283. https://doi.org/10.1016/0022-4804(78)90092-6
32. Kountouras J, Billing BH, Scheuer PJ (1984) Prolonged bile duct obstruction: a new experimental model for cirrhosis in the rat. Br J Exp Pathol 65(3):305–311
33. Chen CY, Shiesh SC, Lin XZ (1999) Biliary sludge and pigment stone formation in bile duct-ligated Guinea pigs. Dig Dis Sci 44(1): 203–209. https://doi.org/10.1023/a:1026639225023
34. Van Campenhout S, Van Vlierberghe H, Devisscher L (2019) Common bile duct ligation as model for secondary biliary cirrhosis. Methods Mol Biol 1981:237–247. https://doi.org/10.1007/978-1-4939-9420-5_15
35. Rose C, Butterworth RF, Zayed J, Normandin L, Todd K, Michalak A, Spahr L, Huet PM, Pomier-Layrargues G (1999) Manganese deposition in basal ganglia structures results from both portal-systemic shunting and liver dysfunction. Gastroenterology 117(3):640–644. https://doi.org/10.1016/s0016-5085(99)70457-9
36. Bosoi CR, Parent-Robitaille C, Anderson K, Tremblay M, Rose CF (2011) AST-120 (spherical carbon adsorbent) lowers ammonia levels and attenuates brain edema in bile duct-ligated rats. Hepatology 53(6):1995–2002. https://doi.org/10.1002/hep.24273
37. Rodrigo R, Jover R, Candela A, Compañ A, Sáez-Valero J, Erceg S, Felipo V (2005) Bile duct ligation plus hyperammonemia in rats reproduces the alterations in the modulation of soluble guanylate cyclase by nitric oxide in brain of cirrhotic patients. Neuroscience 30(2):435–443. https://doi.org/10.1016/j.neuroscience.2004.09.027
38. Jover R, Rodrigo R, Felipo V, Insausti R, Sáez-Valero J, García-Ayllón MS, Suárez I, Candela A, Compañ A, Esteban A, Cauli O, Ausó E, Rodríguez E, Gutiérrez A, Girona E, Erceg S, Berbel P, Pérez-Mateo M (2006) Brain edema and inflammatory activation in bile duct ligated rats with diet-induced hyperammonemia: a model of hepatic encephalopathy in cirrhosis. Hepatology 43(6): 1257–1266. https://doi.org/10.1002/hep.21180
39. García-Ayllón MS, Cauli O, Silveyra MX, Rodrigo R, Candela A, Compañ A, Jover R, Pérez-Mateo M, Martínez S, Felipo V, Sáez-Valero J (2008) Brain cholinergic impairment in liver failure. Brain 131(Pt 11):2946–2956. https://doi.org/10.1093/brain/awn209
40. Cameron GR, Karunaratne WAE (1936) Carbon tetrachloride cirrhosis in relation to liver regeneration. J Pathol Bacteriol 42:1–21
41. Pérez TR (1983) Is cirrhosis of the liver experimentally produced by CCl4 and adequate model of human cirrhosis? Hepatology 3(1):112–120. https://doi.org/10.1002/hep.1840030118
42. Lee GP, Jeong WI, Jeong DH, Do SH, Kim TH, Jeong KS (2005) Diagnostic evaluation of carbon tetrachloride-induced rat hepatic cirrhosis model. Anticancer Res 25(2A):1029–1038
43. Yamamoto H, Sugihara N (1987) Blood ammonia levels and hepatic encephalopathy induced by CCl_4 in rats. Toxicol Appl Pharmacol 91(3):461–468. https://doi.org/10.1016/0041-008x(87)90067-6
44. Leone P (2020) Papel de la inflamación periférica y la neuroinflamación en el deterioro cognitivo y motor en ratas con diferentes grados de daño hepático. Efectos del tratamiento con rifaximina
45. Balzano T, Leone P, Ivaylova G, Castro MC, Reyes L, Ramón C, Malaguarnera M, Llansola M, Felipo V (2021) Rifaximin prevents T-lymphocytes and macrophages infiltration in cerebellum and restores motor incoordination in rats with mild liver damage. Biomedicines 9(8):1002. https://doi.org/10.3390/biomedicines9081002
46. Leone P, Mincheva G, Balzano T, Malaguarnera M, Felipo V, Llansola M (2022) Rifaximin improves spatial learning and memory impairment in rats with liver damage-associated neuroinflammation. Biomedicines 10(6):1263. https://doi.org/10.3390/biomedicines10061263
47. Tuominen I, Fuqua BK, Pan C, Renaud N, Wroblewski K, Civelek M, Clerkin K, Asaryan A, Haroutunian SG, Loureiro J,

Borawski J, Roma G, Knehr J, Carbone W, French S, Parks BW, Hui ST, Mehrabian M, Magyar C, Cantor RM, Ukomadu C, Lusis AJ, Beaven SW (2021) The genetic architecture of carbon tetrachloride-induced liver fibrosis in mice. Cell Mol Gastroenterol Hepatol 11(1):199–220. https://doi.org/10.1016/j.jcmgh.2020.08.010
48. Talcott MR, Akers W, Marini RP (2015) Chapter 25—Techniques of experimentation. In: Laboratory animal medicine, 3rd edn. Academic Press. https://doi.org/10.1016/B978-0-12-409527-4.00025-0
49. Fitzhugh OG, Nelson AA (1948) Liver tumors in rats fed Thiourea or Thioacetamide. Science 108(2814):626–628. https://doi.org/10.1126/science.108.2814.626
50. Dashti H, Jeppsson B, Hägerstrand I, Hultberg B, Srinivas U, Abdulla M, Bengmark S (1989) Thioacetamide- and carbon tetrachloride-induced liver cirrhosis. Eur Surg Res 21(2):83–91. https://doi.org/10.1159/000129007
51. Müller A, Machnik F, Zimmermann T, Schubert H (1988) Thioacetamide-induced cirrhosis-like liver lesions in rats—usefulness and reliability of this animal model. Exp Pathol 34(4):229–236. https://doi.org/10.1016/s0232-1513(88)80155-5
52. Muñoz Torres E, Paz Bouza JI, López Bravo A, Abad Hernández MM, Carrascal ME (1991) Experimental thioacetamide-induced cirrhosis of the liver. Histol Histopathol 6(1):95–100
53. Laleman W, Vander Elst I, Zeegers M, Servaes R, Libbrecht L, Roskams T, Fevery J, Nevens F (2006) A stable model of cirrhotic portal hypertension in the rat: thioacetamide revisited. Eur J Clin Investig 36(4):242–249. https://doi.org/10.1111/j.1365-2362.2006.01620.x
54. Saran T, Hilgier W, Urbańska EM, Turski WA, Albrecht J (2004) Kynurenic acid synthesis in cerebral cortical slices of rats with progressing symptoms of thioacetamide-induced hepatic encephalopathy. J Neurosci Res 75(3):436–440. https://doi.org/10.1002/jnr.10862
55. Schmandra TC, Bauer H, Petrowsky H, Herrmann G, Encke A, Hanisch E (2001) Effect of fibrin glue occlusion of the hepatobiliary tract on thioacetamide-induced liver failure. Am J Surg 182(1):58–63. https://doi.org/10.1016/s0002-9610(01)00659-6
56. Singh S, Trigun SK (2010) Activation of neuronal nitric oxide synthase in cerebellum of chronic hepatic encephalopathy rats is associated with up-regulation of NADPH-producing pathway. Cerebellum 9(3):384–397. https://doi.org/10.1007/s12311-010-0172-y
57. Méndez M, Méndez-López M, López L, Aller MA, Arias J, Cimadevilla JM, Arias JL (2008) Spatial memory alterations in three models of hepatic encephalopathy. Behav Brain Res 188(1):32–40. https://doi.org/10.1016/j.bbr.2007.10.019
58. Wallace MC, Hamesch K, Lunova M, Kim Y, Weiskirchen R, Strnad P, Friedman SL (2015) Standard operating procedures in experimental liver research: thioacetamide model in mice and rats. Lab Anim 49(1 Suppl):21–29. https://doi.org/10.1177/0023677215573040
59. Sathyasaikumar KV, Swapna I, Reddy PVB, Murthy CRK, Roy KR, Gupta D, Senthilkumaran B, Reddanna P (2007) Co-administration of C-phycocyanin ameliorates thioacetamide-induced hepatic encephalopathy in Wistar rats. J Neurol Sci 252:67–75
60. Rahman TM, Hodgson HJ (2000) Animal models of acute hepatic failure. Int J Exp Pathol 81:145–157
61. Andres T, Blei MD, Judy R, Cho MD (1990) Brain edema in acute liver failure. Insight from experimental studies. Rev Investig Clin 42(Suppl):127–130
62. Tuñón MJ, Alvarez M, Culebras JM, González-Gallego J (2009) An overview of animal models for investigating the pathogenesis and therapeutic strategies in acute hepatic failure. World J Gastroenterol 15(25):3086–3098. https://doi.org/10.3748/wjg.15.3086
63. Newsome PN, Plevris JN, Nelson LJ, Hayes PC (2000) Animal models of fulminant hepatic failure: a critical evaluation. Liver Transpl 6:21–31
64. Keppler D, Lesch R, Reutter W, Decker K (1968) Experimental hepatitis induced by D-galactosamine. Exp Mol Pathol 9(2):279–290. https://doi.org/10.1016/0014-4800(68)90042-7
65. Sielaff TD, Hu MY, Rollins MD, Bloomer JR, Amiot B, Hu WS, Cerra FB (1995) An anesthetized model of lethal canine galactosamine fulminant hepatic failure. Hepatology 21:796–804
66. Basile AS, Mullen K. Preclinical models of hepatic encephalopathy. Curr Protoc Neurosci. Chapter 9: Unit9. 2001.3. https://doi.org/10.1002/0471142301.ns0903s08
67. Yang Y, Zhai H, Wan Y, Wang X, Chen H, Dong L, Liu T, Dou G, Wu C, Yu M (2021) Recombinant human HPS protects mice and

nonhuman primates from acute liver injury. Int J Mol Sci 22(23):12886. https://doi.org/10.3390/ijms222312886

68. Zimmermann C, Ferenci P, Pifl C, Yurdaydin C, Ebner J, Lassmann H, Roth E, Hörtnagl H (1989) Hepatic encephalopathy in thioacetamide-induced acute liver failure in rats: characterization of an improved model and study of amino acidergic neurotransmission. Hepatology 9(4):594–601. https://doi.org/10.1002/hep.1840090414
69. Rogers DC, Fisher EM, Brown SD, Peters J, Hunter AJ, Martin JE (1997) Behavioral and functional analysis of mouse phenotype: SHIRPA, a proposed protocol for comprehensive phenotype assessment. Mamm Genome 8:711–713
70. Matkowskyj KA, Marrero JA, Carroll RE, Danilkovich AV, Green RM, Benya RV (1999) Azoxymethane-induced fulminant hepatic failure in C57BL/6J mice: characterization of a new animal model. Am J Phys 277(2):G455–G462. https://doi.org/10.1152/ajpgi.1999.277.2.G455
71. Keppler DO, Pausch J, Decker K (1974) Selective uridine triphosphate deficiency induced by D-galactosamine in liver and reversed by pyrimidine nucleotide precursors. Effect on ribonucleic acid synthesis. J Biol Chem 249(1):211–216
72. Czekaj P, Król M, Limanówka L, Skubis-Sikora A, Kolanko E, Bogunia E, Hermyt M, Michalik M, Sikora B, Prusek A, Grajoszek A, Pająk J (2022) Dynamics of acute liver injury in experimental models of hepatotoxicity in the context of their implementation in preclinical studies on stem cell therapy. Front Biosci (Landmark Ed) 27(8):237. https://doi.org/10.31083/j.fbl2708237
73. Stachlewitz RF, Seabra V, Bradford B, Bradham CA, Rusyn I, Germolec D, Thurman RG (1999) Glycine and uridine prevent D-galactosamine hepatotoxicity in the rat: role of Kupffer cells. Hepatology 29(3): 737–745. https://doi.org/10.1002/hep.510290335
74. Popken RJ, Kropveld D, Oosting J, Chamuleau RA (1983) Quantitative analysis of EEG power spectra in experimental hepatic encephalopathy. Neuropsychobiology 9(4): 235–243. https://doi.org/10.1159/000117971
75. Wright G, Soper R, Brooks HF, Stadlbauer V, Vairappan B, Davies NA, Andreola F, Hodges S, Moss RF, Davies DC, Jalan R (2010) Role of aquaporin-4 in the development of brain oedema in liver failure. J Hepatol 53(1):91–97. https://doi.org/10.1016/j.jhep.2010.02.020
76. Ganai AA, Husain M (2018) Genistein alleviates neuroinflammation and restores cognitive function in rat model of hepatic encephalopathy: underlying mechanisms. Mol Neurobiol 55(2):1762–1772. https://doi.org/10.1007/s12035-017-0454-1
77. Ede RJ, Gove CD, Hughes RD, Marshall W, Williams R (1987) Reduced brain Na+, K+-ATPase activity in rats with galactosamine-induced hepatic failure: relationship to encephalopathy and cerebral oedema. Clin Sci (Lond) 72(3):365–371. https://doi.org/10.1042/cs0720365
78. Lo WD, Ennis SR, Goldstein GW, McNeely DL, Betz AL (1987) The effects of galactosamine-induced hepatic failure upon blood-brain barrier permeability. Hepatology 7(3):452–456. https://doi.org/10.1002/hep.1840070307
79. Farghali H, Kgalalelo Kemelo M, Wojnarová L, Canová NK (2016) In vitro and in vivo experimental hepatotoxic models in liver research: applications to the assessment of potential hepatoprotective drugs. Physiol Res 65(Suppl 4):S417–S425. https://doi.org/10.33549/physiolres.933506
80. Engelmann C, Sheikh M, Sharma S, Kondo T, Loeffler-Wirth H, Zheng YB, Novelli S, Hall A, Kerbert AJC, Macnaughtan J, Mookerjee R, Habtesion A, Davies N, Ali T, Gupta S, Andreola F, Jalan R (2020) Toll-like receptor 4 is a therapeutic target for prevention and treatment of liver failure. J Hepatol 73(1):102–112. https://doi.org/10.1016/j.jhep.2020.01.011
81. Hori T, Chen F, Baine AMT, Gardner LB, Nguyen JH (2011) Fulminant liver failure model with hepatic encephalopathy in the mouse. Ann Gastroenterol 24(4):294–306
82. Sakai H, Park SS, Kikkawa Y (1992) Differential oxidase activity of hepatic and pulmonary microsomal cytochrome P-450 isozymes after treatment with cytochrome P-450 inducers. Biochem Biophys Res Commun 187:1262–1269
83. Gunawan BK, Liu ZX, Han D, Hanawa N, Gaarde WA, Kaplowitz N (2006) c-Jun N-terminal kinase plays a major role in murine acetaminophen hepatotoxicity. Gastroenterology 131:165–178
84. Sakaida I, Kayano K, Kubota M, Mori K, Takenaka K, Yasunaga M, Okita K (1992) Protective effect of deferoxamine for acetaminophen induced liver injury.

Gastroenterol Jpn 27(3):426. https://doi.org/10.1007/BF02777767

85. Isobe-Harima Y, Terai S, Miura I, Segawa M, Murata T, Itamoto K, Taura Y, Shinoda K, Sakaida I (2008) A new hepatic encephalopathy model to monitor the change of neural amino acids and astrocytes with behaviour disorder. Comparative Study Liver Int 28(1): 117–125. https://doi.org/10.1111/j.1478-3231.2007.01589.x
86. Wojnarová L, Canová NK, Farghali H, Kučera T (2015) Sirtuin 1 modulation in rat model of acetaminophen-induced hepatotoxicity. Physiol Res 64(Suppl 4):S477–S487. https://doi.org/10.33549/physiolres.933205
87. Gammal SH, Basile AS, Geller D, Skolnick P, Jones EA (1990) Reversal of the behavioral and electrophysiological abnormalities of an animal model of hepatic encephalopathy by benzodiazepine receptor ligands. Hepatology 11(3):371–378. https://doi.org/10.1002/hep.1840110307
88. Hsu CY, Lee FY, Huo TI, Chan CY, Huang HC, Lin HC, Chang CC, Teng TH, Wang SS, Lee SD (2010) Lack of therapeutic effects of gabexate mesilate on the hepatic encephalopathy in rats with acute and chronic hepatic failure. J Gastroenterol Hepatol 25(7): 1321–1328. https://doi.org/10.1111/j.1440-1746.2010.06235.x
89. Avraham Y, Grigoriadis N, Poutahidis T, Vorobiev L, Magen I, Ilan Y, Mechoulam R, Berry E (2011) Cannabidiol improves brain and liver function in a fulminant hepatic failure-induced model of hepatic encephalopathy in mice. Br J Pharmacol 162(7): 1650–1658. https://doi.org/10.1111/j.1476-5381.2010.01179.x
90. Brattin WJ, Glende EA Jr, Recknagel RO (1985) Pathological mechanisms in carbon tetrachloride hepatotoxicity. J Free Radic Biol Med 1:27–38
91. Chang HM, Liao YW, Chiang CH, Chen YJ, Lai YH, Chang YL, Chen HL, Jeng SY, Hsieh JH, Peng CH, Li HY, Chien Y, Chen SY, Chen LK, Huo TI (2012) Improvement of carbon tetrachloride-induced acute hepatic failure by transplantation of induced pluripotent stem cells without reprogramming factor c-Myc. Int J Mol Sci 13(3):3598–3617. https://doi.org/10.3390/ijms13033598
92. Sepehrinezhad A, Shahbazi A, Sahab Negah S, Joghataei MT, Larsen FS (2021) Drug-induced-acute liver failure: A critical appraisal of the thioacetamide model for the study of hepatic encephalopathy. Toxicol Rep 8:962–970. https://doi.org/10.1016/j.toxrep.2021.04.011. eCollection 2021
93. Bélanger M, Côté J, Butterworth RF (2006) Neurobiological characterization of an azoxymethane mouse model of acute liver failure. Neurochem Int 48:434–440
94. Shimojima N, Eckman CB, McKinney M, Sevlever D, Yamamoto S, Lin W, Dickson DW, Nguyen JH (2008) Altered expression of zonula occludens-2 precedes increased blood brain barrier permeability in a murine model of fulminant hepatic failure. J Investig Surg 21:101–108
95. Gupta H, Suk KT, Kim DJ (2021) Gut microbiota at the intersection of alcohol, brain, and the liver. J Clin Med 10(3):541. https://doi.org/10.3390/jcm10030541
96. Wolstenholme JT, Duong NK, Brocato ER, Bajaj JS (2024) Gut-liver-brain Axis and alcohol use disorder: treatment potential of fecal microbiota transplantation. Alcohol Res 44(1):01. https://doi.org/10.35946/arcr.v44.1.01. eCollection
97. Giménez-Garzó C, Salhi D, Urios A, Ruíz-Sauri A, Carda C, Montoliu C, Felipo V (2015) Rats with mild bile duct ligation show hepatic encephalopathy with cognitive and motor impairment in the absence of cirrhosis: effects of alcohol ingestion. Neurochem Res 40(2):230–240. https://doi.org/10.1007/s11064-014-1330-2
98. Adachi Y, Moore LE, Bradford BU, Gao W, Thurman RG (1995) Antibiotics prevent liver injury in rats following long-term exposure to ethanol. Gastroenterology 108(1):218–224. https://doi.org/10.1016/0016-5085
99. Yan AW, Fouts DE, Brandl J, Stärkel P, Torralba M, Schott E, Tsukamoto H, Nelson KE, Brenner DA, Schnabl B (2011) Enteric dysbiosis associated with a mouse model of alcoholic liver disease. Hepatology 53(1): 96–105. https://doi.org/10.1002/hep.24018
100. Hadjihambi A, Konstantinou C, Klohs J, Monsorno K, Le Guennec A, Donnelly C, Cox IJ, Kusumbe A, Hosford PS, Soffientini U, Lecca S, Mameli M, Jalan R, Paolicelli RC, Pellerin L (2023) Partial MCT1 invalidation protects against diet-induced non-alcoholic fatty liver disease and the associated brain dysfunction. J Hepatol 78(1):180–190. https://doi.org/10.1016/j.jhep.2022.08.008

Chapter 7

Neurobehavioral Tests in Preclinical Studies of Hepatic Encephalopathy

Bilal El-Mansoury, Omar El Hiba, Abdelaati El Khiat, and Arumugam Jayakumar

Abstract

Hepatic encephalopathy (HE) is a complex neuropsychiatric disorder that arises as result of liver dysfunction. HE encompasses a wide range of neurological and neuropsychiatric abnormalities ranging from subclinical changes to coma. It can be divided into two broad categories based on severity: overt HE (OHE) and covert HE (CHE). The exact mechanisms underlying HE are not fully elucidated. However, the use of animal models has contributed not only the understanding the pathophysiology of the disease but also the development of current available treatments. For instance, behavioral testing in preclinical studies of HE offers a robust framework for exploring the neurobehavioral aspects of this condition and helps not only in unraveling the underlying mechanisms of HE but also serve as essential tools in the development and assessment of novel therapeutic strategies. Hence, the use of neurobehavioral experiments has helped in the investigation of new drugs and plays an important role in preclinical studies. This chapter summarizes the commonly used behavioral tests in preclinical studies of HE with focus on rats and mice studies.

Key words Hepatic encephalopathy, Neuropsychiatric disorder, Behavioral tests, Preclinical studies

1 Introduction

Hepatic encephalopathy (HE) is a complex neuropsychiatric disorder caused by impaired liver function and/ or portosystemic shunts [1–3]. HE encompasses a wide range of neurological and neuropsychiatric abnormalities ranging from subclinical changes to coma, in advanced stages of liver disease [4, 5]. Indeed, HE causes a broad range of nonspecific neurological and neuropsychiatric symptoms [6]. Only psychometric tests measuring working memory, attention, psychomotor speed, and visuospatial ability, as well as electrophysiological and other functional brain measures, are affected by HE in its most mild form [7, 8]. As HE worsens, relatives of the patient may describe personality changes, including apathy, impatience, and disinhibition [9], along with noticeable changes in

Tiziano Balzano (ed.), *Experimental and Clinical Methods in Hepatic Encephalopathy Research*, Neuromethods, vol. 219, https://doi.org/10.1007/978-1-0716-4354-9_7,

consciousness and motor abnormalities. Also, patients may experience gradual loss of sense of space and time, inappropriate conduct, an acute confused state characterized by agitation or somnolence, stupor, and, in the most extreme case, coma [10]. Of note, the beginning of asterixis or disorientation is the first indication of overt HE, according to the ISHEN consensus [11].

According to the underlying cause, HE is divided into three main types: Type A HE, associated with acute liver failure (ALF), Type B, associated with portosystemic shunting without liver parenchyma lesions and Type C HE, resulted from chronic liver disease or liver cirrhosis [12–15]. The hallmark of a Type A HE in animal models is the development, within hours or days following ALF induction, of overt neurological symptoms, such as loss of righting reflex (precoma) and loss of corneal reflex (coma). Moreover, cerebral edema, which is frequently seen in the majority of animal models of Type A HE, is depicted by intracranial hypertension, another HE marker linked to ALF [16]. Both Type B and Type C HE, which are chronic types of HE, do not exhibit overt or obvious neurological signs, such as reflex deficits or noticeable behavioral changes; hence, neurobehavioral tests are used to evaluate such disturbances. Neuropathologically, chronic HE is characterized by the presence of an abnormal astrocytes shape known as Alzheimer type 2 astrocytes [15]. Although the mechanisms underlying the development of each type remain elusive, animal studies have greatly contributed to the enhancement of the understanding of the disease's pathogenesis [17] as well as the development of new therapeutics.

Even though it is widely accepted in clinical practice that all forms of HE may reverse completely following the treatment of liver disease, this assumption is no longer accurate as some patients still exhibit persistent neurological symptoms following liver transplantation (LT), and interestingly, compelling evidence comes from recent studies reporting neurodegeneration in HE [18]. Therefore, new treatment approaches for HE mainly focus on a pathogenic component that is causing the observed neurological damage are urgently required [16]. The use of animal models of HE has helped not only in the understanding of the disease but also the development of the current available therapies. Therefore, in order to develop new treatments for HE, preclinical models of the disease are essential to evaluate the drugs therapeutic effects as well as the potential adverse effects. This chapter aimed to review the use of the neurobehavioral experiments in preclinical studies of HE with focus on the most used animal species: rats and mice.

2 Material

2.1 Feeding Behavior: Food Intake Measurement

Food intake measurement is a suitable test in order to check the psychophysiological states (loss of appetite or anorexia) of animals with HE. Of note, the used of a single type of food during all the experiments duration is essential.

2.2 Locomotor Behavior Assessment

HE is associated with impaired locomotor behavior which can be exhibited in various forms. For example, non-comatose individuals with HE may exhibit motor system anomalies such as hypertonia, hyperreflexia, and a positive Babinski sign [18], while reports of seizures in HE are extremely rare [19–21]. Moreover, extrapyramidal dysfunction is frequently observed in patients with Parkinsonian-like tremor, hypomimia, muscular rigidity, bradykinesia, hypokinesia, monotone and slowness of speech, and dyskinesia with reduced voluntary movements [22]. Furthermore, asterixis or "flapping tremor" is often present in the early middle stages of HE and precedes stupor or coma. Actually, it is a negative myoclonus that causes a loss of postural tone rather than a tremor [18]. Notably, each person may not have motor symptoms of HE or they may not develop concurrently, making it challenging to stage the severity of HE [18]. Several neurobehavioral tests are used to study locomotor behavior in animals with HE particularly in rodents. These include open field test, rotarod, parallel bars, beam-balance test, triple horizontal bars, static rods, and actimeters.

2.2.1 Open Field Test

To test the presence of motor deficits in rats or mice with HE for example after the administration of a given treatment, locomotor activity can be assessed in the open field (OF) test [23]. Briefly, the OF consists of a square enclosure with a surface area of 1 m^2 and 30 cm in height (for rats; Fig. 1a), the bottom surface is subdivided into 16 identical squares (Fig. 1c). This test is also used in mice by dividing the OF arena to four quadrants (Fig. 1b) [24].

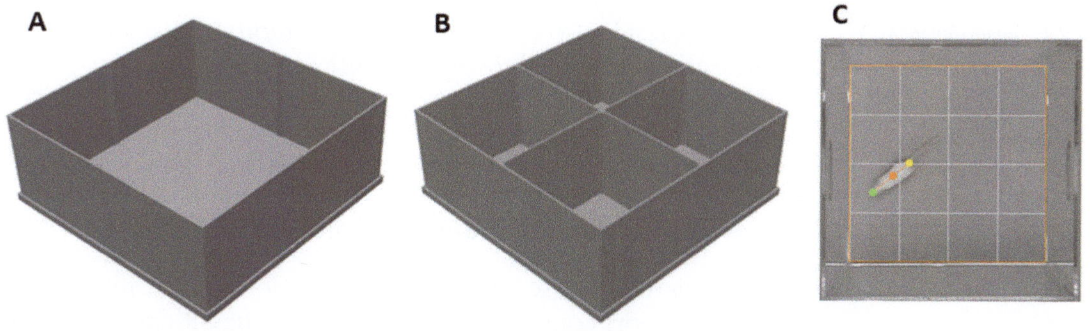

Fig. 1 Illustration of the open field test. (**a**) OF in rats, (**b**) open field in mice, and (**c**) shows the measurement of locomotor activity using a video tracking software

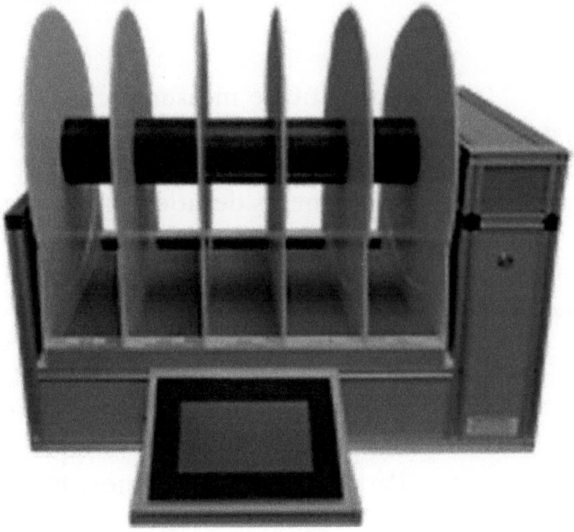

Fig. 2 Illustration of the rotarod test

2.2.2 Actimeters

Locomotor activity can be also assessed using infrared sensory units (motion detectors), which permit quantification of the animal's movements. It consists of an individual cage on a sensory unit linked to a movement recorder that is attached to a personal computer [25]. The actimeter is usually set over up to 48 h to quantify the activity over the circadian cycle. Several actimeter types are now available such as the body-implantable actimeter (nanotag) which enables to measure home-cage activity despite housing the mice in groups [26].

2.2.3 Rotarod Test

Motor coordination and balance of animals (rats and mice) with HE can be assessed using rotarod test. It is a test particularly suitable to assess the motor coordination in rodents [27]. The rotarod apparatus is made up of a multi-track drum, driven in rotation around its axis (rotating rod 75 mm diameter), by an electric motor with adjustable speed, and equipped with a mechanical fall detector and a timer for each track (Fig. 2) [27, 28].

2.2.4 Beam-Balance Test

Beam walk test is also widely used to study motor coordination and balance and can be used to assess the effect of drugs on rats or mice with HE. The device comprises of a wooden beam that is 150 cm long and positioned 60 cm above the ground horizontally (Fig. 3). To lessen the impact of falling rats, a safety box with cushions is positioned beneath the beam. For training mice, the 12-mm-wide square beam and the 28-mm-wide square beam are suggested [29].

2.2.5 Triple Horizontal Bars

This test is widely used to study motor coordination in mice. A wooden support column at either end holds the 38-cm-long brass bars 49 cm above the bench surface (Fig. 4). Three bars with different diameters are used (2-, 4-, and 6-mm). The recommended

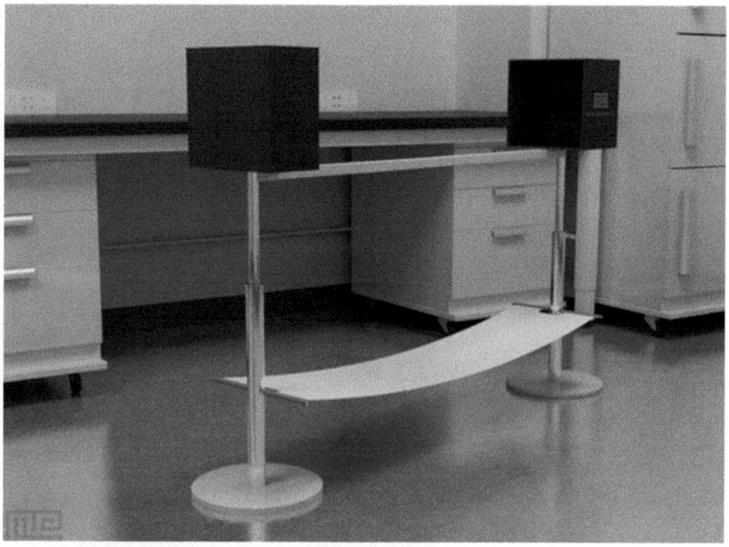

Fig. 3 Photograph of the beam walk test

Fig. 4 Photograph of the triple horizontal bars test

bar to use is 2 mm, yet many mice achieve the highest possible scores with this. Since mice find it difficult to grasp the larger diameter bars, an attempt has been made to improve the test by including two more larder bars [30].

2.2.6 Static Rods Test

Static Rods is adapted for measuring motor coordination in mice and rods of different diameters are used to refine the test. A G-clamp secures five hardwood dowels or rods, each measuring 60 cm in length and with diameters of 35, 28, 22, 15, and 9 mm. The ending line is marked by a mark 10 cm from the end of the rod near the bench. The rods are 60 centimeters above the ground [30] (Fig. 5).

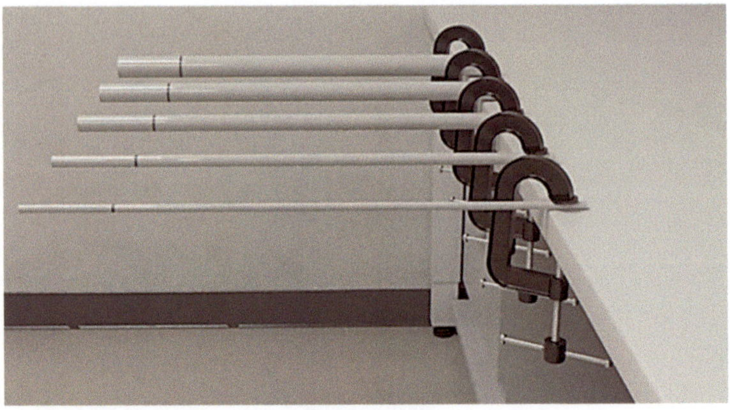

Fig. 5 Photo of the Static rods test in mice

Fig. 6 Photo of the parallel bars test in mice

2.2.7 Parallel Bars

This test is also used to study motor coordination in mice. In short, wooden supporting columns at the ends of two parallel steel bars, each measuring one meter in length and four millimeters in diameter, are placed 30 millimeters apart (on their centers) and are 60 cm above the floor (Fig. 6).

2.3 Neurological Score Assessment in ALF

Several rating measures have been established for acute HE based on reported behavioral abnormalities, reflex impairment, and/or the presence of ataxia [13, 31, 32]. Thus, this test can be applied to test the effect of a drug on the neurological function in mice with acute HE.

2.4 Assessment of Memory Function

Memory impairment is among the complaints of patients with HE, and it is exhibited by most animal models of the disorder. Several neurobehavioral tests can be used to evaluate memory function in preclinical practice. These include Morris water maze test, Y-maze test, T-maze test, social recognition memory, and novel object recognition test among others.

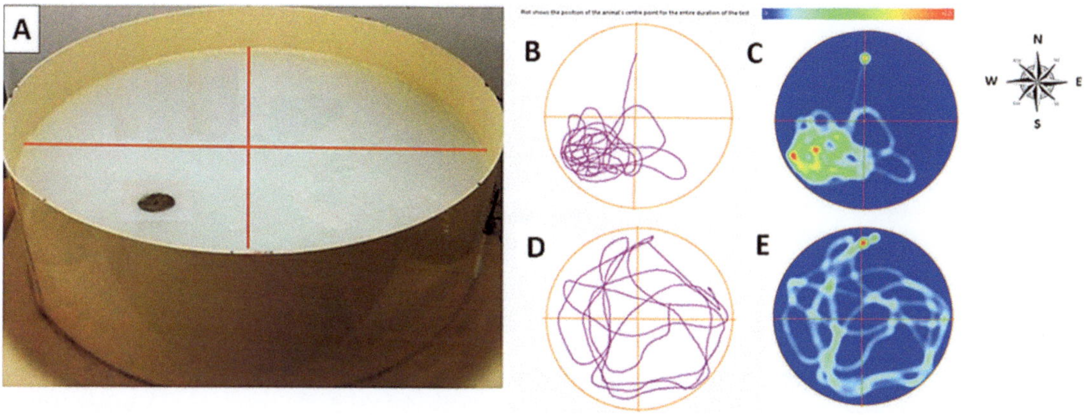

Fig. 7 Illustration of Morris water maze test. (**a**) Morris water maze apparatus, (**c**) heat map of a normal animal with intact memory, (**b**) heat map of an animal with memory impairment, (**b**) track plot of an animal with intact memory, (**d**) track plot of an animal with memory impairment

2.4.1 Morris Water Maze

Morris water maze test is adapted to assess spatial learning and memory in rodents [33]. The maze consisted of a circular water tank (rats; 140 cm in diameter and 60 cm in height) filled with 2/3 water maintained at 25 ± 2 °C and at a level of 40 cm above the floor of tank (Fig. 7) [28]. Four roughly equal quadrants (NE, NW, SE, and SW) make up the pool. This test is also widely used in mice with adapted diameters. A circular platform with a diameter of 10 cm is concealed 1 cm below the water's surface, and nontoxic coloring (milk powder) is added to opacify the water. There are both exterior and internal visual clues in the surroundings. The animal is incentivized to locate the platform and swiftly flee the water by the unpleasant nature of the water, which is the basis of the test.

2.4.2 Radial Arm-Maze Task

The radial maze (RAM) test adapted for long term-spatial memory assessment [34, 35]. The RAM was created initially for rats, while it is recently adapted to mice as well [34]. The radial maze is situated 60 cm above the ground and is made up of eight radial arms that are equally spaced out from the central arena (Fig. 8) [36]. Each arm measures 50 cm in length and 10 cm in width. At the distal end of each arm is a feeding well where rewards are placed, such as sugar-filled milk or sucrose pellets [35, 37]. There is a 3 cm lip around the margins of each arm and the center arena to keep rats from dropping out of the labyrinth. A range of extra-maze stimuli, such as chairs, bookcases, and posters, surround the maze [36].

2.4.3 T-Maze Test

The T-maze test can be used to evaluate short working memory in rodents with hepatic encephalopathy after the administration of a drug. It is well known that this test effectively detects short memory impairment in rodents [38, 39]. The T-maze apparatus is

Fig. 8 Illustration of radial maze test

Fig. 9 Illustration of T-maze apparatus

composed of two lateral arms and one central arm (Fig. 9) (for each arm in rat: 50 cm long, 10 cm in width, with walls of 30 cm heigh) [39]. The test is also used in mice with adapted diameters.

2.4.4 Y-Maze Test

This test can be also used to assess spatial learning and short memory in rodents [28]. The apparatus is composed of three spaced arms (identical arms made from plexiglass or wood) (120°) with a 50 cm length and 20 cm height (Fig. 10). The three arms of the maze are clearly named and identified as "A", "B," and "C."

2.4.5 Novel Object Recognition Test

The novel object recognition test (NOR) is a well-established for testing recognition memory in rodents. The test can be performed in the open field, using two similar objects in color, shape, and appearance which are different from the third one (Fig. 11). Of note, the three objects used can be made from heavy wood to prevent them from being moved by animals.

2.4.6 Social Recognition Memory Test

The test can be done in the Open Field apparatus to test social memory in rats or mice (Fig. 12). It consists of placing the tested rats in the open field with two juvenile rats; a familiar and non-familiar rat. The test based on the fact that a normal rat prefers

Fig. 10 Illustration of Y-maze apparatus

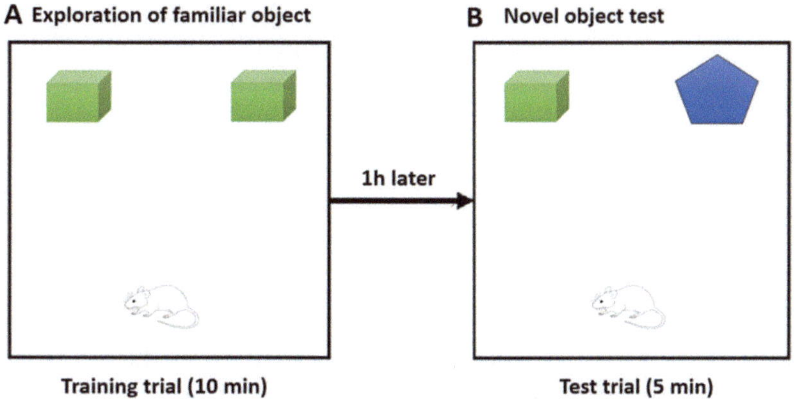

Fig. 11 Schematic representation of novel object recognition procedure

to interact with new rat and investigate new environment rather that familiar one which is an innate response of rodents. This test can be also performed in mice by dividing the open field into four quadrants.

2.5 Anxiety- and Depressive- Like Behaviors

Anxiety and depressive like behaviors are also observed in HE patients and animal models of the disorder. Therefore, the use of behavioral experiments can help to check the presence of these symptoms in preclinical studies. Several tests are used to evaluate the presence of these symptoms. These include elevated plus maze test, dark light box, social interaction test, tail suspension test, and forced swim test.

2.5.1 Elevated Plus Maze Test

The elevated plus maze (EPM) test is widely used to assess anxiety state in rodents with HE. It is one of the most popular tests used to evaluate anxiety in rodents [40]. The purpose of this validated test is to identify the brain regions and mechanisms responsible for anxiety-related behavior as well as to evaluate the anti-anxiety

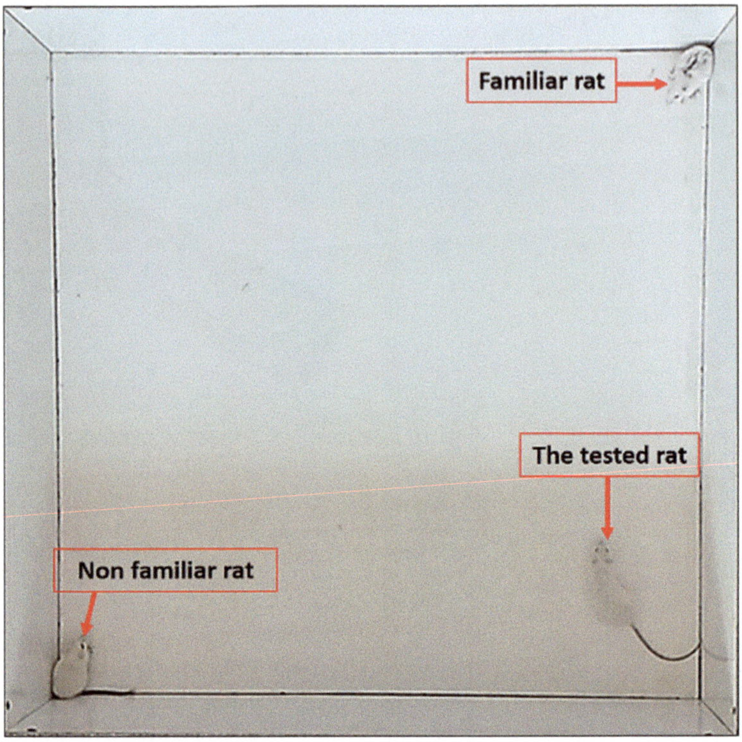

Fig. 12 Photograph of social recognition memory test

effects of pharmaceutical medicines [41, 42]. Briefly, in rats, the maze consisted of two open arms without sidewalls (length 50 cm × wide 10 cm) and two closed (50 cm × 10 cm × 30 cm) arms, and elevated to a height of 50 cm above the floor and placed in a moderately lighted room (Fig. 13) [43]. The test can be also used in mice with adapted apparatus diameters.

2.5.2 *Dark-Light Box Test*

The light/dark test can be used to determine whether a medication will have anxiogenic or anxiolytic effects. It is predicated on rodents' natural aversion to highly lit spaces as well as their impulsive exploration activity in response to mild stressors like new environments and light [44]. The Dark-light box is made up of two plastic compartments that are connected by a tunnel (Fig. 14). One dark compartment with a lid and a white compartment with overhead lighting.

2.5.3 *Social Interaction Test*

Social behavior could be considered a mirror of anxiety, especially in mammals [45]. It is well documented that excessive and inappropriate anxiety is likely to negatively affect an individual's social life [46, 47]. The social interaction test is carried out using the Open Field apparatus (length 100 cm × height 30 cm) (Fig. 15). This assessment of social interest/behavior has been extensively validated and used [48, 49].

Fig. 13 Illustration of elevated plus maze apparatus

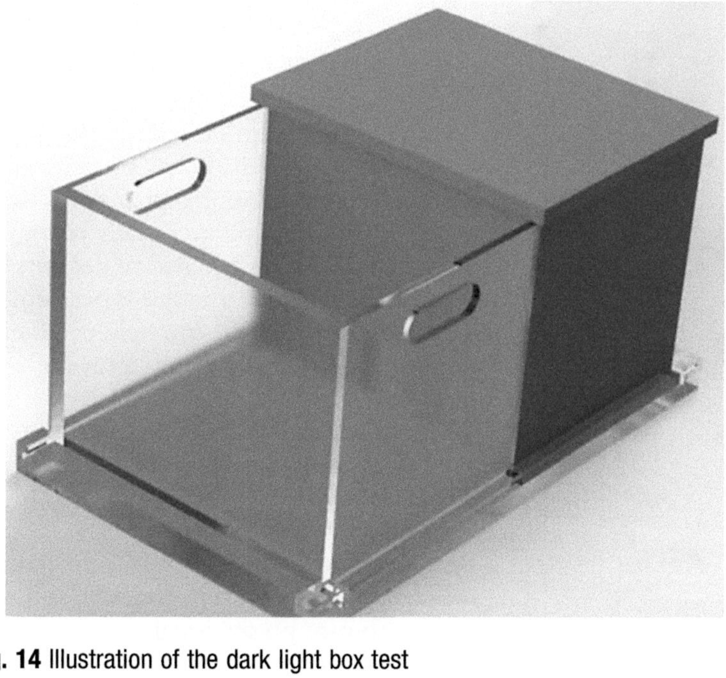

Fig. 14 Illustration of the dark light box test

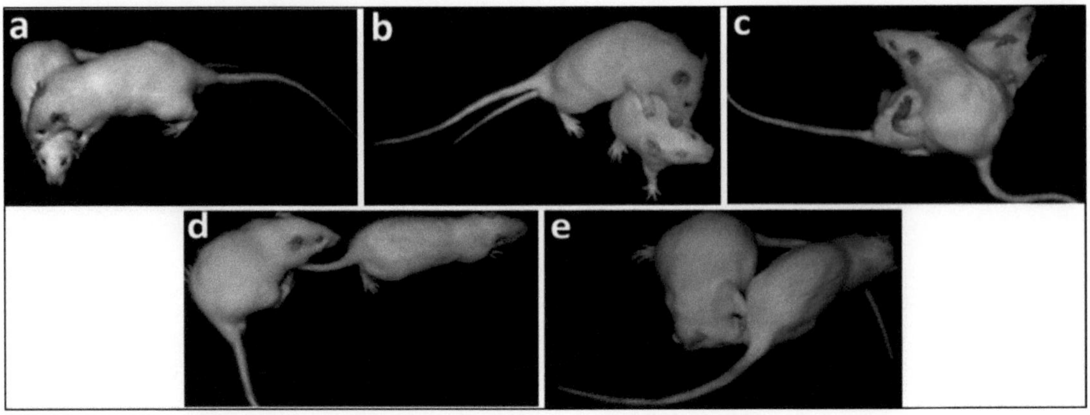

Fig. 15 Photos illustrating social behavioral patterns in rats. (**a**) pouncing, (**b**) wrestling, (**c**) pinning, (**d**) following, and (**e**) genital investigation

2.5.4 Open Field Test (Anxiety)

The open field (OF) results can be also exploited for testing animals' anxiety state. This test is employed in research on the neurological underpinnings of anxiety as well as in the search for new therapeutic targets and anxiolytic substances [50]. The OF test exploits the natural aversion of rodents to exposed field [51]. Open field apparatus consisted of a square arena (50 × 50 cm), with a white floor divided into 16 squares, enclosed by continuous, 30-cm-high walls made of white wood. In this test, the 16 squares adjacent to the wall represent the periphery of the arena, while the other 9 squares represent an exposed field, or "central area."

2.5.5 Forced Swim Test

The forced swim test (FST) is used to assess depressive like behaviors in rodents with HE. It is among the most widely utilized tests for examining rodent behavior that is similar to depression. Based on the idea that an animal placed in a water-filled container will first try to get out but will eventually become immobile, maybe indicating some degree of behavioral despair, which is the cornerstone of the test [52]. Briefly, the test is performed in a transparent Plexiglas cylinder (diameter: 20 cm, height: 40 cm) containing 30 cm of tap water maintained at a temperature of 24 °C ± 1 (Fig. 16).

2.5.6 Tail Suspension Test

The tail-suspension test can be also used to evaluates depression-like behavior especially in mice but it is also adapted for rats [53–55] (Fig. 17).

2.6 Pain Assessment

Pain is among the complaints of patients with cirrhosis and its management is of great concern and consists a clinical challenge for health care professionals [56–58]. Of note, impaired temperature perception was reported in patients with HE [59]. Therefore, to assess sensitivity to thermal pain, the tail flick test and the hot

Fig. 16 Schematic representation of mobility and immobility in the forced swim test

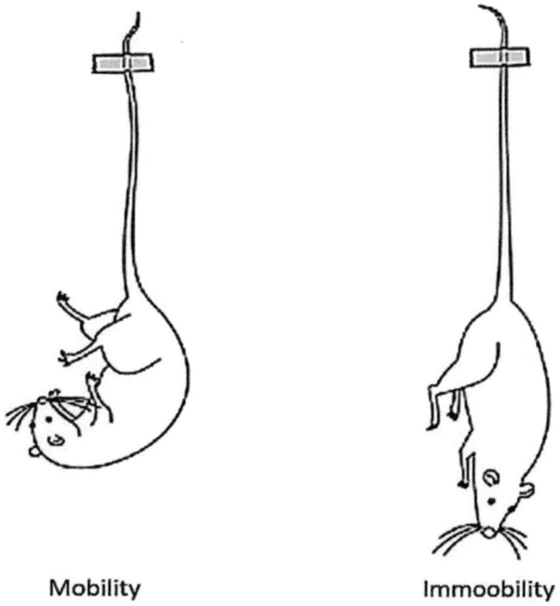

Fig. 17 Schematic representation of immobility and mobility in the tail suspension test

plate test can be used to evaluate the animal's reaction to pain. For centrally acting analgesics, the tail-flick and hot plate procedures are both commonly utilized.

2.6.1 Hot Plate Test

This test is based on pain reflexes due to contact of the rat's pads with a heated surface (Fig. 18). The rodent is placed in an open area during the experiment, and the floor is made of a plate that is heated to a consistent temperature of 55.4 °C (for rats).

Fig. 18 Photo of the hot plate test

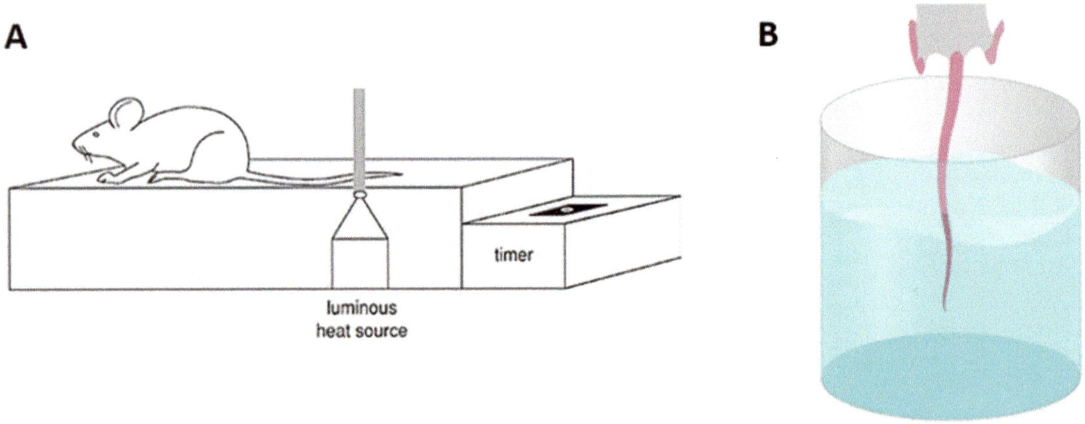

Fig. 19 Illustration of tail-flick tests

2.6.2 Tail-Flick Test

The tail-flick test is a popular method for assessing the antinociceptive impact of drugs. It was first described by D'Amour and Smith in 1941. The tail-flick assay uses heat as a noxious stimuli, similar to hot plate test [60]. When the tail is subjected to unpleasant heat stimulation, a basic nociceptive reflex reaction is produced, which is a tail flick away from the heat source (Fig. 19a). This is a spinally mediated flexion reflex, as opposed to the complex activity of licking the hind paw in the hot-plate test. Tail-flick device equipped with an automated timer or a human stopwatch. The warm water test is an additional substitute for this test tail-flick test (Fig. 19b) [61].

3 Methods

3.1 Food Intake Measurement

Briefly, each rat of each group is placed in an individual cage, to get used to the isolation (habituation phase). For the test, a quantity of the food is measured (e.g., 40 g for rats and 15 g for mice) and placed in the cage without bedding in the presence of a single rat with *ad libitum* access to water. The nocturnal food intake of 12 h (8 p.m. to 8 a.m.) is then measured after drying the rest of the food under the hood to remove any possible traces of urine or water on the food likely to overestimate the weight. Finally, the food intake ratio is calculated according to the following equation:

$$\text{Food intake ratio (FIR)} = (\text{Quantity of food ingested/body weight at the time of the study})^{*}\ 100$$

3.2 Open Field Test

The habituated is carried out to the apparatus 24 h before the test. The test is carried out by placing each animal in the center of the OF, then the movements of the animal are freely filmed for 5 min by a camera placed at the top of the device. The videos are then examined and the following parameters are evaluated including distance moved and the number of boxes crossed; the box is considered to have been covered when the animal crosses it with all four legs. These parameters are calculated for the entire test duration [23].

3.3 Rotarod Test

Prior to testing, the animals undergo two days of training (first day at a speed of 8 rpm, second day at a speed of 10 rpm). While in the test day, each rat or mouse is placed on the axis which rotates at a gradual speed of ten rotations per minute, and the animals whose motor coordination has deteriorated slipped off the rod into a tray 10 cm below, which serves as the basis for the test's concept. Each animal is given a test duration of 60 s, and the endurance time is calculated by timing the animal's positioning on the moving rod until it falls. An animal's endurance time of 60 s or more is considered normal response [5]. Any decrease in endurance time is interpreted as a sign of declining motor coordination.

3.4 Beam-Balance Test

A training session is performed at 24–48 h before the test which consist of placing the animal on the beam for a 60 s trial. Permit the rat to fall into the safety box if it is unable to balance on its own. Start the beam with a mouse at the beginning, start timing when the animal is firmly in place, and measure the latency to cross the beam (up to 60 s) [28, 62]. Briefly, set the timer, place the animal on the beam, monitor the rat intently for 60 s, note the score on the worksheet, and then return the rat back in its home cage for a quick rest (3–5 min). Repeat the process with each animal for a total of three tests [63].

3.5 Triple Horizontal Bars

Briefly, hold the mouse by the tail for a short while and allow it to use only its forepaws to grab the horizontal bar at the center. Of note, start with the narrow bar (2 mm) because it is the easiest to grasp by the animals and give each mice a 30 s maximum duration. If the experimenter did not place the mouse poorly and the mouse falls before 5 s, try again up to three times to see if you can acquire a score more than 5 s [64]. Use the best result as the datum. A mouse scores just 1 in if it continuously cannot sustain itself for longer than 5 s. If the triple-bar version is being utilized, the mouse can be tested on thicker bars if it receives a score of 5 on the first 2 mm bar. A brief break can be taken while testing one or more additional mice. The final score is the cumulated total, and the scoring method for the 4- and 6-mm bars is the same as for the 2 mm bar. For example, a mouse that hits the 2 mm bar and then falls off the 4 mm bar after 13 s receives a cumulative score which is calculate-d as: 5 + 3 = 8 [30]. Score the trials on the horizontal bars as follows:

Falling between 1 and 5 s = 1

Falling between 6 and 10 s = 2

Falling between 11 and 20 s = 3

Falling between 21 and 30 s = 4

Falling after 30 s = 5

Placing one forepaw on a bar support without falling = 5

3.6 Parallel Bars

With its longitudinal axis perpendicular to the bars, position the mouse in the middle of the two bars, with its front paws on one and its hind paws on the other. The scoring is comparable to that of the stationary rods. Keep track of the mouse's orientation time until it reaches one of the end supports and the time it takes to turn 90 degrees from the initial position. Moreover, record the mouse turning upside down. With regard to the static rods test, the maximum period is allowed if orientation and transit are not accomplished without the rods turning upside down. In less than 5 s, mice in relatively good health seldom fall; when they do, it is usually because of inappropriate positioning. Discard the outcome and test the mouse once more [30].

3.7 Neurological Score Assessment in ALF

Briefly, the results of five reflex tests—the ataxia test, the pinna reflex, the corneal reflex, the tail flexion test, the escape response, and the righting reflex—are used. Each of the aforementioned five tests, as well as the ataxia test, receives a semi-quantitative evaluation that determines the neurological score, and each is assigned a score from 0 to 2 (2 = reflexes intact, no ataxia; 1 = weak or delayed reflex, minor ataxia, 0 = no reflex evident, significant ataxia). The summation of these six reflexes and ataxia gives a neurological score between 0 and 12 [16]. There is a lack of application of this scale in rat models with acute HE.

1. Pinna reflex: applying a cotton applicator to the auditory meatus and observing movement of the head or ear retraction.
2. Corneal reflex: consists of measuring the blink reaction after lightly touching the cornea with a cotton swab.
3. Tail flexion: consists of using a forceps to pinch the tail and assessing the flexion of the animal's tail.
4. Escape response: consists of pinching the mouse's tail by forceps and observe the subsequent movement of the animal away from stimuli.
5. Righting reflex: consists of putting the mice on their backs and determining how long it will take them to right themselves.
6. Ataxia: Mice are placed on the lid of a wire cage, and the number of foot falls through the lid is counted.

3.8 Morris Water Maze

For each test, the animal is placed in the water, facing the wall and has 60 s to find the platform, and once found it, the animal is lift on the platform for 20 s to memorize the environment. The experimenter gently moves the rat through the platform if it cannot find it. The test is performed in 6 days. The first day (training session) consist of habituating the animals with the swimming arena and the presence of a hidden platform by placing each animal at starting point of all four quadrants randomly. From the second to the fifth day (learning stage), the animal learn that the platform is located in the south-west (SW) quadrant. The sixth day (the test day), each animal is placed at a determined starting point of one of the four quadrants [65], and given 60 s time duration in the absence of the platform. Each test is recorded by a camera above the swimming arena and the videos were later analyzed using the sophisticated ANY-maze video tracking software (Stoelting, Wood Dale, IL, USA) [28].The analyzed parameters are pathlength (the distance to target the platform), escape latency (the latency to find the platform), the mean speed of swimming, and the time spent in the SW quadrant [33]. During the first training day, the path length, speed, and escape latency are measured to rule out any physical or visual anomalies that can impede the learning and recalling processes. From day 2 to 5, all the trials are recorded which indicates learning process. On the sixth day (last day), however, time in zone is noted as a memory indicator (Fig. 7) [33].

3.9 Radial Arm-Maze Task

During the initial days of training, every animal spends a total of 10 min in the apparatus to acquaint itself with the maze and the reward before being put back in its cage [35]. Only the arms' food wells were baited on the sixth day. Each animal underwent one trial every day; they are each placed individually on the central platform and given 8 min to investigate, or until they had reached every arm. By counting the number of errors (all re-entries into the arms, up to a maximum of 16 arm entries) throughout each trial, choice accuracy is evaluated. After 15 days, the animals achieved an asymptotic performance (<0.5 error/rat/day) [36]. After the camera records the behaviors, the videos are examined by a skilled researcher, blinded to the treatments [35].

3.10 T-Maze Test

The test consists of two quick tests that are administered to each animal in rapid succession. In the second test, the rodent is believed to show spontaneous alternation, wherein it chooses a new arm instead of going back to the one it visited before, indicating that it memorized its first choice [66–68]. The investigated parameter is the percentage (%) of the left and right arms' spontaneous alternation based on which arm is visited during the previous session. For statistical analysis, these data are converted into scores: 0% indicates no alternation, and 100% indicates a proper alternation in each test. Typically, control animals get about 80% in average of spontaneous alternation [68].

3.11 Y-Maze Test

Each animal is positioned in arm "A" (start arm) during the test, with its nose pointed toward the center. All animals are given a 10-min period to investigate each of the maze's three arms as part of the test for evaluating SWM. Since rodents are naturally curious to explore new places, they typically return straight to the arm they have visited after visiting another one (arm), and a normal animal should exhibit a tendency to enter a less recently visited arm. To determine the percentage of alternations, the number of arm entries and the number of correct alternations are counted. For every group, the percentage (%) of alternation behavior is determined [69, 70]. In addition, the distance travelled to novel arm (path length) and heatmap are recorded using ANY-maze video tracking software (Fig. 7) [28].

3.12 Novel Object Recognition Test

Briefly, three sessions make up the test: one for habituation to the OF, one for training (exploration of the two similar objects), and one for testing. Each animal is placed alone in the OF without any objects for 5 min to investigate it during the habituation phase.

Then, each animal is individually exposed to two identical objects placed in the OF apparatus at an equal distance during the training session, which lasts for 10 min of exploration. Regarding the test, an hour later, every animal is separately exposed to one of the two old objects, and the other object is changed with a new one (novel) and for 5 min exploration [71]. In the absence of external restrictions or reinforcement, mice and rats typically show an innate predisposition to explore new objects for a notably longer period of time than they do familiar ones. Exploring the novel object thus shows the use of recognition and learning memory.

The criteria (biting, sniffing, contact with forepaws or nose) that signify item investigation or engagement are recorded. For every group, the discrimination ratio and the average exploration duration are determined [71]. If an animal spends longer time interacting with a new object, memory is thought to be intact [72], and the discrimination ratio (DR), which measures how much more time is spent examining the new object than the old one during the entire exploration period, is used to quantify the performance of the NOR task.

3.13 Social Recognition Memory Test

Before undertaking the test, each animal (rats or mice) is habituated individually by placing it in the device for 4 min (first habituation), the second habituation consists of the acquisition of social memory by introducing a juvenile rat for 4 min (the juvenile rat then becomes familiar) (Fig. 12). The test is done 60 min later. The adult animal is re-exposed to two juvenile rats simultaneously (1 familiar and 1 unfamiliar). The rat's activity is recorded by camera for 5 min. Therefore, the following parameters are measured: the percentage of time spent interacting with the familiar rat compared to that of interacting with the unfamiliar rat. A selective recognition memory is considered intact if we note a longer investigation time for the (unfamiliar) juvenile rat in the third trial compared to that of the familiar juvenile rat [73]. This test can be also performed in mice by dividing the open field unto four quadrants.

3.14 Elevated Plus Maze Test

The animals are positioned one at a time in the center of the maze, facing a closed arm. During a 5-min observation session, the number of entries and the amount of time spent in the closed and open arms are noted [43]. Arm entries are defined as entry of all four paws into an arm [74]. Time spent in the open and closed arms as well as the number of entries to the closed and opened arms are calculated.

3.15 Dark-Light Box Test

Each animal is observed individually for 5 min starting in the dark compartment facing the tunnel [75]. The times spent in the light box, tunnel, and dark compartment as well as the entrances into each of these three areas are measured [76].

3.16 Social Interaction Test

Briefly, during a first phase of habituation, each animal is placed individually for 5 min exploration in the experimental set-up. In the second phase, an unfamiliar juvenile rat male (150 g or less in body weight) is placed in the same device, as far away as possible from the tested rat [48, 49]. The animal's behavior is freely filmed for 5 min by a camera placed at the top of the device and the percentage of time spent on social behavior (Fig. 15) (move toward, anogenital sniffing, close following) and nonsocial behavior (resting, moving away, grooming, exploration of the open field) is calculated.

3.17 Open Field Test (Anxiety)

The test is initiated by placing a single animal (rat or mouse) in the middle of the arena and letting it move freely for 5 min. The animal's behavior is continuously filmed by a video camera placed over the apparatus. Then, videos are examined and the time spent in the peripheral arena and the number of entries into the central arena are counted [51]. An animal that spends less time in the periphery of the arena and pass through the center frequently is considered less anxious.

3.18 Forced Swim Test

FST consisted of training and test sessions performed at a 24-h interval. The habituation is performed 24 h before the test and consists of placing each animal individually in the water-filled cylinder for 15 min. Twenty-four hours following the training (the day of the test), rats are allowed to swim under the same conditions for 5 min and their movements are video-recorded. The total immobility time over a 5-min period is measured (the rat is considered immobile if it floats in the water, in an upright position, and only makes small movements to keep its head above water) [77, 78]. The primary evaluation criterion is the duration of immobility, which is characterized as the minimal motions required to maintain head above water or, alternatively, the times when swimming is not actively demonstrated [52]. The entire amount of time that the animal's body is immobile, with the exception of the little movements required to maintain its head above water, is referred to as the immobility time [52]. While a shorter period of immobility is a sign of antidepressant effect, a longer period is linked to conduct suggestive of despair [52].

3.19 Tail Suspension Test

Briefly, the tail ends were fixed with adhesive and the animals are suspended down about 10 cm above the ground (Fig. 17) [77], and each animal is suspended by the tail in the suspension box for 5 min, and immobility time is recorded by a video camera [53]. During the behavioral analysis, the immobility time (no movement of limbs and body) of each rat is calculated as indication of depression-like behavior.

3.20 Hot Plate Test

Two behavioral components can be measured in terms of reaction time or latency (in seconds), namely licking the rat's paw pads and/or paw jumping. Both are considered a response. A special feature of this test is that it can only be performed once for each animal [79].

3.21 Tail-Flick Test

In this experiment, the noxious heat stimulus is warm water rather than light. The distal half of the tail of a mildly restrained animal (a mouse or rat) is submerged in circulating water that is thermostatically controlled at 49 °C in this modified tail-flick assay. A stop watch is used to assess the tail's latency to vigorously flex in response to the heat stimulus [60]. Tail-flick latency for each animal (rat or mice) is calculated as the average of three consecutive measurements.

4 Notes

The neurobehavioral experiments are invaluable for gaining a comprehensive understanding of HE by assessing changes in behavior and cognitive function, and serve as essential tools during preclinical studies of HE in developing effective treatments, and ultimately improving patient outcomes. Additionally, they will help not only in investigating the therapeutic effects of tested drugs, but also evaluating their potential side effects to ensure that they are safe for use. Thus, findings from neurobehavioral studies in animal models are often translated to human clinical studies.

The neurobehavioral tests must be carried out by trained experimenters in order to reduce variability and increase the reliability of the results. Additionally, during all the experiment duration, handling methods should be applied in order to minimize stress that might interfere with the experiment's results. Additionally, before behavioral testing, animals should be placed at the room tests 30 min before the experiments to get acclimatized with the new environment. Finally, after each behavioral assessment trial, the apparatus should be cleaned with 70% ethanol.

Although the significant role of neurobehavioral experiments, the absence of standardized protocols for neurobehavioral testing in HE models can lead to inconsistent results. Moreover, differences in experimental design, such as animal species, sex, and age, can affect outcomes. Additionally, different animal models of HE may yield different results. This variability can complicate the interpretation and comparison of data among studies.

In a nutshell, HE still a complex neuropsychiatric disorder of great concerns and the development of new therapies is urgently needed. Hence, behavioral tests in preclinical studies of HE play an important role in exploring the neurobehavioral aspects of this condition and the potential therapeutic effects of treatments under investigation. Therefore, these tests serve as essential tools in the development and assessment of new therapies. Future research should continue to refine these methodologies and expand the repertoire of behavioral assays to further understanding of HE and improve patient outcomes.

References

1. Jayakumar AR, Rao KVR, Norenberg MD (2015) Neuroinflammation in hepatic encephalopathy: mechanistic aspects. J Clin Exp Hepatol 5:S21–S28
2. El-Mansoury B et al (2024) Short working memory impairment associated with hippocampal microglia activation in chronic hepatic encephalopathy. Metabolites 14(4):193
3. El Khiat A et al (2023) P14 altered locomotor behavior associated with Purkinje cells and astrocytes dysfunctions in rat with acute hepatic encephalopathy involving the oxidative stress and neuroinflammation. Off J Am Coll Gastroenterol ACG 118(9S):S11
4. El-Mansoury B et al (2023) P8 cognitive impairment associated with cortical and hippocampal microglia activation and altered synaptic function in a rat model of chronic hepatic encephalopathy. Off J Am Coll Gastroenterol ACG 118(9S):S7
5. Tong XY et al (2024) Age and sex in the development of hepatic encephalopathy: role of alcohol. Biology (Basel) 13(4):228
6. Ferenci P, Lockwood A, Mullen K, Tarter R, Weissenborn K, Blei AT (2002) Hepatic encephalopathy—definition, nomenclature, diagnosis, and quantification: final report of the working party at the 11th world congresses of gastroenterology, Vienna, 1998. Hepatology 35(3):716–721
7. Gitlin N, Lewis DC, Hinkley L (1986) The diagnosis and prevalence of subclinical hepatic encephalopathy in apparently healthy, ambulant, non-shunted patients with cirrhosis. J Hepatol 3(1):75–82
8. Lockwood AH (2000) 'What's in a name?' Improving the care of cirrhotics. J Hepatol 32(5):859–861
9. Wiltfang J, Nolte W, Weissenborn K, Kornhuber J, Rüther E (1998) Psychiatric aspects of portal-systemic encephalopathy. Metab Brain Dis 13:379–389
10. Weissenborn K (1998) Diagnosis of encephalopathy. Digestion 59(Suppl. 2):22–24
11. Bajaj JS et al (2011) The design of clinical trials in hepatic encephalopathy–an International Society for Hepatic Encephalopathy and Nitrogen Metabolism (ISHEN) consensus statement. Aliment Pharmacol Ther 33(7): 739–747
12. El-Mansoury B, El Hiba O, Jayakumar AR (2023) Physiology and function of glial cells in health and disease. IGI Global
13. DeMorrow S, Cudalbu C, Davies N, Jayakumar AR, Rose CF (2021) 2021 ISHEN guidelines on animal models of hepatic encephalopathy. Liver Int 41(7):1474–1488
14. El Hiba O, El-Mansoury B, Norenberg M, Jayakumar AR (2024) Astroglia and acute hepatic encephalopathy: the mechanisms of cytotoxic brain edema. In: Physiology and function of glial cells in health and disease. IGI Global, pp 243–260
15. El-Mansoury B, El Hiba O, El Khiat A, Jayakumar AR (2024) Hepatic encephalopathy as a gliopathy: the mechanisms of Alzheimer Type II astrocytosis. In: Physiology and function of glial cells in health and disease. IGI Global, pp 223–242
16. Demorrow S, Cudalbu C, Davies N, Jayakumar AR, Rose CF (2021) 2021 ISHEN guidelines on animal models of hepatic encephalopathy. Liver Int 41(2021):1474–1488. https://doi.org/10.1111/liv.14911
17. Lima LCD, Miranda AS, Ferreira RN, Rachid MA, e Silva ACS (2019) Hepatic encephalopathy: lessons from preclinical studies. World J Hepatol 11(2):173–185. https://doi.org/10.4254/wjh.v11.i2.173
18. Lynch A (2022) Hepatic encephalopathy. Small Anim Crit Care Med 5 (March):506–509. https://doi.org/10.1016/B978-0-323-76469-8.00096-4
19. Delanty N, French JA, Labar DR, Pedley TA, Rowan AJ (2001) Status epilepticus arising de novo in hospitalized patients: an analysis of 41 patients. Seizure 10(2):116–119
20. Eleftheriadis N, Fourla E, Eleftheriadis D, Karlovasitou A (2003) Status epilepticus as a manifestation of hepatic encephalopathy. Acta Neurol Scand 107(2):142–144
21. Prabhakar S, Bhatia R (2003) Management of agitation and convulsions in hepatic encephalopathy. Indian J Gastroenterol 22:S54–S58

22. RD A (1953) The neurological disorder associated with liver disease. Metab Toxic Dis Nerv Syst:198–237
23. Braissant O, Rackayová V, Pierzchala K, Grosse J, McLin VA, Cudalbu C (2019) Longitudinal neurometabolic changes in the hippocampus of a rat model of chronic hepatic encephalopathy. J Hepatol 71(3):505–515. https://doi.org/10.1016/j.jhep.2019.05.022
24. Smimih K et al (2023) Sensory motor function disturbances in mice prenatally exposed to low dose of ethanol: a neurobehavioral study in postnatal and adult stages. Neurol Int 15(2): 580–594
25. López L et al (2002) Cytochrome oxidase activity of the suprachiasmatic nucleus and pineal gland in rats with portacaval shunt. Exp Neurol 173(2):275–282
26. Funabashi D, Wakiyama Y, Muto N, Kita I, Nishijima T (2022) Social isolation is a direct determinant of decreased home-cage activity in mice: a within-subjects study using a body-implantable actimeter. Exp Physiol 107(2): 133–146
27. Monville C, Torres EM, Dunnett SB (2006) Comparison of incremental and accelerating protocols of the rotarod test for the assessment of motor deficits in the 6-OHDA model. J Neurosci Methods 158(2):219–223. https://doi.org/10.1016/j.jneumeth.2006.06.001
28. Dhanda S, Sandhir R (2015) Role of dopaminergic and serotonergic neurotransmitters in behavioral alterations observed in rodent model of hepatic encephalopathy. Behav Brain Res 286:222–235. https://doi.org/10.1016/j.bbr.2015.01.042
29. Carter RJ, Morton J, Dunnett SB (2001) Motor coordination and balance in rodents. Curr Protoc Neurosci 15(1):8–12
30. Deacon RMJ (2013) Measuring motor coordination in mice. J Vis Exp 75:1–8. https://doi.org/10.3791/2609
31. Matkowskyj KA, Marrero JA, Carroll RE, Danilkovich AV, Green RM, Benya RV (1999) Azoxymethane-induced fulminant hepatic failure in C57BL/6J mice: characterization of a new animal model. Am J Physiol 277(2):G455–G462
32. McMillin M et al (2014) Gli1 activation and protection against hepatic encephalopathy is suppressed by circulating transforming growth factor β1 in mice. J Hepatol 61(6):1260–1266
33. Vorhees CV, Williams MT (2006) Morris water maze: procedures for assessing spatial and related forms of learning and memory. Nat Protoc 1(2):848–858
34. Shimizu T, Nayar SG, Richardson WD (2003) An appetitive spatial working memory task for mice, using a semiautomated 8-arm radial maze, 03 November 2023, PROTOCOL (Version 1) available at Protocol Exchange. https://doi.org/10.21203/rs.3.pex-2431/v1
35. Bellettini-Santos T et al (2023) Vitamin D3 improves spatial memory and modulates cytokine levels in aged rats. Metab Brain Dis 38(4): 1155–1166. https://doi.org/10.1007/s11011-022-01152-9
36. Poorheidari G, Mashhadi Akbar Boojar M (2023) Scopolamine induced non-cognitive effects in the radial-arm maze: alleviation by Physostigmine. Adv Pharmacol Ther J 3(1): 9–21. https://doi.org/10.18502/aptj.v3i1.12487
37. Cassity S et al (2024) Cholinergic modulation of rearing in rats performing a spatial memory task. Eur J Neurosci 59(9):2240–2255
38. Ramos-Moreno T, Castillo CG, Martínez-Serrano A (2012) Long term behavioral effects of functional dopaminergic neurons generated from human neural stem cells in the rat 6-OH-DA Parkinson's disease model. Effects of the forced expression of BCL-XL. Behav Brain Res 232(1):225–232
39. Deacon RMJ, Rawlins JNP (2006) T-maze alternation in the rodent. Nat Protoc 1(1): 7–12
40. Rodgers RJ, Dalvi A (1997) Anxiety, defence and the elevated plus-maze. Neurosci Biobehav Rev 21(6):801–810. https://doi.org/10.1016/S0149-7634(96)00058-9
41. Walf AA, Frye CA (2007) The use of the elevated plus maze as an assay of anxiety-related behavior in rodents. Nat Protoc 2(2):322–328. https://doi.org/10.1038/nprot.2007.44
42. Benammi H, El O, Romane A, Gamrani H (2014) A blunted anxiolytic like effect of curcumin against acute lead induced anxiety in rat: involvement of serotonin. Acta Histochem 116(5):920–925
43. Drapier D et al (2007) Effects of acute fluoxetine, paroxetine and desipramine on rats tested on the elevated plus-maze. Behav Brain Res 176(2):202–209. https://doi.org/10.1016/j.bbr.2006.10.002
44. Bourin M, Hascoe M (2003) The mouse light / dark box test. 463:55–65. https://doi.org/10.1016/S0014-2999(03)01274-3
45. Amaral DG (2002) The primate amygdala and the neurobiology of social behavior: implications for understanding social anxiety. Biol Psychiatry 51(1):11–17
46. Wu T, Luo Y, Broster LS, Gu R, Luo Y (2013) The impact of anxiety on social decision-

making: behavioral and electrodermal findings. Soc Neurosci 8(1):11–21

47. Brown EJ, Turovsky J, Heimberg RG, Juster HR, Brown TA, Barlow DH (1997) Validation of the social interaction anxiety scale and the social phobia scale across the anxiety disorders. Psychol Assess 9(1):21
48. Kent S et al (1992) Different receptor mechanisms mediate the pyrogenic and behavioral effects of interleukin 1. Proc Natl Acad Sci 89(19):9117–9120
49. Schoemaker RG, Smits JFM (1994) Behavioral changes following chronic myocardial infarction in rats. Physiol Behav 56(3):585–589
50. Kraeuter A-K, Guest PC, Sarnyai Z (2019) The open field test for measuring locomotor activity and anxiety-like behavior. In: Pre-clinical model: techniques and protocols, pp 99–103
51. Carola V, D'Olimpio F, Brunamonti E, Mangia F, Renzi P (2002) Evaluation of the elevated plus-maze and open-field tests for the assessment of anxiety-related behaviour in inbred mice. Behav Brain Res 134(1–2): 49–57. https://doi.org/10.1016/S0166-4328(01)00452-1
52. Önel T, Arıcıoğlu F, Yıldırım E, Zortul H, Yaba A (2023) The effect of maternal separation stress-induced depression on ovarian reserve in Sprague Dawley Rats: the possible role of imipramine and agmatine through a mTOR signal pathway. Physiol Behav 269:114270
53. Dehdar K et al (2023) Allergen induces depression-like behavior in association with altered prefrontal-hippocampal circuit in male rats. Neuroscience 524:21–36
54. Dunn AJ, Swiergiel AH (2005) Effects of interleukin-1 and endotoxin in the forced swim and tail suspension tests in mice. Pharmacol Biochem Behav 81(3):688–693
55. Frenois F et al (2007) Lipopolysaccharide induces delayed FosB/DeltaFosB immunostaining within the mouse extended amygdala, hippocampus and hypothalamus, that parallel the expression of depressive-like behavior. Psychoneuroendocrinology 32(5):516–531
56. Wright G, Jalan R (2007) Management of hepatic encephalopathy in patients with cirrhosis. Best Pract Res Clin Gastroenterol 21(1): 95–110
57. Rodenbaugh D, Vo CT, Redulla R, McCauley K (2020) Nursing management of hepatic encephalopathy. Gastroenterol Nurs 43(2): E35–E47
58. Bloom A et al (2020) Can adequate analgesia be achieved in patients with cirrhosis without precipitating hepatic encephalopathy? A prospective study. Clin Exp Hepatol 6(3):243–252
59. Brenner M et al (2015) Patients with manifest hepatic encephalopathy can reveal impaired thermal perception. Acta Neurol Scand 132(3):156–163
60. Bannon AW, Malmberg AB (2007) Models of nociception: hot-plate, tail-flick, and formalin tests in rodents. Curr Protoc Neurosci Chapter 8(October):1–16. https://doi.org/10.1002/0471142301.ns0809s41
61. Janssen PA, Niemegeers CJ, Dony JG (1963) The inhibitory effect of fentanyl and other morphine-like analgesics on the warm water induced tail withdrawl reflex in rats. Arzneimittelforschung 13:502–507
62. Hausser N, Johnson K, Parsley MA, Guptarak J, Spratt H, Sell SL (2018) Detecting behavioral deficits in rats after traumatic brain injury. JoVE (J Vis Exp) 131:e56044
63. Hausser N, Johnson K, Parsley MA, Guptarak J, Spratt H, Sell SL (2018) Detecting behavioral deficits in rats after traumatic brain injury. J Vis Exp January:1–11. https://doi.org/10.3791/56044
64. Abdelmohcine A et al (2023) Hyperammonemia induces gut microbiota dysbiosis and motor coordination disturbances in mice: new insight into gut-brain axis involvement in hepatic encephalopathy. Acta Neurobiol Exp (Wars) 83(2):203–215
65. Bromley-Brits K, Deng Y, Song W (2011) Morris water maze test for learning and memory deficits in Alzheimer's disease model mice. JoVE (J Vis Exp) 53:e2920
66. Bontempi B, Whelan KT, Risbrough VB, Lloyd GK, Menzaghi F (2003) Cognitive enhancing properties and tolerability of cholinergic agents in mice: a comparative study of nicotine, donepezil, and SIB-1553A, a subtype-selective ligand for nicotinic acetylcholine receptors. Neuropsychopharmacology 28(7):1235–1246
67. El Khiat A et al (2019) Kinetic deterioration of short memory in rat with acute hepatic encephalopathy: involvement of astroglial and neuronal dysfunctions. Behav Brain Res 367 (March):201–209. https://doi.org/10.1016/j.bbr.2019.03.046
68. Draoui A, El Hiba O, Abbaoui A, El Fari R, Aitihya M, Gamrani H (2019) Differential impairment of short working and spatial memories in a rat model of progressive Parkinson's disease onset: a focus on the prodromal stage. Brain Res Bull 150:307–316
69. Fukuda T, Sato Y, Takiguchi M, Yamamoto T, Murasawa H, Pawlak A (2020) Dehydroeffusol rescues amyloid β 25–35 -induced spatial working memory deficit. Plant Foods Hum Nutr 75:279–282

70. Onaolapo OJ, Onaolapo AY, Mosaku TJ, Akanji OO, Abiodun OR (2012) Elevated plus maze and Y-maze behavioral effects of subchronic, oral low dose monosodium glutamate in Swiss albino mice. J Pharm Biol Sci 3(4):21–27
71. Bevins RA, Besheer J (2006) Object recognition in rats and mice: a one-trial non-matching-to-sample learning task to study'recognition memory'. Nat Protoc 1(3):1306–1311
72. Fernandes VS et al (2012) Repeated treatment with a low dose of reserpine as a progressive model of Parkinson's disease. Behav Brain Res 231(1):154–163
73. Bernal-Mondragón C, Rivas-Arancibia S, Kendrick KM, Guevara-Guzmán R (2013) Estradiol prevents olfactory dysfunction induced by A-β 25-35 injection in hippocampus. BMC Neurosci 14. https://doi.org/10.1186/1471-2202-14-104
74. Pellow S, Chopin P, File SE, Briley M (1985) Validation of open:closed arm entries in an elevated plus-maze as a measure of anxiety in the rat. J Neurosci Methods 14(3):149–167. https://doi.org/10.1016/0165-0270(85)90031-7
75. Fuss J, Ben Abdallah NMB, Hensley FW, Weber K-J, Hellweg R, Gass P (2010) Deletion of running-induced hippocampal neurogenesis by irradiation prevents development of an anxious phenotype in mice. PLoS One 5(9): e12769
76. Binder E, Droste SK, Ohl F, Reul JMHM (2004) Regular voluntary exercise reduces anxiety-related behaviour and impulsiveness in mice. 155:197–206. https://doi.org/10.1016/j.bbr.2004.04.017
77. Huang X, Yang C, Huang M (2023) Protective mechanism of the EZH2/microRNA-15a-5p/CXCL10 axis in rats with depressive-like behaviors. J Chem Neuroanat 132(97):102283. https://doi.org/10.1016/j.jchemneu.2023.102283
78. Yankelevitch-yahav R, Franko M, Huly A, Doron R (2015) The forced swim test as a model of depressive-like behavior. March:1–7. https://doi.org/10.3791/52587
79. Le Bars D, Gozariu M, Cadden SW (2001) Animal models of nociception. Pharmacol Rev 53(4):597–652. https://doi.org/10.1002/9783527611942.ch9

Chapter 8

Human Liver Spheroids from Peripheral Blood: A New Insight Approach for Hepatic Encephalopathy Research

Zorica A. Becker-Kojić, Anne-Kathrin Schott, Ivan Zipančić, Lucie Picard, and Vicente Hernández-Rabaza

Abstract

Hepatic encephalopathy (HE) is a nervous system disorder produced by liver failure. HE patients display a broad range of symptoms, involving motor, emotional, and cognitive disorders. HE severe cases end in patient death. This disease is a worldwide affection, and which number is increasing in parallel to liver pathologies associated with occidental lifestyles. The therapeutic HE approaches are limited by liver transplantation offers and by the absence of human hepatocyte research models. To address this technical limitation, we present a method to obtain human liver spheroids from unmanipulated peripheral blood. Human liver spheroids are obtained by using blood-derived pluripotent stem cells (BD-PSCs), previously generated from unmanipulated peripheral blood by activation of human membrane-bound GPI-linked protein, following differentiation into human hepatocytes. Human liver spheroids possess better physiological responses than 2D cultures, including a more realistic biologically microenvironment, and more accurate prediction of disease state and in vivo-like drug response. This chapter provides the guidelines for discovering our method, introduces a promising new source of human hepatocyte spheroids, and offers a new technical HE research tool.

Key words Liver, Hepatic encephalopathy, Blood, BD-PSCs, Spheroids

1 Introduction

Hepatic encephalopathy (HE) is a disease which displays a gradient of motor, emotional and cognitive alterations, evolving, in the worst cases, into the death of the patient [1]. HE represents a worldwide health problem affecting millions of persons. HE symptoms cover a broad range of symptoms from slight motor disturbances, such as drive alterations, to severe cognitive deficits [2]. HE can be initiated by several pathologies, including virus infection,

Supplementary Information The online version contains supplementary material available at https://doi.org/10.1007/978-1-0716-4354-9_8.

Tiziano Balzano (ed.), *Experimental and Clinical Methods in Hepatic Encephalopathy Research*, Neuromethods, vol. 219, https://doi.org/10.1007/978-1-0716-4354-9_8,

liver alcoholic-associated pathologies and liver cancer, which all can evolve into cirrhosis and liver failure, which finally is the physiological director that triggers HE [3, 4]. The incapacity of liver detoxification and the consequent accumulation of toxic metabolites into the bloodstream, mainly ammonium, is the main cause of HE [5]. Additionally, the immune system, blood pressure, and metabolic functions are also altered and contribute to the clinical symptoms [6, 7]. Finally, hepatic insufficiency is translated into nervous system lesions, affecting motor, cognitive, and emotional capacities [8–10], which require liver treatments [11, 12].

The therapeutical approaches must adapt to the grade of liver lesion and metabolic derangements induced by HE, once liver fibrosis and necrosis achieve a massive level, transplantation is the only recognized treatment [13]. Currently, livers demand highly overcomes the offer, and the patients and families must face this hard reality. Moreover, an expanding occidental lifestyle, building on sedentary and unhealthy feeding habits, and abuse of recreative drugs, draw an increase in HE prevalence in the coming years. New therapeutic and research approaches are demanded to address the expanding and serious health problems associated with HE [14].

A three-dimensional (3D) spheroid culture system represents a promising research model to analyze cellular and molecular behaviors, as well as test pharmacology effects [15]. Hepatic 3D cultures provide a technological opportunity to advance research and therapeutic innovation in HE [16].

1.1 Hepatic 3D Cultures, Between the Two-Dimensional (2D) Cell Culture and Complex Organ Approaches

Growing cells in 2D has been the most very well-established method for culturing human cells for many years. This method is simple, using straightforward protocols that are easy to perform and analyze. Cells are growing in monolayer, flat stretched and polarized in multi-well cell culture plates, with the addition of appropriate growth surfaces and media. Reagents and consumables are widely available and inexpensive.

Unfortunately, 2D cell cultures harbor several limitations, most importantly, they do not reflect on the complexity of the natural structure of tissue, like cell-cell and cell-environment 3D interactions. The expression pattern of 2D cell cultures differs from that of in vivo models. These limitations have driven the development of alternative models that better mimic the natural surroundings, physiology, and structure of biological tissues. A 3D culture system is more expensive than a 2D with fewer commercially available representatives. The most important advantage is that it mimics in vivo interactions and shows expression patterns that are similar to in vivo models [17].

In 3D culture, cells often grow to form spheroid cells. 3D cell cultures arrange themselves during proliferation into sphere-like formations [18]. 3D cultures can be prepared using a support known as a scaffold which allows growth in all directions. Examples of scaffolds include hydrogels, polymeric hard materials, and hydrophilic glass fiber. In scaffold-based, cells are embedded into a matrix, which mimics extracellular matrix (ECM), with hard polymer scaffolds providing the physical support of specialized high-fiber tissue such as skin, tendons, or bones. Scaffold-free approaches also exist, and these rely on the self-aggregation of cells. One of the most well-known scaffold-free approaches is the hanging drop method; a technique that can be used both with and without specialized plates. In this method, cells are encouraged to naturally form cellular aggregates, facilitated by gravitational and surface tension forces. In addition, bioreactors are increasingly used for large-scale cell culture. Examples of these include general bioreactors that are independent of a CO_2 incubator, to alternative bioreactors such as orbital shakers, spinner flasks, and rotating wall vessels. All of which employ motion to prevent cells from adhering to surfaces, keeping cells and spheroids in solution.

Organoids can be defined as self-assembled 3D tissue structures that are derived from primary tissue or stem cells [19]. They are a mini simplified version representing an organ, that is produced in vitro and three dimensions. Organoids mimic the functional, structural, and biological complexity of an organ. Organoids are formed from human embryonic (hESCs), human induced pluripotent stem cells (hiPSCs), stem cells, or mesenchymal cells from adult tissues, all due to their capability to self-organize in 3D culture condition owing to their self-renewal and differentiation capacities [20]. With rapidly growing and improving organoid technology particularly with 3D bioprinting technology at high speed in a 96-well plate, it can be expected that this technology can be used in personalized diagnostic, tissue engineering, gene, and cell therapies as well as in regenerative medicine [21].

3D cell cultures offer scientists a model platform for a wide range of applications due to better and more accurate insight into normal and diseased physiological function, disease progression, and drug toxicity, including liver diseases [22].

1.2 Hepatocytes

Human hepatocytes are the main parenchymal cells of the liver and constitute 80% of the liver mass. The function of hepatocytes is very important including protein synthesis, lipid metabolism (synthesis of cholesterol, phospholipids, and bile salt), protein storage, carbohydrate metabolism, detoxification, and excretion of exogenous and endogenous substances. Hepatocytes are commonly used in cell, biological, and biopharmaceutical research to explore the mechanism of drug metabolism and even predict in vivo drug responses. For that purpose, hepatocytes are isolated from animal or human tissue and kept in 2D culture as a sandwich with an additional layer of ECM. The use of primary human hepatocytes in 2D culture brings some important limitations: accessibility and differentiation of cellular profile. Primary human hepatocyte availability is limited and once in 2D cultures hepatocytes rapidly differentiate reducing or even losing the proper hepatocyte profile [23]. Hepatocytes can be differentiated from PSCs like hESCs or iPSCs [24] and then used for the derivation of 3D hepatic cultures [25].

As mentioned earlier, liver 3D cultures offer a more realistic cellular model than 2D cultures. 3D spheroids provide a microenvironment which mimics better that 2D the natural cellular condition, as indicated by 3D cellular orientation and proximity, which are both better than in 2D cultures, additionally, the ECM production is also higher in 3D cultures [17], the last fact is very relevant due to the role that the ECM plays in tissue behaviors.

1.3 A Solution Between the 2D and Complex Organs Approaches

Additionally, besides the 3D cellular cultures advantages, the spheroids also represent a valuable, and necessary, alternative to complex organ research [15]. Liver tissue regeneration and complex liver organoid development is a current challenging goal, which will grow from 3D hepatocyte spheroid models. Hepatic spheroids provide the microenvironment to test the behaviors of different cell types and targeted molecular pathways, becoming essential to accelerate liver regeneration innovation [26]. Besides regeneration research, the 3D cultures can play a relevant role in regenerative medicine, potentially being a clinical alternative, especially under specific circumstances with high mortality rates, such as acute liver diseases (around 80% of mortality rate), in this scenario, the spheroids could partially recover liver function patients providing more time until a suitable transplant can be done [27].

The roles that human liver spheroids can play in cellular culture and liver regeneration are both probably the most relevant applications, but not the only ones. Additional spheroid potential functions, such as extracellular vesicles (EVs) sources or animal model alternatives, are also remarkable.

The EVs research is a promising biomedical approach [28, 29]. Hepatocyte spheroids provide a cellular model to collect EVs induced by several treatments [30]. Finally, human spheroids represent an alternative to animal model research. Specie limitations and ethical issues can both be minimized by human cellular cultures. Rodent animal models, the most common research models, have biological and molecular differences from humans, which means limitations in the extrapolation of conclusions and applications [31]. Moreover, the use of animal models tends to be reduced and controlled by ethical committee regulations, trying to reduce and replace animal use by alternative methods. In this line, human spheroids represent a solid alternative in molecular, cellular, and toxicology studies.

Sum the liver human spheroids represent a relevant tool in biomedical science, even so, human liver hepatocyte availability is limited. Likewise, the method to generate human hepatocytes through hepatic differentiation using hESCs or iPSCs is restricted by several factors including some technical like low yield of fully differentiated cells, small-scale cell culture, and immunological rejection problems as well [32, 33]. However, the possibility of generating new hepatocytes and 3D cultures from blood samples provides a new method that can reduce all the previous limitations. Furthermore, opens the possibility to create personal genetic cultures from autologous samples.

Considering all the listed properties, we have developed a protocol for human hepatic spheroids generation by using blood-derived pluripotent stem cells (BD-PSCs) and grew them in ultra-low attachment plates. We have prepared spheroids containing 4000 to 1×10^6 cells. To characterize the spheroids, light microscopy and immunofluorescence (IF) analyses were used. Hepatic function was tested by expression assessment of P450 cytochrome family, which are involved in the cellular detoxification process, specifically, P450 3A4 (CYP3A4) and 2E1 (CYP2E1) enzymes [25].

2 Materials

2.1 Fluorescence Microscope for Live Staining and Immunofluorescence (IF) Analysis

To image BD-PSC-derived liver spheroids for live staining and IF of 2D cell cultures, we used an inverted microscope (CKX53 equipped with cell imaging software cellSens and fluorescence light source U-RFL-T, Olympus, Germany). The stage was equipped with a universal holder to accommodate microtiter plates (OLS OMNI Life Science, Germany) as well as multidishes (Nunclon 4-Well dish, Sigma Aldrich Germany). The objective with 4× magnification was chosen for live staining and objectives with 20× and 40× magnification for 2D cell cultures. The brightfield pictures of cells/spheroids were directly taken before the corresponding fluorescence analysis. The exposure settings were made automatic. Fluorophores were excited using the filters for B-excitation (Goat anti-Mouse IgG DyLight 488, Antibodies online Germany) and G-excitation (Rabbit anti-chicken IgG Texas red and Goat anti-Mouse IgG Cy3, Antibodies online, Germany). Primary anti-human antibodies were purchased as follows: alpha-fetoprotein, CYP3A4 and CYP2E1 from Proteintech, Germany; albumin, hepatocyte nuclear factor 4 alpha and transthyretin from Sigma Aldrich, Germany; cytokeratin 18 from Agilent Dako, Germany.

2.2 Cell Culture Media Preparation

Five hundred milliliters of hepatoblast medium was prepared using knockout serum replacement (KSR)/dimethyl sulfoxide (DMSO) medium consisting of 76.4% knockout DMEM (KO-DMEM, Thermo Fisher Scientific Germany), 20% KSR (Fisher Scientific Germany), 0.5% L-glutamine, 1% non-essential amino acids (NEAA), 0.1% β-mercaptoethanol (all from Thermo Fisher

Scientific Germany), 1% DMSO (Sigma Aldrich Germany), and 1% penicillin-streptomycin (Pen/Strep, BioChrom GmbH Germany). Five hundred milliliters of hepatocyte maturation medium was prepared using 1% L-glutamine, 10 μM hydrocortisone 21-hemisuccinate sodium salt (HCC, Biomol Germany), and 1% Pen/Strep. Medium from the stock was aliquoted and fresh hepatocyte growth factor (HGF, Thermo Fisher Scientific) and oncostatin M (OSM, Thermo Fisher Scientific) were added at final concentrations of 10 ng/mL and/or 20 ng/mL for each medium change. Oncostatin M is a cytokine belonging to the interleukin 6 group of cytokines, important for hematopoiesis as well as for liver development.

2.3 Preparation of Glass Coverslips for the Generation of Human Hepatocytes

To facilitate the microscopical and immunophenotypically analysis of hepatocyte re-differentiation, we used glass coverslips (Langenbrinck Germany) that were prepared as follows. For cleaning the glass coverslips (14 mm) were incubated in nonionic detergent for 10 min and washed in deionized water till no bubbles remained. Glass coverslips were further incubated in 1 M HCl for 30 min (adapted from Marchenko et al. [33]). Deionized water was used for the final wash of glass coverslips and this procedure was repeated at least 3×. Coverslips were left overnight at room temperature to dry and sterilized in a suitable container in an autoclave.

3 Methods

Ethical approvals were obtained (ACA CELL Biotech GmbH/ 25b-5482.2-64-1) for performing these experiments and informed consent was signed by all donors before blood extraction in compliance with institutional guidelines. This protocol was previously published in Jove [25], and the present communication has been approved by the Jove editorial, including text and image use.

3.1 Preparing Mononuclear Cells (MNCs) from Human Peripheral Blood (PB)

To prepare the human leukocytes from peripheral blood (PB), we used blood from healthy donors, extracted by trained medical personnel, following the standard protocol using density gradient media according to the published protocol [34].

Thirty milliliters of PB was extracted from the veins of healthy donors using a vacutainer with heparin according to standard protocol.

Peripheral blood mononuclear cells (PBMNCs) were isolated by Ficoll preparation according to the protocol published by Becker-Kojić et al. [34]. The interphase layer between plasma and Ficoll containing the MNCs was isolated, by pipetting. Isolated cells were washed using sterile phosphate buffer saline (PBS) by centrifugation at 300× *g* and this procedure was repeated three times.

The number of cells was counted using the Neubauer chamber.

3.2 Dedifferentiation of MNCs Upon Activation with Human GPI-Anchored Glycoprotein

Dedifferentiation is a transient process by which the cells regress from a mature cell phenotype, by losing their phenotypic specializations and returning to an earlier stage, increasing cell potency, and gaining the feature to re-differentiate into alternative specialized cell types. This process is contrary to differentiation where differences in gene expression, morphology, and physiology arise in cells making its function increasingly specialized. In our laboratory, we have developed a robust dedifferentiation method of human blood progenitor cells, which is based on the activation of the human membrane GPI-linked protein ACA by antibody cross-linking at the membrane of blood progenitor cells [35, 36]. Membrane signaling of ACA-protein via PLCy/PI3K/Akt/mTOR/PTEN upregulates WNT, B-catenin, Notch, Bmi-1 and HoxB4 genes that promote self-renewal leading to the generation of human pluripotent stem cells, capable of re-differentiation into cells belonging to all three germ layers. ACA-specific antibody was used for initiation of the signaling mechanism behind the GPI-membrane protein.

6×10^6 mononuclear cells in PBS/1% bovine serum albumin (BSA) in a polystyrene tube (15 mL) were incubated with the specific antibody for 30 min at 37 °C according to Becker-Kojić et al. [34].

Cells were centrifuged at 300× *g* at room temperature and the incubation medium (PBS/BSA) was replaced with Iscove's modified Dulbecco's medium supplemented with 10% fetal bovine serum (FBS). Culture medium did not have any of the growth factors usually used for the cultivation of stem cells.

We used polystyrene tubes to culture the cells in 5% CO_2 -incubator at 37 °C.

The human cells were grown for 8–10 days as described by Becker- Kojić et al. [34]. On day 5 (D5), 1–2 mL of Iscove's medium supplemented with 10% FBS should be added to each 15 mL tube.

3.3 Sorting of Newly Generated Dedifferentiated Cells

Membrane GPI-linked protein ACA is expressed in all terminally non-differentiated (progenitor) or stem cells in the blood. Only these cells can be reprogrammed by the nucleus-to-membrane ACA-signaling network [37]. To obtain the population of stem cells after reprogramming, the mature CD45-positive cells should be removed.

After 8–10 days of culturing, the reprogrammed cells in suspension are centrifuged at room temperature for 10 min at 300× *g* and the resulting supernatant is aspirated with a sterile pipette according to Becker-Kojić et al. [34].

The pellet obtained by centrifugation was resuspended in 90 μL of cold PBS buffer (PBS pH 7.2, 0.5% BSA, and 2 mM EDTA) and 10 μL of CD45 positive nano-sized magnetic beads was added to the resuspended cultured cells and incubated on ice for 15 min.

The cell suspension was washed with 2 mL of PBS buffer and centrifuged at 300× *g* for 10 min at room temperature. The cells were resuspended in 500 μL of PBS buffer.

The column was washed with 500 μL of pre-cooled PBS buffer and placed in the magnetic field using a magnetic stand.

The cells were placed onto the column and washed twice with 500 μL of PBS buffer. The flow-through, which contained the CD45 negative cells, was collected. Iscove's medium supplemented with 10% FBS was added to the collected reprogrammed cells.

The number of reprogrammed cells was counted in a Neubauer chamber.

3.4 Coating Cell Culture Plates with Biolaminin for 2D Hepatic Differentiation of BD-PSCs

Laminin is a major component of the basement membrane, which represents a continuous sheet of extracellular matrix that forms an interface between epithelial, muscle or neuronal cells and their adjacent stroma. Basement membrane plays an essential role in tissue organization by influencing cell adhesion, migration, and proliferation. Human recombinant laminins are indispensable xeno-free adhesion proteins that support cell viability and hepatocyte function in vitro [38].

Using a sterile pair of tweezers, autoclaved coverslips were placed in 4-well plates and exposed to UV light for 30 min to ensure sterile conditions.

A 120 μL aliquot of 5 μg/mL biolaminin of human recombinant biolaminin (LN111) was added to each glass coverslip. The coated glass coverslips were left overnight at 4 °C.

The excess of biolaminin was removed and 200 μL of hepatocyte differentiation medium was put in each well, as described below.

3.5 Culturing Hepatic Cells Differentiated from BD-PSCs

Differentiation of human ESCs or iPSC toward hepatocytes provides a very important platform for studying liver diseases, drug discovery, and drug induced toxicity. The shortage of human primary hepatocytes and their feature to dedifferentiate in vitro (in culture conditions) losing their hepatic function, as well as an increased therapeutic and pharmaceutical request for functional hepatocytes, has urged the development of protocols for derivation of human hepatocytes using PSCs [16]. Several protocols for in vitro differentiation of human pluripotent stem cells into hepatocyte-like cells are developed and these cells are currently used to test drug efficacy and toxicity as well as modelling various chronic and inherited human liver diseases [39].

Therefore, we have developed our own two-stage protocol using our blood-derived pluripotent stem cells (BD-PSCs). Firstly, BD-PSCs were differentiated into endoderm/hepatic progenitor cells at D4, following differentiation to hepatoblasts (hepatic progenitor cells) at D8 with typical polygonal morphology, and finally, maturation to hepatocytes at D15–24 as published in Schott et al. [25].

3.5.1 Details

3 × 10^5 BD-PSCS cells were put into each well of a 4-well plate coated with biolaminin and placed in the incubator at 37 °C and 5% CO_2.

The cells were cultured for 5 days in KSR/DMSO hepatoblast medium that supports endodermal differentiation, and every second-day medium was changed.

At D5 medium was switched to hepatocyte maturation medium and the cells were cultured for an additional 20 days in the incubator at 37 °C and 5% CO_2. Medium change was every second day.

3.6 Immunofluorescence Analysis of Newly Generated 2D Liver Cell Cultures

Immunophenotyping for 2D liver cultures was performed by using typical liver progenitor markers. Alpha-fetoprotein (AFP) is the most abundant plasma glycoprotein that is found in human fetuses. The adult organism expresses a very low concentration of this protein; therefore, it is considered a marker for hepatocyte progenitor cells [40]. Transthyretin (TTR) is a 55 kDa homotetramer and a major thyroid-hormone binding protein that transports the thyroid hormone thyroxine (T4) from the bloodstream to the brain [23, 41]. This protein is expressed at the first stage (D4 to D8) of hepatic differentiation of BD-PSCs. Following hepatic differentiation at D15, its expression decreases, while the expression of albumin (ALB) increases. Human serum albumin is produced mainly in the liver and represents the most abundant protein in blood plasma. Its function is to transport hormone fatty acids and other compounds and to buffer blood pH maintaining oncotic pressure. Expression of ALB is most critical for hepatic differentiation [42]. We also examined the expression of hepatocyte nuclear factor 4 alpha (HNF-4α), a transcription factor indispensable for the expression of liver-specific genes that play a role in the development of the liver. This protein appeared at D4 during the hepatic differentiation of BD-PSCs, and rose throughout the differentiation process reaching a strong and stable expression during maturation time at D15-D24.

3.6.1 Details

1. BD-PSCs cells were cultured according to the differentiation method described above for 4, 8, 15, and 24 days and culture media was removed.
2. Cells were incubated with pre-warmed fixative consisting of 4% paraformaldehyde in PBS for 10 min. The fixative was discarded, and the cells washed twice with PBS for 5 min each. Triton X-100 at the concentration of 0.1% was added carefully, and cells were permeabilized for 5 min and washed twice with PBS.
3. Blocking solution made of PBS and 5% BSA was added, and 4-well plates placed on a rocker plate for 1 h at room temperature.

4. Primary antibodies were diluted in dilution buffer 1% BSA/PBS as follows: albumin (ALB) 1:50, alpha-fetoprotein (AFP) 1:250, cytokeratin 18 (CK18) 1:50, hepatocyte nuclear factor 4 alpha (HNF4α) 1:1000 and transthyretin (TTR) 1:50. 50 μL of antibody dilution was used per well.
5. The cells were incubated for 1 h at room temperature, and the antibody solution was discarded. Cells were then washed for 5 min. This step was repeated three times.
6. The following secondary antibodies were prepared in dilution buffer: rabbit anti-chicken IgG (Texas red) 1:1000, goat anti-mouse IgG (488) 1:1000, and goat anti-rabbit (488) 1:500. 50 μL of antibody dilution was used per well and the cells were incubated for 30 min at room temperature.
7. The cells were washed three times with PBS for 5 min each and the coverslips were mounted with mounting media containing DAPI for microscopic analysis.

3.7 3D Spheroid Hepatic Differentiation

3D culture of human cells gained great scientific interest due to their better resemblance to the cell environment, mimicking in vivo tissue organization and cell-cell interactions. 3D hepatic cultures originating from PSCs maintain patient-specific phenotypes and functions for weeks in a culture, making them useful in translational pharmacology and toxicology. 3D hepatic culture can help to understand the molecular mechanism of metabolic and infectious diseases, such as HE, malaria, and viral hepatitis [22]. Moreover, the knowledge about 3D hepatic structures facilitates the development of organoids, a miniaturized in vitro structure that resembles the liver and can be used to study its biology. Liver organoids can be used to study development, disease, as well as drug discovery in the laboratory. There is a hope that such models can be established to study disease development and potential drugs and are potentially applicable in the field of regenerative medicine, such as for the cure of HE [43]. We developed our two-stage protocols to develop human hepatocytes from BD-PSCs by using the same media as described for 2D liver differentiation. Growing the cells in ultra-low attachment plates that enable the hepatic differentiated cells to self-assemble shortens the time for full differentiation to mature hepatocytes as compared to 2D cultures [39].

3.7.1 Details

BD-PSCs were counted using a Neubauer chamber.

BD-PSC suspension was centrifuged at 300× *g* for 10 min at room temperature, the supernatant removed, and the cells resuspended in KSR/DMSO medium at 2 × 10^6 cells/mL.

The cells were passed through a 40 μm cell strainer to ensure a single-cell suspension and to remove any additional debris.

BD-PSCs were counted again using a Neubauer chamber to prepare a sufficient volume for each cell seeding density to dispense the required volume per well. We prepared a gradient with top seeding of 1×10^6 cells to a low seeding density of 4000 cells.

One hundred microliters of KSR/DMSO medium was dispensed in 96-well ultra-low attachment plates and 100 μL of cell suspension was added in seeding dilution.

The ultra-low attachment plate was placed in the incubator at 37 °C and 5% CO_2 for 5 days. Fifty percent of the medium was changed 3–4 days after seeding to be sure that the spheroids were sufficiently compact.

3.8 Live Staining of Liver Spheroids

To test the newly formed spheroids for their capability of hepatocyte-specific function, we analyzed through fluorescent live staining the expression of cytochrome P450 3A4 (CYP3A4) and 2E1 (CYP2E1) enzymes. They belong to the cytochrome P450 family and play important roles in cellular and drug metabolism through the process of detoxification [44]. The presence of cytoskeletal protein Cytokeratin 18 (CK18) is of interest as well, as it is expressed in the liver as one of the major components of intermediate filament [45]. The expression of ALB and AFP as mentioned above for 2D liver culture was examined as well.

1. The culture media was removed in a manner to not touch the newly formed spheroids, and a freshly prepared solution of PBS containing 0.1% Triton X-100 was added for 5 min to allow the cells to be permeabilized.
2. The spheroids were then washed with medium for 5 min, this step was repeated twice. The pipetting was done carefully to not manipulate the spheroids.
3. The primary anti-human antibodies were diluted in PBS with 1% BSA as follows: ALB (1:50), AFP (1:250), CK18 (1:50), CYP2E1 (1:200), and CYP3A4 (1:200) and placed for 1 h in 5% CO_2 incubator at 37 °C.
4. After the incubation time the excess of antibody solution was removed by washing the spheroids with medium, this was done three times.
5. The dilutions of the corresponding secondary antibodies goat anti-mouse IgG (Cy3), goat anti-mouse IgG (488) and rabbit anti-chicken IgG (Texas red) at 1:1000 and 1:500 for goat anti-rabbit (488) in PBS in 1% BSA were prepared. The incubation time was 20 min in the incubator at 37 °C and 5% CO_2.
6. Three washing steps with the medium were carried out and the ultra-low attachment plate was placed for 30 min in the incubator at 37 °C and 5% CO_2 before performing fluorescence microscopy analysis.

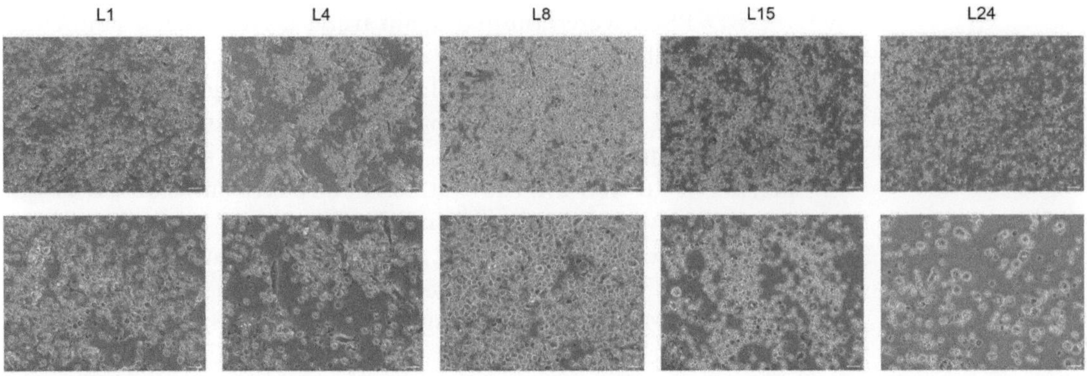

Fig. 1 Differentiation of BD-PSCs to hepatic-like cells. Representative micrographs of morphological changes throughout hepatic differentiation of BD-PSCs showing endodermal L4, or polygonal shape L8 morphology finally reaching

3.8.1 Method Comments

This method describes the differentiation of human BD-PSCs into endoderm/hepatic progenitor cells and hepatocytes by applying a two-step protocol. Morphological changes during the hepatic differentiation process are shown in Fig. 1. BD-PSCs differentiate into hepatocytes going through three different stages. The first stage represents the differentiation into endodermal cells L4, the second, differentiation to hepatic progenitor cells (hepatoblast) L8, exhibiting a typical polygonal morphology, and the third, the maturation to hepatocytes L15-L24.

Immunofluorescence analysis was performed to confirm the hepatic differentiation of BD-PSCs as presented in Fig. 2. Strong expressions of endoderm/human liver progenitor markers, like alpha-fetoprotein (AFP), a major plasma protein in fetal serum considered as a marker for hepatocytes' precursor, and transthyretin (TTR), a major thyroid-hormone binding protein involved in transporting thyroxine from the bloodstream to the brain, are both found in the cells at the first stage of the hepatic differentiation process at L4 to L8. However, their expression decreases at L15, while the expression of albumin (ALB), the most abundant plasma protein produced mainly by the liver and utterly critical for hepatic differentiation, as well as hepatocyte nuclear factor 4 alpha (HNF-4α), rises throughout the differentiation time L4-L15 reaching a strong and stable expression during the maturation time L15-L24. Cytokeratin 18 (CK18) is a cytoskeletal protein, one of the major components of intermediate filament expressed in the liver [45]. The results show that, as expected, CK18 expression correlates with mature hepatocytes (L15-L24), and it is not expressed in hepatocyte progenitor cells.

The well-defined protocol for hepatocyte differentiation in 2D cultures enables the engineering of hepatic 3D cultures starting with BD-PSCs. We demonstrate here that spontaneous aggregation of these cells in low attachment plates containing hepatocyte

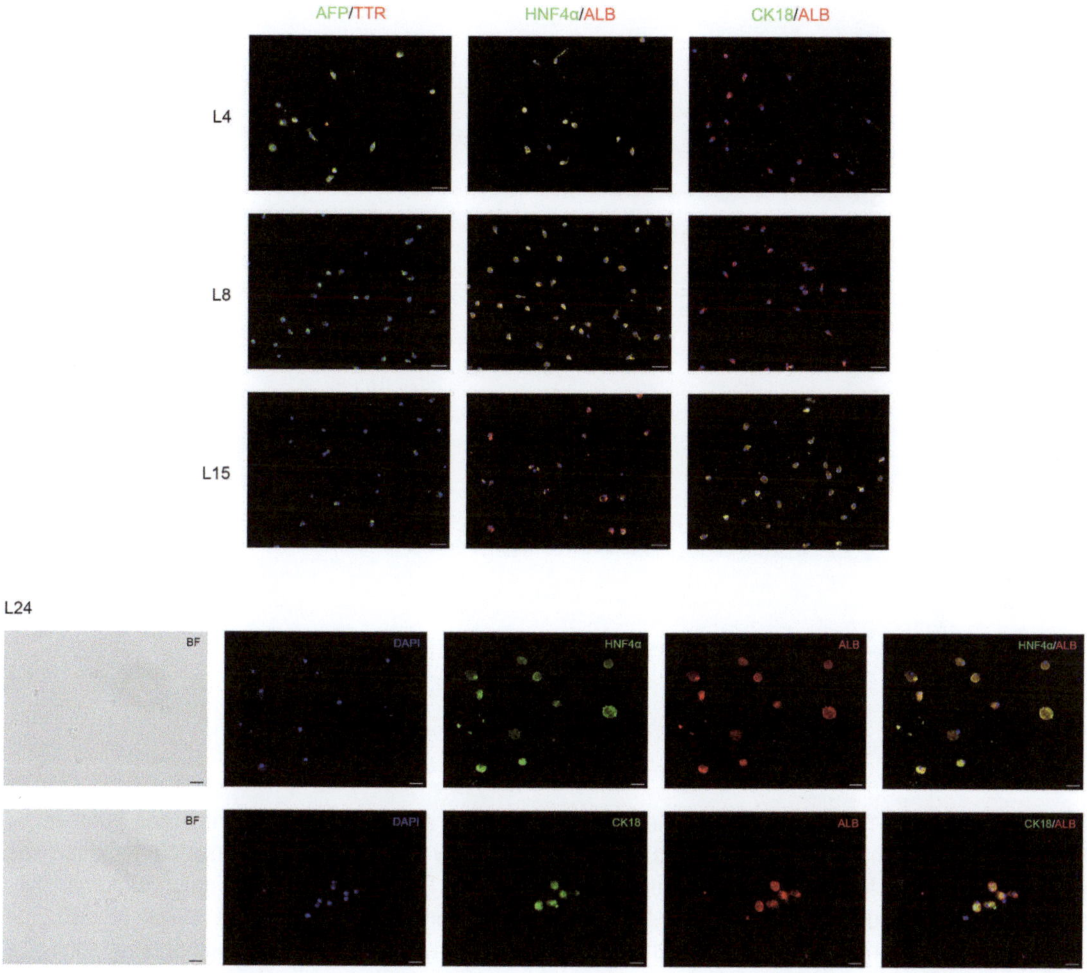

Fig. 2 Immunofluorescence analysis of BD-PSCs re-differentiation towards hepatic cells. Endoderm/hepatocytes progenitor and hepatocytes specific markers are expressed during liver differentiation of BD-PSCs in 2D cultures. On days L4 to L8, micrographs show decreased expression of endoderm/hepatic progenitor AFP and TTR while their expression disappeared from L8-L24. Expression of hepatocytes ALB and HNFα markers arise at L4 and increases during maturation, whereas the expression of CK18 appeared first at L15, reaching the maximum at L24. Scale bars for graphs L4-L15: 50 μm and for L24: 20 μm. Control is presented in Supplementary Fig. 1

induction/maturation medium initiates spheroid formation. The growth track was followed by imaging cells at L2, L4, and L7 (Fig. 3a). There is a consistent correlation between spheroid volume and the variable number of cells, as presented in Fig. 3b.

The liver is an organ in which most of the drugs in the human body get metabolized. Cytochrome P450 is a superfamily of enzymes (monooxygenases) that are of pivotal importance in the processes of drug and cellular metabolism, detoxification of xenobiotics, and homeostasis [44]. To assess the potential functional activity of BD-PSCs derived hepatic spheroids, we analyzed the

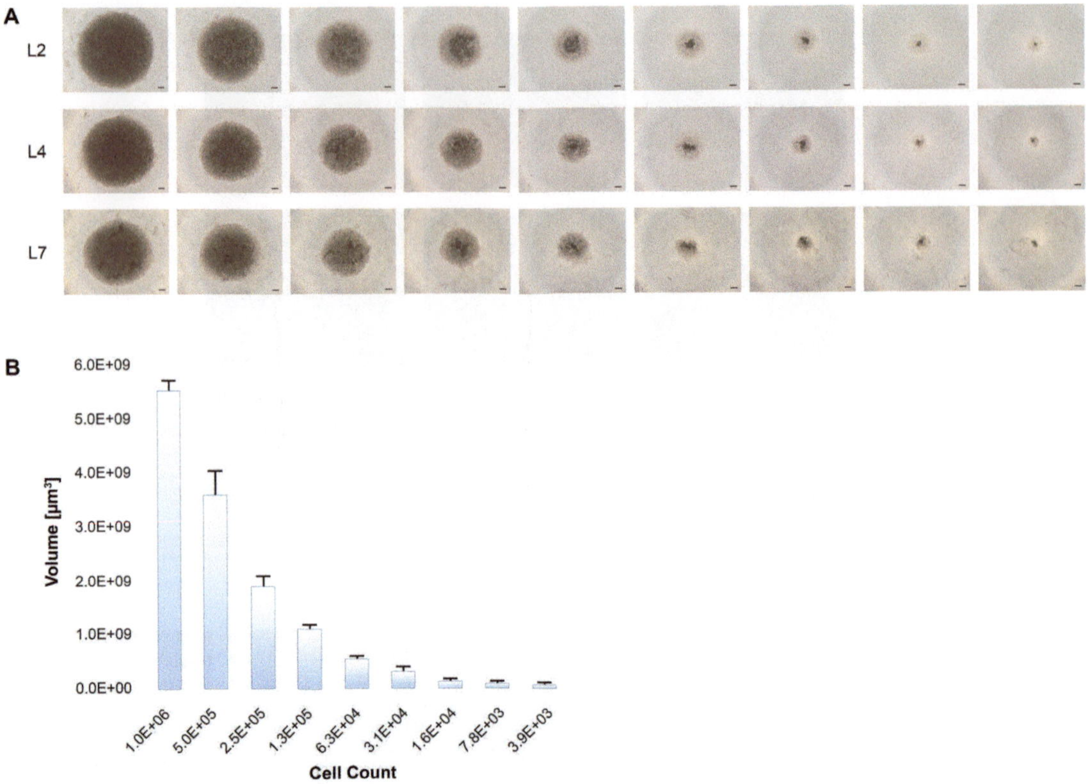

Fig. 3 Formation of 3D spheroids upon hepatic differentiation of BD-PSCs. (**a**) Variable cell numbers of BD-PSCs starting with 1×10^6 to 4000 cells were seeded into low attachment plates, and differentiation was performed according to the two-stage procedure as stated in the Protocol. The generation of 3D human liver spheroids was imaged at different time points; shown are representative phase contrast images at each time during the culture period. Scale bar: 200 µm. (**b**) Diameters of at least four hepatic spheroids for each size were measured at L4 using microscope imaging software and volumes were calculated. Error bars show standard deviation

expression of drug-metabolizing enzymes like CYP3A4 and CYP2E1, members of CYP3 and CYP2 families. Most of the drugs that are used today including codeine, cyclosporin A, erythromycin, acetaminophen, and diazepam, as well as many steroids and carcinogens, are metabolized due to the activity of the CY3A4 enzyme. CYP2E1 is involved in the metabolism of endogenous substrates like ethylene glycol, benzene, and carbon tetrachloride, and highlights mutagenic compound metabolism, such as with nitrosamine. The spheroids that are formed and differentiated according to the protocol at D14, live stained with antibodies to these two enzymes, reveal the potential hepatic functional activity of BD-PSCs derived spheroids (Fig. 4).

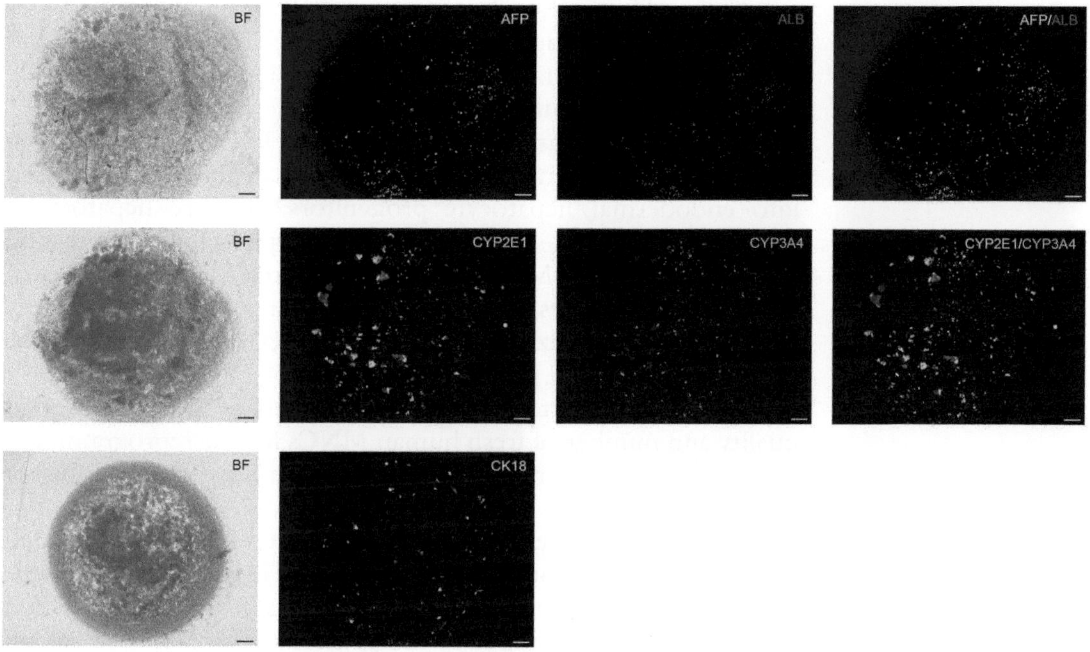

Fig. 4 *CYP3A4* and *CYP2E1*. Hepatocyte functional markers are expressed in BD-PSCs-derived liver spheroids. BD-PSCs were differentiated into hepatocytes. Direct immunofluorescence analysis was performed on live cells at L14 using antibodies to ALB, AFP, CK18, and CYP2E1 and CYP3A4, members of the cytochrome P450 family. Scale bar: 200 μm

3.8.2 HE Opportunities

HE ranges from minimal, mild to severe cases, (the official nomenclature is under review [1, 2]). Liver failure triggers HE development, drawing the mentioned broad spectrum of symptoms. HE therapeutic advances are limited by the liver transplantable availability and hepatocyte research models. The most severe cases require liver transplantation, but the health systems cannot answer the current demand, and the common animal models or the 2D human hepatocyte cultures cannot mimic a final realistic human hepatic model.

Primary human hepatocytes (PHH) are often used to study liver toxicity. These cells can be maintained in culture for a short time retaining their specific functions. Also, the number of cells available from a single donor is limited; in addition, these cells cannot be easily expanded in the culture; therefore, the shortage of donor PHH remains the main obstacle for hepatotoxicity studies. PSCs represent a renewal source of human tissues and can be used for the generation of 3D hepatic cultures [18].

Liver 3D culture systems show multiple advantages when compared to 2D. Shorter differentiation time and accurate mimicking of in vivo processes enable more precise studies on drug-induced toxicity, better prediction of liver liability, and are more cost-effective [46]. The liver spheroid cultures due to their autologous feature could be a great advantage over primary human hepatocytes (PHH) circumventing the disadvantages related to their use and may become a gold standard for application in testing drug toxicity and has a potential future use in regenerative medicine.

We have demonstrated here that BD-PSCs generated from steady-state peripheral blood can be successfully differentiated into endodermal/hepatocyte progenitors/ mature hepatocytes, with steady albumin secretion and phenotypic stability expressing hepatocyte markers. Moreover, the engineered 3D human hepatocyte spheroid cultures demonstrate the potential functional activity by expressing the enzymes that belong to cytochrome P450, like CYP3A4 and CYP2E1.

The most important step in the protocol is to obtain a good quality and number of fresh human MNCs for the reprogramming process. The use of frozen MNCs results in a reduced number of reprogrammed cells.

We have engineered human liver spheroids using activated PBMNC cultures with and without applying immunomagnetic sorting that separates reprogrammed cells from mature blood cells. The slight difference in using these two methods relies on the higher density of 3D structure when using purified reprogrammed cells. The expression of hepatocyte markers remains consistent in both cell culture preparations.

4 Conclusion

Due to the limited availability of PHH, this method represents a biologically relevant alternative to autologous fresh hepatocytes. Regarding the study of HE, this in vitro method emerges as a promising methodology to improve the analysis strength of human samples. The possibility of using hepatocyte spheroids from BD-PSCs opens a new technical approach to address the current research limitations in HE research. It also opens up a possibility for application in regenerative medicine. Our laboratory has developed a methodology to provide new research opportunities to HE experimental groups.

Acknowledgements

This work was supported by ACA CELL Biotech GmbH Heidelberg, Germany. The authors like to thank to Journal of Visualized Experiments (JoVE) for its generosity in allowing to spread and sharing of the methodology and data.

Disclosures The data presented in this chapter was previously published in the Journal of Visualized Experiments (styled JoVE), in detail, Schott, AK, Zipančić, I, Hernández-Rabaza, V, Becker-Kojić, ZA. Human Liver Spheroids from Peripheral Blood for Liver Disease Studies. J. Vis. Exp. 2023; (191). The current publication was officially approved by JoVE.

Additionally, the corresponding author declares that she is a patent holder related to a Novel human GPI-linked protein. She co-founded and works with ACA CELL Biotech GmbH. The other authors declare there are no conflicts of interest.

References

1. Weissenborn K (2019) Hepatic encephalopathy: definition, clinical grading and diagnostic principles. Drugs 79:5–9. https://doi.org/10.1007/S40265-018-1018-Z/TABLES/2
2. Häussinger D, Dhiman RK, Felipo V et al (2022) Hepatic encephalopathy. Nat Rev Dis Prim 8(1):43. https://doi.org/10.1038/s41572-022-00366-6
3. Hirode G, Vittinghoff E, Wong RJ (2019) Increasing burden of hepatic encephalopathy among hospitalized adults: an analysis of the 2010–2014 national inpatient sample. Dig Dis Sci 64:1448–1457. https://doi.org/10.1007/S10620-019-05576-9/METRICS
4. Sepanlou SG, Safiri S, Bisignano C et al (2020) The global, regional, and national burden of cirrhosis by cause in 195 countries and territories, 1990-2017: a systematic analysis for the global burden of disease study 2017. Lancet Gastroenterol Hepatol 5:245–266. https://doi.org/10.1016/S2468-1253(19)30349-8
5. Jayakumar AR, Norenberg MD (2018) Hyperammonemia in hepatic encephalopathy. J Clin Exp Hepatol 8:272–280. https://doi.org/10.1016/J.JCEH.2018.06.007
6. Vaquero J, Chung C, Cahill ME, Blei AT (2003) Pathogenesis of hepatic encephalopathy in acute liver failure. Semin Liver Dis 23: 259–269. https://doi.org/10.1055/S-2003-42644
7. Häussinger D, Schliess F (2008) Pathogenetic mechanisms of hepatic encephalopathy. Gut 57:1156–1165. https://doi.org/10.1136/GUT.2007.122176
8. Wright G, Sharifi Y, Jover-Cobos M, Jalan R (2014) The brain in acute on chronic liver failure. Metab Brain Dis 29:965. https://doi.org/10.1007/S11011-014-9553-0
9. Fabrellas N, Moreira R, Carol M et al (2020) Psychological burden of hepatic encephalopathy on patients and caregivers. Clin Transl Gastroenterol 11:e00159. https://doi.org/10.14309/CTG.0000000000000159
10. Ochoa-Sanchez R, Tamnanloo F, Rose CF (2021) Hepatic encephalopathy: from metabolic to neurodegenerative. Neurochem Res 46:2612–2625. https://doi.org/10.1007/S11064-021-03372-4
11. Cordoba J, Ventura-Cots M, Simón-Talero M et al (2014) Characteristics, risk factors, and mortality of cirrhotic patients hospitalized for hepatic encephalopathy with and without acute-on-chronic liver failure (ACLF). J Hepatol 60:275–281. https://doi.org/10.1016/J.JHEP.2013.10.004
12. Sundaram V, Shah P, Mahmud N et al (2020) Patients with severe acute-on-chronic liver failure are disadvantaged by model for end-stage liver disease-based organ allocation policy. Aliment Pharmacol Ther 52:1204–1213. https://doi.org/10.1111/APT.15988
13. Burra P, Burroughs A, Graziadei I et al (2016) EASL clinical practice guidelines: liver transplantation. J Hepatol 64:433–485. https://doi.org/10.1016/J.JHEP.2015.10.006
14. Acharya C, Bajaj JS (2021) Hepatic encephalopathy and liver transplantation: the past, present, and future toward equitable access. Liver Transpl 27:1830–1843. https://doi.org/10.1002/LT.26099
15. Fennema E, Rivron N, Rouwkema J et al (2013) Spheroid culture as a tool for creating 3D complex tissues. Trends Biotechnol 31: 108–115. https://doi.org/10.1016/j.tibtech.2012.12.003
16. Messina A, Luce E, Hussein M, Dubart-Kupperschmitt A (2020) Pluripotent-stem-cell-derived hepatic cells: hepatocytes and organoids for liver therapy and regeneration. Cells 9. https://doi.org/10.3390/CELLS9020420
17. Ryu NE, Lee SH, Park H (2019) Spheroid culture system methods and applications for

mesenchymal stem cells. Cells 8:1620. https://doi.org/10.3390/CELLS8121620
18. Chen S, Wang J, Ren H et al (2020) Hepatic spheroids derived from human induced pluripotent stem cells in bio-artificial liver rescue porcine acute liver failure. Cell Res 30:95–97. https://doi.org/10.1038/S41422-019-0261-5
19. Corrò C, Novellasdemunt L, Li VSW (2020) A brief history of organoids. Am J Physiol Cell Physiol 319:C151–C165. https://doi.org/10.1152/AJPCELL.00120.2020
20. Tang XY, Wu S, Wang D et al (2022) Human organoids in basic research and clinical applications. Signal Transduct Target Ther 7(1):168. https://doi.org/10.1038/s41392-022-01024-9
21. Dey M, Ozbolat IT (2020) 3D bioprinting of cells, tissues and organs. Sci Rep 10. https://doi.org/10.1038/S41598-020-70086-Y
22. Ingelman-Sundberg M, Lauschke VM (2022) 3D human liver spheroids for translational pharmacology and toxicology. Basic Clin Pharmacol Toxicol 130(Suppl 1):5–15. https://doi.org/10.1111/BCPT.13587
23. Riede J, Wollmann BM, Molden E, Ingelman-Sundberg M (2021) Primary human hepatocyte spheroids as an in vitro tool for investigating drug compounds with low clearance. Drug Metab Dispos 49:501. https://doi.org/10.1124/DMD.120.000340
24. Lv Y, Rao Z, Liu L et al (2024) The efficient generation of functional human hepatocytes from chemically induced pluripotent stem cells. Cell Prolif 57. https://doi.org/10.1111/CPR.13540
25. Schott AK, Zipančić I, Hernández-Rabaza V, Becker-Kojić ZA (2023) Human liver spheroids from peripheral blood for liver disease studies. J Vis Exp. https://doi.org/10.3791/64703
26. Rossi G, Manfrin A, Lutolf MP (2018) Progress and potential in organoid research. Nat Rev Genet 19:671–687. https://doi.org/10.1038/S41576-018-0051-9
27. Hussein M, Pasqua M, Pereira U et al (2023) Microencapsulated hepatocytes differentiated from human induced pluripotent stem cells: optimizing 3D culture for tissue engineering applications. Cells 12:865. https://doi.org/10.3390/CELLS12060865/S1
28. Van Niel G, D'Angelo G, Raposo G (2018) Shedding light on the cell biology of extracellular vesicles. Nat Rev Mol Cell Biol 19(4):213–228. https://doi.org/10.1038/nrm.2017.125
29. Thietart S, Rautou PE (2020) Extracellular vesicles as biomarkers in liver diseases: a clinician's point of view. J Hepatol 73:1507–1525. https://doi.org/10.1016/J.JHEP.2020.07.014
30. Bell CC, Hendriks DFG, Moro SML et al (2016) Characterization of primary human hepatocyte spheroids as a model system for drug-induced liver injury, liver function and disease. Sci Rep 6:1–13. https://doi.org/10.1038/srep25187
31. Nevzorova YA, Boyer-Diaz Z, Cubero FJ, Gracia-Sancho J (2020) Animal models for liver disease – a practical approach for translational research. J Hepatol 73:423–440. https://doi.org/10.1016/J.JHEP.2020.04.011
32. Vallverdú J, García M, de la Torre RA, Mannaerts I et al (2021) Directed differentiation of human induced pluripotent stem cells to hepatic stellate cells. Nat Protoc 16:2542–2563. https://doi.org/10.1038/S41596-021-00509-1
33. Xie X, Zhou X, Liu T et al (2022) Direct differentiation of human embryonic stem cells to 3D functional hepatocyte-like cells in alginate microencapsulation sphere. Cells 11. https://doi.org/10.3390/CELLS11193134
34. Becker-Kojić ZA, Schott AK, Zipančić I, Hernández-Rabaza V (2021) GM-free generation of blood-derived neuronal cells. J Vis Exp 2021:e61634. https://doi.org/10.3791/61634
35. Becker-Kojić ZA, Ureña-Peralta JR, Zipančić I et al (2013) Activation by ACA induces pluripotency in human blood progenitor cells. Bull Exp Biol Med 155:552–567. https://doi.org/10.1007/S10517-013-2196-Z
36. Becker-Kojić ZA, García-Verdugo JM, Schott A-K et al (2022) Membrane-to-nucleus Signaling in human blood progenitor cells reveals an efficient GM-free reprogramming to pluripotency. https://doi.org/10.5772/INTECHOPEN.108950
37. Becker-Kojić ZA, Ureña-Peralta JR, Saffrich R et al (2013) A novel human glycoprotein ACA is an upstream regulator of human hematopoiesis. Bull Exp Biol Med 155:536–551. https://doi.org/10.1007/S10517-013-2195-0
38. Takayama K, Nagamoto Y, Mimura N et al (2013) Long-term self-renewal of human ES/iPS-derived hepatoblast-like cells on human laminin 111-coated dishes. Stem Cell Reports 1:322–335. https://doi.org/10.1016/J.STEMCR.2013.08.006

39. Graffmann N, Scherer B, Adjaye J (2022) In vitro differentiation of pluripotent stem cells into hepatocyte like cells – basic principles and current progress. Stem Cell Res 61. https://doi.org/10.1016/J.SCR.2022.102763
40. Crandall BF, Lau HL (1981) Alpha-fetoprotein: a review. Crit Rev Clin Lab Sci 15:127–185. https://doi.org/10.3109/10408368109105870
41. Magalhães J, Eira J, Liz MA (2021) The role of transthyretin in cell biology: impact on human pathophysiology. Cell Mol Life Sci 78:6105–6117. https://doi.org/10.1007/S00018-021-03899-3
42. Spinella R, Sawhney R, Jalan R (2016) Albumin in chronic liver disease: structure, functions and therapeutic implications. Hepatol Int 10:124–132. https://doi.org/10.1007/S12072-015-9665-6
43. Guan Y, Enejder A, Wang M et al (2021) A human multi-lineage hepatic organoid model for liver fibrosis. Nat Commun 12. https://doi.org/10.1038/S41467-021-26410-9
44. Zhao M, Ma J, Li M et al (2021) Cytochrome P450 enzymes and drug metabolism in humans. Int J Mol Sci 22. https://doi.org/10.3390/IJMS222312808
45. Korver S, Bowen J, Pearson K et al (2021) The application of cytokeratin-18 as a biomarker for drug-induced liver injury. Arch Toxicol 95:3435. https://doi.org/10.1007/S00204-021-03121-0
46. Kammerer S (2021) Three-dimensional liver culture systems to maintain primary hepatic properties for toxicological analysis in vitro. Int J Mol Sci 22. https://doi.org/10.3390/IJMS221910214

Chapter 9

Stereological Estimation of Purkinje Cell Number in Nissl-Stained Sections of Mouse Cerebellum

Tiziano Balzano

Abstract

Hepatic encephalopathy (HE) is a neurocognitive syndrome associated with liver disease, presenting a spectrum of neuropsychiatric abnormalities ranging from mild cognitive impairment to severe disorientation, coma, and potentially death. Traditionally considered a gliopathy affecting astrocytes without neuronal damage, recent evidence suggests a neurodegenerative component in HE pathogenesis. Studies have shown cerebellar neurodegeneration in HE and chronic liver disease, highlighting the need for robust methods to estimate cerebellar neuron numbers to better understand HE's neurological impact.

Unbiased stereology, the gold standard for quantifying biological objects within three-dimensional structures from two-dimensional tissue images is crucial in this context.

This protocol documents a biological application of computerized stereology to estimate the total number of Purkinje neurons in the mouse cerebellum, thereby enhancing our understanding of neurodegenerative processes in HE.

Key words Hepatic encephalopathy, Neurodegeneration, Stereology, Cerebellum, Purkinje neurons

1 Introduction

Hepatic encephalopathy (HE) is a potentially reversible neurocognitive syndrome that occurs in patients with acute or chronic liver disease. It encompasses a wide spectrum of neuropsychiatric abnormalities, ranging from subclinical changes (such as mild cognitive impairment) to marked disorientation, confusion, coma, and, in severe cases, death. Historically, HE has been considered a gliopathy, primarily affecting the functioning of astrocytes without impacting neuronal architecture [1]. However, recent evidence points to an evolving perspective, with increasing data suggesting a neurodegenerative component in the pathogenesis of HE [2].

Several studies involving post-mortem human brain sections and animal models have documented cerebellar neurodegeneration in HE and chronic liver disease [3–7]. Consequently, robust methods for estimating the absolute number of cerebellar neurons in

Tiziano Balzano (ed.), *Experimental and Clinical Methods in Hepatic Encephalopathy Research*, Neuromethods, vol. 219,
https://doi.org/10.1007/978-1-0716-4354-9_9,

brain sections from animal models and patients with HE would enhance our understanding of the impact of neurological insults and the progression of neurodegenerative processes in HE. For this purpose, unbiased stereology stands as the gold standard method for quantifying biological objects within three-dimensional structures, such as defined brain regions, from two-dimensional tissue images viewed under the microscope [8]. The key components of this approach are unbiased (systematic-random) sampling of anatomically defined structures and the subsequent quantification of biological objects within these defined spaces.

Thus, this protocol aims to document a biological application of computerized stereology to estimate the total number of Purkinje neurons in the mouse cerebellum.

2 Materials

2.1 Nissl Staining

2.1.1 Equipment for Nissl Staining

1. Gelatin-Coated Microscope Slides.
2. Coverslips.
3. Orbital agitator.
4. Water bath.
5. Pencil.
6. Slide racks and dip trays.
7. Optical microscope.

2.1.2 Reagents for Nissl Staining

1. Sodium phosphate (0.1 M, pH 7.4).
2. 0.1% cresyl violet solution.
3. Differentiation solution (95% ethyl alcohol).
4. Distilled water.
5. Ethanol (70%, 95%, 100%).
6. Xilene.
7. Permanent mounting medium.

2.2 Stereology

2.2.1 Hardware Stereo Investigator

1. Olympus BX53 microscope.
2. Digital camera Lumenera LT665R.
3. Prior H117 ProScan Motorized Stage (Prior Scientific, Rockland, MA, USA).
4. Controller with encoder ProScan III (Model: 500-H31XYZEF, Prior Scientific, Rockland, MA, USA).
5. ProScan III joystick (Model: P-PS3J100, Prior Scientific, Rockland, MA, USA).
6. Olympus U-LH100–3 Microscope Light Source Halogen 12v 100w.

2.2.2 Hardware Windows Computer

1. Processor: Intel® Xeon® W-2245 CPU @ 3.90 GHz on 16 Cores.
2. Installed Memory (RAM): 64 GB.
3. Video Card: NVIDIA GeForce RTX A4000.
4. Video Memory: 16 GB.

2.2.3 Software

1. Stereo Investigator version 2023.1.2 (Micro Bright Field, St. Albans VT).

3 Methods

3.1 Nissl Staining of Mice Cerebella

3.1.1 Mounting the Sections

1. Mount thin sections on gelatin-coated slides, using sodium phosphate to slide the sections into position on the slide. Further information on tissue organization and selection can be found in **Note 1**.
2. Label them with pencil on the etched end and leave them to dry over night at room temperature to allow tissue to adhere better to the slides.
3. When the sections are dry, place them in a slide rack, ready to be stained.

3.1.2 Staining

1. Incubate sections in 70% ethanol overnight.
2. Rinse quickly in distilled water.
3. Stain in agitation in 0.1% cresyl violet solution for 5 min at 45 ° C.
4. Rinse quickly in distilled water.
5. Dehydrate in 70% ethanol 1 min.
6. Dehydrate in 95% ethanol 1 min.
7. Incubate sections with chloroform for 10 min (in agitation).
8. Differentiate in 95% ethyl alcohol for 1–5 min (check microscopically for best result).
9. Dehydrate in 100% ethanol for 1 min.
10. Clear in xylene 3 times for 5 min.
11. Mount with permanent mounting medium.

3.2 Unbiased Stereological Estimation of the Total Number of Purkinje Neurons in the Mouse Cerebellum Using the Optical Fractionator

The Optical Fractionator employs thick sections and extrapolates the total cell count from a subset of cells sampled using a Systematic Randomly Sampled (SRS) set of unbiased virtual counting spaces, covering the entire area of interest with consistent spacing along the X, Y, and Z axes. Initiating a new data file typically commences with positioning the reference point, a crucial step for preserving precise alignment of sections and traces when you work on a data file in multiple sessions. After establishing the reference point for your section, proceed to the Optical Fractionator Workflow accessible from the main menu, and initiate a new subject (to access a previous session, please *see* **Note 2**). Follow the workflow instructions provided below:

Step 1. Setup the section: enter serial section information. It includes five sub-sections:

- *Number of sections to count:* The initial step before conducting the Optical Fractionator probe is to establish the section fraction by assessing the quantity of available sections. In this protocol, the stereological analysis of Purkinje cells involves sampling six sections, each spaced at intervals of 480 μm. This equates to selecting 1 out of every 12 40-μm-thick sections, resulting in a total count of six sections. Further information on tissue organization and selection can be found in **Note 1**.
- *Section cut thickness:* Record the tissue cut thickness immediately after sectioning, before any artifacts such as shrinkage occur. For this protocol, coronal sections of 40 μm thickness were used.
- *Section evaluation interval:* Specify the interval of the serially cut sections that will be sampled. To sample every other section, enter 2; to sample every third section, enter 3, and so on. In this protocol, as previously defined, one out of every twelve sections were sampled.

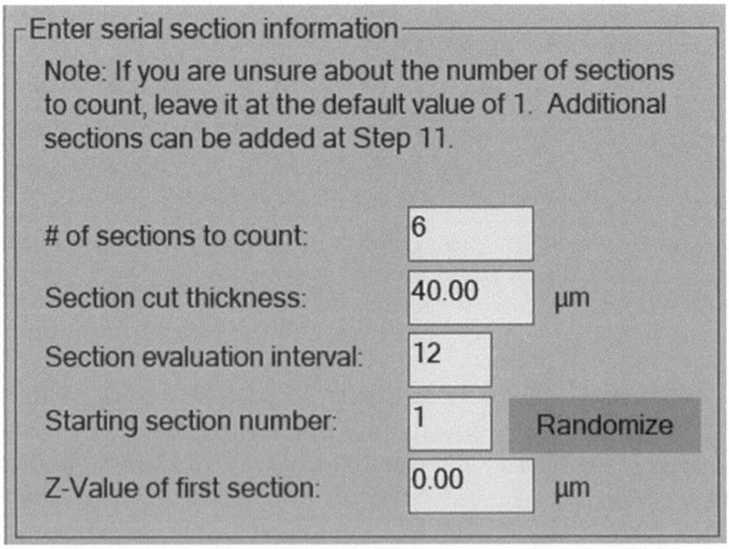

Fig. 1 Illustrative example of step 1 box, showing serial section parameters used for this protocol

- *Starting section number:* Enter the number of the first section you will start from in your serial sections. The first sampling section should be randomly selected, but by setting it to "1," sampling will start from the first section.
- *Z-Value of first section:* This value is automatically adjusted based on the starting section number entered earlier; it reflects an interval of the cut thickness.

All employed values used for this protocol are shown in Fig. 1.

Step 2. Set microscope to low magnification: select the low-magnification objective to trace the regions of interest. For this protocol, a 2× objective has been used to trace the entire cerebellum. Therefore, set the microscope to 2× objective and select "2× Mag Lens" from the menu.

Step 3. Trace the region(s) of interest: Using the low-magnification objective set in Step 2, define and clearly outline the regions to be examined (in this case, the entire cerebellum). Ensure accurate tracing of boundaries for precise analysis. Begin with section 1 and trace all regions of interest using a designated contour (*see* Fig. 2). Repeat this step for all sections. Additional information about this step is provided in **Note 3**.

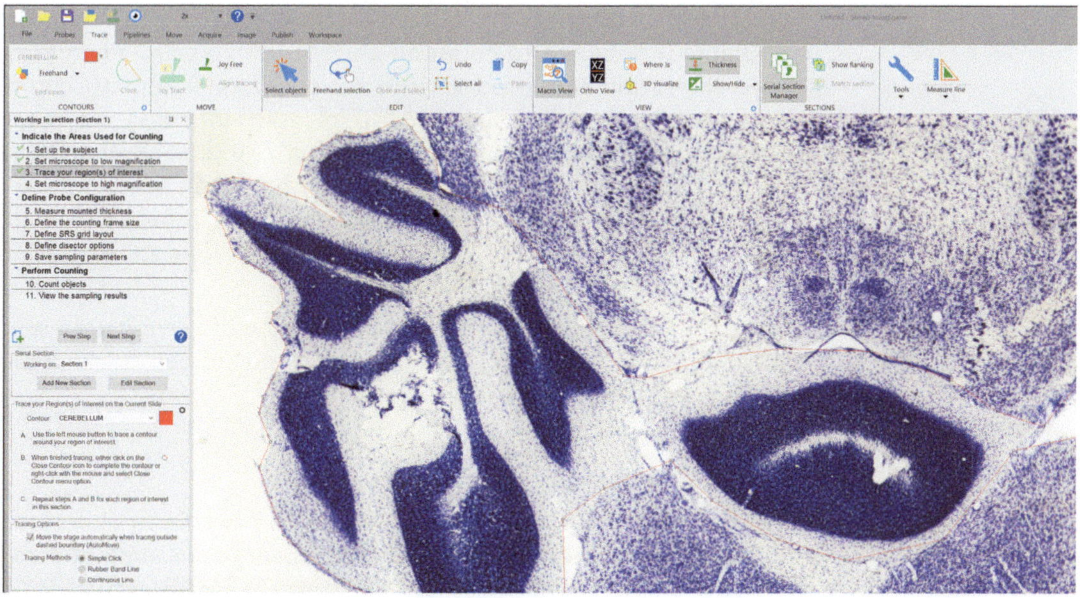

Fig. 2 Example of region of interest tracing. The entire volume of the cerebellum was used as the reference volume to estimate Purkinje neurons. The region of interest was outlined using a 2× objective

Step 4. Set the microscope to high magnification: select a higher objective from the "SELECT HIGH MAG LENS" drop-down menu that is suitable for imaging Purkinje neurons. For this protocol, a 40× objective has been used. Thus, change the objective to 40× and select "40× Mag lens" from the dropdown menu.

Step 5. Measure mounted thickness: it is recommendable to measure thickness at every sampling site for greater accuracy (*see* **Note 4**). In this case, the measured thickness from the counting sites is averaged and used in the calculations of the height sampling fraction. For this purpose, check the boxes "Measure the mounted thickness while counting," "Refocus to top of section at each site," and "Measure the Mounted Thickness While Counting." Finally, set the Evaluation Interval to "1."

Step 6. Define the counting frame size: The counting frame should be sized to contain, on average, one to five objects. This guideline helps ensure accuracy and consistency in cell counting. According to Napper and West [9] the following frame size of 160 μm × 200 μm (*see* **Note 5**) has been used.

Step 7. SRS grid layout settings: In this step, the SRS grid size determines the proximity of your counting frames. This protocol aims to count approximately 600 cells to achieve the desired level of accuracy for the population estimate. With 6 sections chosen, this requires counting approximately 100 cells per section (6 sections × 100 cells = 600 cells). Each counting frame is designed to count an average of two cells. Therefore, each section will need around 50 counting frames to count 100 cells (2 cells/frame × 50 frames = 100 cells). To meet these requirements, a sampling grid of 750 μm × 750 μm was selected. Note that the sections may vary significantly in their profile areas, which might affect the exact number of cells counted per section. For information on determining your SRS grids settings, *see* **Note 6**.

Step 8. Define disector options: In this step, the disector height and guard zones will be set. Based on previous studies, it is recommended to use the following settings:

- *Disector height:*15 μm.
- *Guard zone height:* 2.

Step 9. Save sampling parameters: Here it is possible to choose to save the parameters for future experiments. All sampling parameters used for this protocol are shown in Fig. 3.

Step 10. Count objects.

- Select region of interest from menu. Press the blue arrow to begin a new probe run; the user will be directed to the first sampling site.

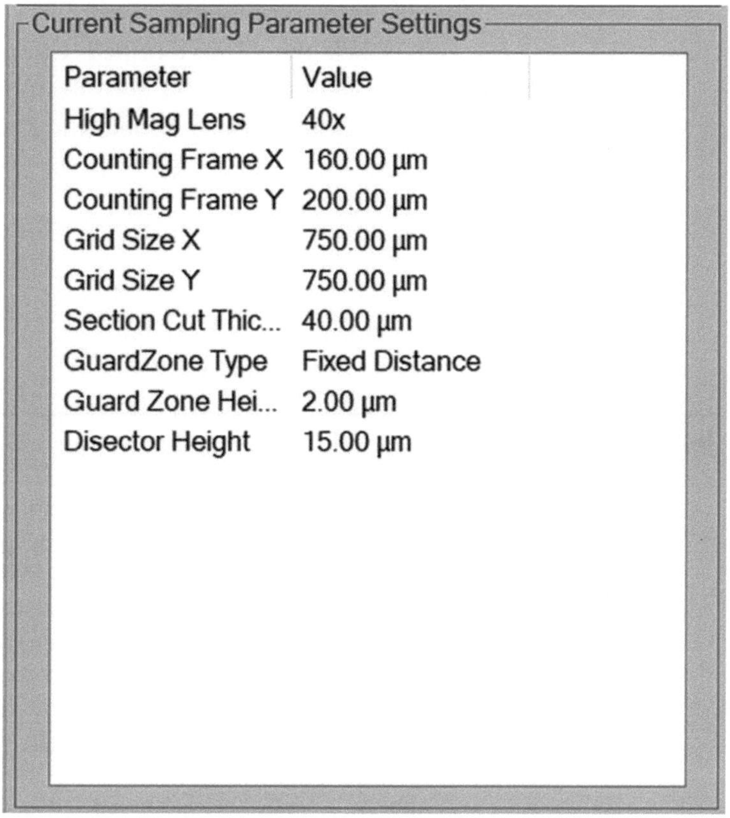

Current Sampling Parameter Settings

Parameter	Value
High Mag Lens	40x
Counting Frame X	160.00 µm
Counting Frame Y	200.00 µm
Grid Size X	750.00 µm
Grid Size Y	750.00 µm
Section Cut Thic...	40.00 µm
GuardZone Type	Fixed Distance
Guard Zone Hei...	2.00 µm
Disector Height	15.00 µm

Fig. 3 Sampling parameters for the stereological estimation of Purkinje cell number in Nissl-stained sections of mouse cerebellum

- The Focus Top of Section dialog box is displayed. Using the fine focus knob, focus to the top of the section and then click OK in the dialog box.
- The Focus Bottom of Section dialog box is displayed. Using the fine focus knob, focus to the bottom of the section and then click OK.
- Select a marker from the Use Marker drop down menu in the workflow panel, or from the marker toolbar.
- Count the cells following pre-established counting rules (*see* **Note** 7). For this protocol, the cell nucleolus was adopted as the counting unit. The main reason for choosing the nucleolus is that because of its relatively small size (around 2 µm in diameter) and it has been demonstrated for Purkinje cells in adult rats that fractionator estimates of the numbers of nucleoli and disector estimates of the numbers of nuclei coincide [10].
- Click the "fast forward" button to go to the next site. Repeat step for all the regions in all the sections.

Step 11. View the sampling results.

- Click the "I'm Finished Counting" button when done. Click the Next Step button to move to **step 11**.
- Click the Display Probe Run List button.
- Select all the probe runs for which you want results from the list.
- Click "View Results."
- Export to Excel (*see* **Note 9**).
- Repeat for each animal.

4 Notes

1. Tissue organization and selection is a critical step in stereology protocols. The sections must be placed in the correct order both when sectioned and when mounted on the slide to ensure accurate determination of the distance between sections. This is essential to meet the stereological assumptions and make precise calculations. In this protocol, sections from 6-week-old C57BL6/C mice were used. To obtain these sections, the cerebellum was isolated from the cerebrum and sliced in the coronal plane using a vibratome at a thickness of 40 μm, resulting in 6 matched series of 12 sections each, approximately. Therefore, to sample six sections spaced at intervals of 480 μm, one section from every two sections from one series should be chosen (this means selecting one out of every twelve 40 μm-thick sections across all six series).
2. To access a previous session, follow the workflow instructions provided below:
   1. Click on Optical Fractionator Menu and look for the Load Project.
   2. Click on this option to open a dialog box that allows you to navigate your computer's file system.
   3. Navigate to the directory where you have saved your previous session file. Stereo Investigator typically saves files with ".dat" extension.
   4. Select the file you want to open.
   5. Once you have selected the file, click Open. The software will load your previous session, including all associated data and settings.
   6. It is always recommendable to check that all your previous work, including images, tracings, and annotations, are correctly loaded. Also ensure that any settings or parameters you were using are intact.

7. Once your session is loaded and verified, you can continue your work from where you left off. Make any necessary adjustments and proceed with your analysis.

3. Purkinje cells are arranged in a monolayer making difficult to determine a reference volume reliably consisting of the Purkinje cell layer alone. For this reason, it is better to use the entire volume of the cerebellum as the reference volume for the Purkinje cell populations. This will not affect the total number of neurons.
4. The Population Estimate by Number-Weighted Section Thickness requires a thickness measurement at each site. Refocus To Top of Section at Each Grid Site (default): This is turned on by default and should always be used since the top of your section will vary from counting site to counting site.
5. It is common for some counting frames to have zero objects or more than five objects. This variation is normal and anticipated. Limiting the number of cells per counting frame is intended to minimize user error and fatigue. When focusing through the tissue to mark cells, tracking a large number of cells can be difficult, especially in densely populated areas. Keeping the cell count per frame low helps ensure accuracy.
6. SRS grid can be alternatively defined by choosing one of the following methods:
   (a) Percentage of region to sample: specify the percentage of the region of interest that you want to sample. This allows for flexible sampling based on the proportion of the area you wish to cover.
   (b) Desired number of sites: enter the desired number of sampling sites in the approximate sites box. This approach is useful when you have a specific number of sites in mind and want the grid size to be adjusted accordingly.

   These alternative methods offer flexibility in determining the SRS grid layout, accommodating different study requirements and preferences.
7. Cell counting involves identifying the boundaries of the cells of interest in three dimensions. The following rules should be followed in the execution of rigorous stereological analysis [11]:
   - Decide on a unique point (the characteristic point) that can be identified for each cell. This can be the top of the cell, i.e., the first part of the cell to come into focus while focusing down through the tissue, or perhaps the top of the nucleus, or even the nucleolus if the cells being examined have only one nucleolus per cell.

- The unique point must be visible for each cell to be counted.
- At each investigated microscopic field, an unbiased counting frame consisting of two exclusion lines and two inclusion lines is placed on the section.
- Select a cell which is a potential candidate for counting and focus on its selected "point" (e.g., "top," nucleolus, etc.) and follow the rules below for counting:
  1. Any point that touches the exclusion line, crosses the exclusion line, or is outside of the counting frame is not counted.
  2. For any point that is still a valid candidate and that is (a) inside the counting frame, (b) touching an inclusion line from either inside or outside the counting frame, or (c) crossing an inclusion line, remains a candidate to be counted.
  3. If an object touches or crosses both an inclusion line and an exclusion line, it is not counted (as per rule #i above).
  4. Restrict your analysis to a virtual space within the section thickness. For any cell that is a valid candidate, count it only if its unique point—in addition to its position relative to the unbiased counting frame (as per rules i, ii, and iii)—comes into focus within this virtual space. In other words, a cell is counted only if its unique point is found within an unbiased virtual counting space (Fig. 4).

8. The purpose of the guard zone is to prevent potential errors in object estimation caused by issues like uneven cutting surfaces or damaged objects near the tissue's surface. It's important to select a total depth for the disector and guard zones that fits within the thickness of your thinnest tissue sections.
9. Interpreting your results sheet:
   1. Several estimates will be calculated for your object of interest:
      (a) *Estimated Population Using User-Defined Section Thickness:* This measure is the least precise and defaults to 0 if no section thickness is manually entered.
      (b) *Estimated Population Using Mean Section Thickness:* This estimate is derived from the average tissue thickness measured across all sampled sites, regardless of whether cells were counted.
      (c) *Estimated Population Using Mean Section Thickness with Counts:* This estimate is based on the average tissue thickness only at sites where cells were marked and counted.

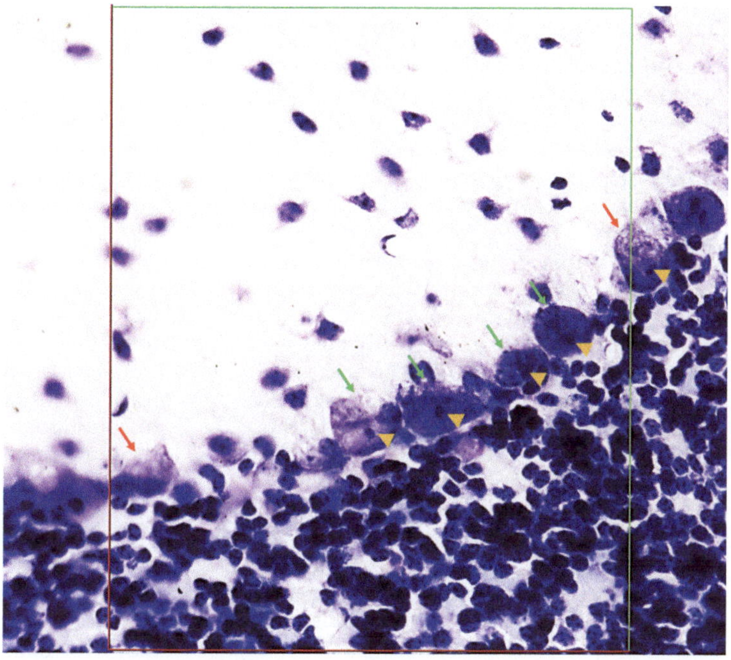

Fig. 4 Unbiased counting frame, counting units, and inclusion criteria. The cell estimation was performed using the 2D unbiased counting frame with left and bottom edges serving as exclusion lines (red), and with the upper and right edges of the frame as inclusion lines (green). Purkinje cell nucleolus was adopted as the counting unit (orange arrowhead). Counting units were counted when completely inside the counting frame or partly inside the frame, provided that they did not touch the exclusion lines. Thus, four cells were counted (green arrows) and two excluded (red arrow), one because the nucleolus was not visible within the unbiased virtual counting space and the other one because the Purkinje nucleolus cell was outside the counting frame

(d) *Estimated Population Using Number-Weighted Section Thickness:* This estimate adjusts for variations in tissue thickness.

2. *Coefficient of Error:* It is recommended to use a Coefficient of Error (Gundersen), where $m = 1$, which should ideally be ≤ 0.1 to ensure accurate sampling. The coefficient of error does not have biological significance (i.e., it does not measure inter- or intra-subject variability) and is solely used to assess the precision of stereological estimates. Mathematically, it is defined as the standard error of the mean of repeated estimates divided by the mean, providing a prediction of accuracy.
3. *Planimetry tab*: Area of each section and volume calculated from this area. This value can be useful if you want to determine population density within a region of interest whose volume is known. Note that this information is biased; for an unbiased estimate of area/volume, use the Cavalieri Estimator.

References

1. Norenberg MD (1987) The role of astrocytes in hepatic encephalopathy. Neurochem Pathol 6:13–33
2. Ochoa-Sanchez R, Tamnanloo F, Rose CF (2021) Hepatic encephalopathy: from metabolic to neurodegenerative. Neurochem Res 46:2612–2625
3. Balzano T, Leone P, Ivaylova G et al (2021) Rifaximin prevents T-lymphocytes and macrophages infiltration in cerebellum and restores motor incoordination in rats with mild liver damage. Biomedicines 9:1002
4. Balzano T, Forteza J, Borreda I et al (2018) Histological features of cerebellar neuropathology in patients with alcoholic and nonalcoholic steatohepatitis. J Neuropathol Exp Neurol 77: 837–845
5. Balzano T, Forteza J, Molina P et al (2018) The cerebellum of patients with steatohepatitis shows lymphocyte infiltration, microglial activation and loss of Purkinje and granular neurons. Sci Rep 8:3004
6. García-Lezana T, Oria M, Romero-Giménez J et al (2017) Cerebellar neurodegeneration in a new rat model of episodic hepatic encephalopathy. J Cereb Blood Flow Metab 37:927–937
7. Tamnanloo F, Chen X, Oliveira MM et al (2024) Excessive intragastric alcohol administration exacerbates hepatic encephalopathy and provokes neuronal cell death in male rats with chronic liver disease. J Neurosci Res 102: e25337
8. West MJ (2012) Introduction to stereology. Cold Spring Harb Protoc 2012:pdb. top070623
9. Napper RM, West JR (1995) Permanent neuronal cell loss in the cerebellum of rats exposed to continuous low blood alcohol levels during the brain growth spurt: a stereological investigation. J Comp Neurol 362:283–292
10. Braendgaard H, Gundersen HJ (1986) The impact of recent stereological advances on quantitative studies of the nervous system. J Neurosci Methods 18:39–78
11. Glaser J, Greene G, Hendricks S (2007) Stereology for biological research: with a focus on neuroscience. MBF Press

Part III

Methods for Exploring Mechanisms and Pathophysiology

Chapter 10

Gut Microbiota-Based Strategies for Targeting Hepatic Encephalopathy

Marina Romaní-Pérez, María del Carmen Cenit, and Ana Agustí

Abstract

Hepatic encephalopathy (HE) is a complex neuropsychiatric syndrome that causes cognitive and motor impairment in patients with liver disease. Recent evidence indicates that the gut microbiota plays a key role in the pathogenesis of HE, suggesting that interventions based on the gut microbiome may have preventive and/or therapeutic benefits. Indeed, strategies targeting the gut microbiota, such as lactulose and rifaximin, probiotics, prebiotics, postbiotics and fecal microbiota transplantation, have shown some success in treating HE by increasing the abundance of potentially beneficial taxa and/or decreasing the abundance of potentially harmful taxa. However, challenges remain in the effective application of microbiome-targeted therapies in the context of HE. These include understanding the precise gut microbiota changes that causally relate to HE, determining the specific microbial species or consortia involved in HE risk and prognosis, assessing the factors involved in the resilience of the HE gut microbiome in some individuals, and evaluating the clinical outcomes of microbiota modifications in the context of the disease. Accordingly, further research is needed to gain greater insight into the complex interplay between the gut microbiome and HE, including identifying the factors that negatively influence the gut-liver-brain axis leading to HE in liver cirrhosis and exploring their relationship with the gut microbiome. Furthermore, studies that integrate multi-omics analysis of microbiota-gut-brain axis parameters are necessary to develop precision models based on specific individual features, such as microbiome profiling before intervention. These models could aid in classifying patients with HE based on their predicted clinical response to specific microbiome-targeted interventions with potential therapeutic effects for HE. In this chapter, we provide details of the most commonly used available techniques, methods, and recommendations for achieving tailored microbiome-based strategies that could aid in the prevention or treatment of HE.

Key words Hepatic encephalopathy, Cirrhosis, Neurology, Gut microbiota, Gut-brain axis, Gut-liver-brain axis, Fecal microbiota transplant

1 Introduction

1.1 The Pathophysiology of Hepatic Encephalopathy

Hepatic encephalopathy (HE) is a complex neuropsychiatric syndrome that causes cognitive, psychiatric, and motor changes in patients with liver disease. Three different forms of HE has been described so far: Type A, which occurs in patients with HE due to acute liver failure; Type B, which occurs in patients with HE with

Tiziano Balzano (ed.), *Experimental and Clinical Methods in Hepatic Encephalopathy Research*, Neuromethods, vol. 219,
https://doi.org/10.1007/978-1-0716-4354-9_10,

significant porto-systemic shunting in the absence of cirrhosis or liver failure [1, 2]; and Type C, which occurs in patients with HE due to chronic liver failure, typically cirrhosis. Type C is further categorized into minimal or sub-clinical HE (MHE) and clinical HE [3]).

Under normal conditions, the liver converts toxic ammonia into urea through the urea cycle. However, impaired liver function can lead to hyperammonemia, which is characterized by high levels of ammonia in blood. This condition can disrupt brain function, leading to neuropsychiatric features such as personality changes, disorientation, depression, changes in sleep-awake patterns, cognitive decline, and motor activity and coordination issues [4, 5]. Hyperammonemia, endotoxins, systemic inflammation, and central neuroinflammation synergistically contribute to the pathophysiology of HE. Recent research has highlighted the crucial role of the gut microbiota in cirrhosis progression [6], implicating its significance in the development of HE. Understanding the interplay between the gut, liver, and brain is thus essential in elucidating the development of HE [7, 8].

1.2 The Gut Microbiota and the Gut-Brain Axis

The gut microbiota is a complex ecosystem of bacteria, viruses, protozoa, archaea, and fungi, with bacteria representing the majority of constituents [9]. It is estimated that the adult human gut microbiota comprises over 1000 distinct bacterial species, totaling over 7000 individual strains. Microorganisms in the gut contribute to human physiology by participating in key biological processes such as the metabolism of nutrients and xenobiotics, as well as the regulation of crucial immune and neuroendocrine functions [10]. Some of these processes are facilitated by by-products of bacterial metabolism such as the short-chain fatty acids (SCFAs) propionate, butyrate, and acetate, among others. SCFAs also regulate gut barrier permeability and function, the inflammatory response, and the metabolic balance across various tissues [11]. It is recognized that changes in the gut microbiota, referred to as dysbiosis, are associated not only with gastrointestinal conditions, but also with extra-gastrointestinal disorders such as obesity, autoimmune arthritis, psychiatric disorders, and also cirrhosis [12–14].

The gut-brain axis refers to the bi-directional communication between the gut and the brain, which is enabled by hormonal, immunological, and neural transmission [15]. Disruptions in this communication have been linked to metabolic and psychiatric disorders, and to comorbid non-psychiatric disorders [16]. Changes in gut microbiota composition and function often coincide with these conditions, which may be cause or a consequence of the disease. These changes can further disrupt the intricate molecular communication between the gut and brain.

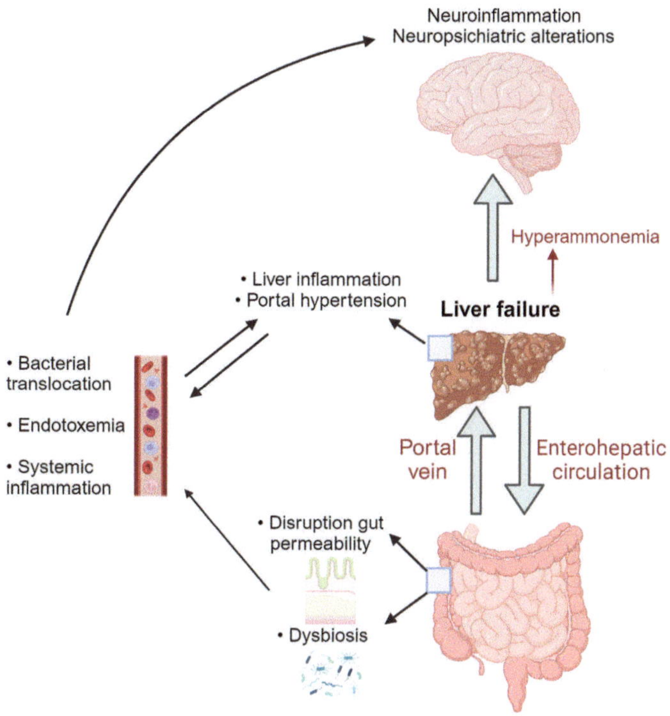

Fig. 1 Schematic representation of the gut-liver-brain axis

1.3 Scientific Evidence of the Link Between the Gut Microbiota and Hepatic Encephalopathy

In recent decades, there have been many efforts to better understand the complex relationship between the gut microbiota and HE. The gut-liver-brain axis (Fig. 1) represents the bi-directional communication between the gut and its microbiota, the liver, and the brain. This complex relationship is shaped by signals that are influenced by diet, genetic predisposition, and environment [17, 18]. HE can be considered as a condition resulting from abnormal interactions between the gut, liver, and brain. Although the underlying mechanisms remain poorly understood, emerging evidence suggests that changes in gut microbiota composition play a critical role in the development of HE [19–21]. Importantly, changes in liver function can also impact the gut microbiota [22]. With regard to gut function, breakdown in gut barrier function can be mediated by several mechanisms, including local infection and bacterial translocation, which can disrupt gut permeability and increase bacterial translocation, inducing both systemic and local inflammation. It also has implications for the composition of the gut microbiota, reducing the abundance of beneficial commensal bacteria and their secondary metabolites, including SCFAs. This creates a permissive environment for harmful substances to enter the bloodstream from the gut, damaging liver and brain function and potentially triggering HE [14, 23], as the liver is connected to the gut though the portal vein system. Decreased liver function induces hyperammonemia, which leads to neuropsychiatric changes [24]. In addition, portal hypertension facilitates bacterial translocation, resulting in elevated serum endotoxin levels that, in turn, exacerbates portal hypertension by inducing systemic and splanchnic vasodilation and by triggering a hepatic inflammatory response via tumor necrosis factor α (TNF-α) [25]. Also, evidence suggests that the increased risk of dysbiosis in patients with cirrhosis, together with the resulting clinical consequences, is mainly attributed to the multiple pathological interactions occurring between the liver and the gastrointestinal tract. Gut motility changes, elevated pH levels, and reduced bile acid levels in the colon of cirrhotic patients might result in the inability to limit the excessive growth of harmful gut bacteria [26]. Patients with cirrhosis present with a notable decrease in monocyte spreading, chemotaxis, and neutrophil activity when compared with healthy subjects [27]. In addition, these patients have an impaired reticuloendothelial system, which reduces the liver's ability to monitor infection by Kupffer cells and sinusoidal endothelial cells. Moreover, the reduced synthesis of hepatic proteins related to innate immunity and pattern recognition impairs the bactericidal capacity of phagocytic cells [28, 29]. This may compromise the integrity of the gut barrier, facilitating bacterial translocation and increasing the risk of intestinal bacterial infection and liver decompensation [30].

1.4 The Role of the Gut Microbiota in the Diagnosis and Treatment of Hepatic Encephalopathy

The diagnosis of HE and MHE is primarily based on clinical evaluation, as there are no definitive biomarkers available to diagnosis the disease. MHE has a subclinical symptomatology and is diagnosed using psychometric and neurophysiological tests [31, 32], whereas HE is characterized by overt clinical manifestations such as asterixis (bilateral tremor) or disorientation [33]. The advent of next-generation sequencing (NGS) has opened new avenues for investigating the gut microbiome in cirrhosis and HE for both diagnostic and therapeutic strategies, and many studies have attempted to identify a distinctive gut microbiome profile capable of distinguishing patients with cirrhosis, with or without MHE or HE [34–37]. While there is evidence of distinct differences in the gut microbiota between healthy individuals and patients with cirrhosis and/or HE [26, 38], a definitive microbiota *profile* that can accurately distinguish between these groups for diagnostic purposes has yet to be established. This is due to two main issues. First, as the gut microbiota is highly dynamic and influenced by many factors including diet, age, lifestyle, alcohol/drug use, and genetic factors, the efficacy of targeted interventions may vary depending on individual characteristics and disease severity. In addition, a high inter-individual variability in the clinical response to different microbiome-targeted interventions has been documented [39–41], with subgroups of patients with the same diagnosed condition responding differently to certain strategies, supporting the existence of specific, non-shared underlying mechanisms between subgroups of patients. Second, differences in sample collection, processing, and data analysis methodologies across trials often introduce large inconsistencies, making it difficult to directly compare results from different trials. The most widely used approach to the management of HE involves treatments that focus on the gut, including the use of lactulose and lactitol, as well as non-absorbable antibiotics such as rifaximin and neomycin [42]. However, the recognition of the gut microbiota as a therapeutic target has led to the exploration of precision probiotics, prebiotics, and fecal microbiota transplant (FMT) to modulate intestinal microbes and therapeutically manage HE [43–47]. These strategies appear to be feasible and are supported by promising results, at least in some patients [38], with improvements in cognitive function, suppression of harmful gut bacteria, and reduced [40] absorption of ammonia.

Although microbiome research has largely focused on bacteria, which make up >99% of the genetic material in the microbiome, it is important to note that viral, fungal, and archaeal components are also present. Notably, these non-bacterial components (including the virome, the mycobiome and other microbiome's fractions), which are under-researched in relation to HE, are beginning to show a key role in physical and mental health. It will therefore be important to further investigate the part that non-bacterial components play in shaping the gut ecosystem during liver dysfunction. In fact, the virome includes the genome of viruses that infect host cells but also the genome of Bacteriophages (Phageome), which are prokaryotic viruses that specifically target bacteria. Of note, bacteriophages, in spite of the challenge for their study, are starting to show a key role for gut microbiome dynamics and human health. Additionally, it will be crucial to examine the impact of external factors on the long-term development of the bacteriome and other gut microbiome interactions, as well as the involvement of these interactions and their related mechanisms in cirrhosis and HE. Specifically, a better understanding of the role of the Phageome in HE could lead to more precise manipulation of the human gut microbiota through the modulation of viruses to improve the prognosis of the disease. Moreover, until recently, research on the gut microbiota has focused on identifying the microbes present in a sample, rather than their functionality. However, it is fundamental to consider the functionality of the gut microbiota when studying disease. Here, we describe research methods and models that we consider are crucial for investigating the role of the gut microbiota in HE, with an aim to develop microbiota-based therapies. This includes a discussion on techniques for the identification of potential probiotics that could help in the prevention or treatment of HE and also those for the evaluation of the preclinical efficacy of selected potential HE probiotics. Of note, these methods can be integrated with other methods that are not described here, such as nuclear magnetic resonance spectroscopy-based metabolomics in liver and brain, to identify gut microbiota-associated metabolic signatures related to HE. This can also help pinpoint potential markers for monitoring the progression of HE and the therapeutic effects of microbiome-based interventions. Additionally, by integrating results from meta-transcriptomics, metabolomics, epigenomics, and metagenomics, *multi*-omics studies could provide valuable insights into the mechanisms that underlie host-microbe interactions in the context of HE, ultimately facilitating the development of tailored microbiome-targeted strategies to combat HE.

2 Materials

2.1 Analysis of the Gut Microbiota Composition and Function

2.1.1 Identification of Microbial Taxa

(a) DNA extraction

AllPrep DNA/RNA Mini Kit, QIAshredder and Proteinase K (Qiagen); Lysozyme (10 mg/mL, Sigma-Aldrich) and mutanolysin (5000 U/mL, Sigma-Aldrich); Glass Beads 0.1 mm diameter, 2-mL sterile vials and caps; Qubit 1X dsDNA High Sensitivity Kit (Thermo Fisher). Instruments: Mini Beadbeater; Qubit® Fluorometer; Vortex; Centrifuge; Lab weighing scale; Channel pipettes (P2, P10, P20, P200, and P1000).

(b) Bacterial 16S V3-V4 amplification for Hi-throughput sequencing and purification, library preparation, amplicon sequencing, and bioinformatic processing.

Eppendorf tubes; Barcoded 16S primers, Nuclease-free PCR grade water, Phusion HF Buffer, dNTPs Mix and Taq Phusion High Fidelity Polymerase; 96-well PCR plate; Agarose, Midori-green, loading buffer, DNA ladder. Channel pipettes (P2, P10, P20, P200, and P1000), Illustra GFX PCR DNA and Gel Band Purification Kit (GE Healthcare) and KAPA Library Quantification Kit. Instruments: 96-well block PCR thermocycler; Microwave oven, Electrophoresis chamber, MiSeq instrument (Illumina).

2.2 Identification of Microbiota-Derived Metabolites: Analysis of SCFAs

0.2 μm blue filters (Corning) and 2 mL syringes, Mili-Q water, eppendorf tubes, 250 μL inserts (Agilent), amber glass 2-mL GC vials (Agilent), silicone/PTFE screw caps (Agilent), a 10 mM stock internal standard (IS, 2-ethylbutyric acid solution), a 100 mM stock fatty acid mix of 45.7 μL of butyric acid, 46.5 μL of iso-butyric acid, 37.4 μL of propionic acid and 28.6 μL of acetic acid. Instruments: a gas chromatography platform with flame ionization detection (GC-FID).

2.3 Animal Models to Study Causality and the Relationships Between the Gut Microbiota and Hepatic Encephalopathy

2.3.1 Fecal Microbiota Transplant

(a) Depletion of the gut microbiota in rats using antibiotics

Conventional or specific pathogen free (SPF) rats with or without HE; antibiotic cocktail: ampicillin and metronidazole (1 g/L each), vancomycin, (500 mg/L), ciprofloxacin HCl (200 mg/L), imipenem (250 mg/L), all resuspended in autoclaved deionized water; stainless steel feeding needles or oral gavage needles for rats—curved/50 mm/18G/2.25 mm; DNA extraction kit (*see* Subheading 2.1.1).

(b) Transferring the gut microbiota from a donor to a recipient animal

Sterile forceps; eppendorf tubes and 15 mL centrifuge tubes; reusable and sterile feeding needle—curved/50 mm/18G/2.25 mm; PBS, cysteine, glycerol; vortex; refrigerated centrifuge; ice and −80 °C freezer.

2.4 Pre-Clinical Trials to Identify Potential Probiotics for Hepatic Encephalopathy

(a) Animals with a complex (non-controlled) gut microbiota: conventionally raised and specific pathogen-free animals

Conventional or SPF animals (mouse or rat) with or without HE; doses of the previously cultured microbial candidate(s): frequently 109 CFU/per mouse in pre-reduced PBS (0.1 mL) and maintained at −80 °C until administration; doses of vehicle: pre-reduced PBS (0.1 mL); stainless steel feeding needles or oral gavages: 22 g × 1″, 1.25 mm-straight; all the required equipment for phenotypic characterization.

(b) Gut microbiota-deficient animal models

A gnotobiotic facility; GF mice with HE; doses of the previously cultured microbial candidate(s): frequently 109 cfu/per mouse in pre-reduced PBS (0.1 mL) and maintained at −80 °C until administration; doses of vehicle: pre-reduced PBS (0.1 mL); stainless steel feeding needles or oral gavages: 22 g × 1″, 1.25 mm-straight; all the required equipment for phenotypic characterization.

3 Methods

3.1 Analysis of the Gut Microbiota Composition and Function

3.1.1 Identification of Microbial Taxa

(a) DNA extraction

For human studies, stool samples from healthy individuals and patients with HE are collected at home and immediately stored at −20 °C. Samples are transported to the hospital in refrigerated bags to prevent sample thawing and are stored at −80 °C until processing. For animal studies, cecal samples are stored at −80 °C until processing.

DNA for gut microbiome analysis is extracted from 0.15 to 0.2 g of frozen stool using the QIAamp Fast DNA Stool Mini Kit (Qiagen) or similar, with the following modifications. Specifically, the extraction protocol we use combines bacterial enzymatic lysis, Proteinase K digestion, mechanical bead beating, debris removal (Qiagen, QIAshredder), and nucleotide isolation (QIAgen AllPrep RNA/DNA Mini Kit). The protocol effectively removes PCR inhibitors from even the most difficult stool sample types using a buffer specifically designed for this purpose. Samples are added to a bead-beating tube for rapid and thorough homogenization to lyse bacterial cells within the fecal material. Total genomic DNA is captured on a silica membrane in a spin column format and then eluted. DNA is stored at −20 °C and is ready for PCR analysis and other downstream applications.

Stool Sample Aliquoting

1. Weigh 150–200 mg of stool in a flow cabinet to prepare an aliquot for DNA isolation. It is important to prepare stool aliquots while keeping the samples frozen at around −80 °C.
2. Place the original sample back in the −80 °C freezer.
3. Place the weighed stool aliquot in a 2 mL microcentrifuge tube on ice.

Enzymatic Lysis and Proteinase K Digestion

4. Add 110 μL of TE buffer containing 20 μL of lysozyme (10 mg/mL, Sigma-Aldrich, L6876), 2 μL of mutanolysin (5000 U/mL, Sigma-Aldrich, M9901) and 15–20 μL Proteinase K (Qiagen) to lyse the cells and release the DNA and RNA in the sample.
5. Mix well by vortexing for 10 s.
6. Incubate at 37 °C for at least 30 min (suggested 1 h) and vortex for 10 s at least every 10 min.

Mechanical Lysis

7. Transfer sample to a 2 mL sterile bead-beating tube filled with 1 mL of 0.1-mm glass beads.
8. Fill the remaining volume of the tube with RLT Plus w/β-ME before adding the screw top cap.
9. Place the tube in a Beadbeater for three successive rounds of 1 min.
10. Remove the tube from the Beadbeater.
11. Heat the sample at 95 °C for 10 min and centrifuge for 5 min at max speed to pellet the debris.
12. Transfer the supernatant from the bead tube to a sterile 1.5 mL tube.
13. Centrifuge for 2 min at max speed to remove any additional debris.

Qiagen AllPrep DNA/RNA Mini Protocol

14. Pipette the supernatant from step 11 into a QIAshredder spin column placed in a 2 mL collection tube and centrifuge for 2 min at maximum speed. Note: incomplete homogenization leads to significantly reduced RNA yields and can cause clogging of the AllPrep DNA and RNeasy spin columns.
15. Transfer the homogenized lysate to the AllPrep DNA spin column placed in a 2 mL collection tube (supplied). Close the lid gently and centrifuge the column for 30 s at ≥8000 × g (≥10,000 rpm). Ensure that no liquid remains on the column membrane after centrifugation. If necessary, repeat the centrifugation until all liquid has passed through the membrane.
16. Place the AllPrep DNA spin column in a new 2 mL collection tube (supplied) and store at room temperature (15–25 °C) or at 4 °C for later DNA purification.

Genomic DNA Extraction and DNA Elution

17. Add 500 μL Buffer AW1 to the AllPrep DNA spin column from step 16. Close the lid gently and centrifuge for 15 s at ≥8000 × g (10,000 rpm) to wash the spin column membrane. Discard the flow through. *Reuse the spin column in the next step. Buffer AW1 is supplied as a concentrate. Ensure that ethanol is added to Buffer AW1 before use.
18. Add 500 μL Buffer AW2 to the AllPrep DNA spin column. Close the lid gently and centrifuge for 2 min at full speed to wash the spin column membrane. Buffer AW2 is supplied as a concentrate. Ensure that ethanol is added to Buffer AW2 before use. The longer centrifugation step dries the spin column membrane, ensuring that no ethanol is carried over during DNA elution. Residual ethanol may interfere with downstream reactions. After centrifugation, carefully remove the AllPrep DNA spin column from the collection tube. If the column contacts the flow-through, empty the collection tube and centrifuge the spin column again for 1 min at full speed.
19. Place the AllPrep DNA spin column in a new 1.5 mL collection tube (supplied). Add 50 μL Buffer 1× low TE or elution buffer directly to the spin column membrane and close the lid. Incubate at room temperature (15–25 °C) for 1 min and then centrifuge for 1 min at ≥8000 × g (10,000 rpm) to elute the DNA.
20. Repeat step 19 to ensure that all the DNA is eluted. To prevent dilution of the first DNA eluate, use a new 1.5 mL collection tube (not supplied) to collect the second DNA eluate. To combine the first and second DNA eluates, reuse

the collection tube from step 19. To achieve a higher DNA concentration, elute with 2 × 50 μL instead of 100 μL Buffer 1× low TE or elution buffer. The final DNA yield, however, may be reduced.

21. After DNA extraction, *check DNA concentration and quality of isolated DNA* using the Qubit dsDNA HS kit (Thermo Fisher Scientific) and the Qubit fluorometer. Briefly, to determine the concentration and quality of isolated DNA using Qubit, add 10 μL of Qubit standards to the appropriate tube, add 1–20 μL of each user sample to the appropriate tube, mix each tube vigorously by vortexing for 3–5 s, allow all tubes to incubate at room temperature for 2 min and then proceed to measure the standards and samples.

22. *Make an aliquot of 20 ng/mL DNA for 16S rDNA amplicon sequencing* and store the remaining purified DNA at −20 °C until needed for downstream applications.

Bacterial 16S V3-V4 Amplification for Hi-Throughput Sequencing

1. Set up triplicate reactions per sample using the aliquot of 20 ng/mL DNA for 16S rDNA amplicon sequencing, according to the scheme described in (Table 1). Prior to setting up the PCR, prepare working dilutions of each dual-barcoded primer stock at 5 μM in elution buffer. Prepare the amount of the PCR master mix you will need (the mix of Nuclease-Free water, Buffer HF, dNTPs Mix and Taq Phusion High Fidelity Polymerase) for all PCR reactions. Furthermore, measure the DNA concentration of each sample using a Nanodrop Spectrophotometer or Qubit fluorometer with the 1× dsDNA HS assay and normalize DNA samples to a concentration of 20 ng/μL in elution buffer or nuclease-free water.

Table 1
Preparation of the aliquot of 20 ng/mL DNA for 16S rDNA amplicon sequencing

Reagent	Stock	Vol	Final concentration
Nuclease-free water	–	23.4 μL	–
Buffer HF	5X	7.0 μL	1X
dNTPs mix	10X	1.5 μL	0.43X
Library mix (dual-barcoded oligo mix)	5 uM	2.0 μL	0.29 μM
Taq Phusion high Fidelity polymerase	2 U/μL	0.1 μL	0.007 U/μL
DNA	20 ng/μL	1.0 μL	0.67 ng/μL
	Total	35.0 μL	

Table 2
PCR cycling conditions

Temperature	Time	Cycles
98 °C	3 min	1
95 °C	20 s	
55 °C	20 s	25
72 °C	20 s	
72 °C	5 min	1

2. Dispense 34 μL of the PCR master mix + barcoded primers to each respective well across the 96-well plate. It is important to note that in this step, samples must be tagged with barcodes to allow later multiplexing during the sequencing process.
3. Add 1 μL (20 ng) of each diluted DNA sample to a single well of the 96-well PCR plate(s) for a total reaction volume of 35 μL (as we run triplicate reactions per sample, the same DNA sample will be added to three PCR wells of the 96-well PCR plate).
4. Mix well by pipetting. Seal the plates tightly with adhesive seal to prevent evaporation during the PCR. Briefly spin the plate in a refrigerated centrifuge (4 °C) to ensure that the sample is at the bottom of each well.
5. Place PCR plate, or plates, on the thermocycler and run the PCR for all reactions according to the following cycling conditions (Table 2).
6. After the PCR, check for the successful production of 16 rDNA amplicons on a 2% agarose gel electrophoresis using 1× TAE running buffer.
7. For the electrophoreses, use 5 μL Midorigreen per 50 mL agarose, mix 6 μL PCR product with 3 μL of 3× loading buffer and run together with samples (PCR amplicons) a positive control, a non-template control (NTC) to check for contamination, and a DNA ladder to estimate the size of PCR amplicons tested.
8. Establish an adequate voltage for the successful separation of 8 V/cm of distance between electrodes. After electrophoresis, visualize the DNA bands and compare their size to the size of the DNA Ladder. Expected amplicons must show a unique band of approximately 480 bp in length.
9. Combine the triplicate sample amplicons in the same tube for amplicon purification.

Purification, Library Preparation, Amplicon Sequencing, and Bioinformatic Processing

Purify the generated amplicons using the Illustra GFX PCR DNA and Gel Band Purification Kit (GE Healthcare) according to the manufacturer's instructions. This is a fast and easy-to-use method with less than 10 min hands on time designed for the purification and concentration of DNA from PCR mixtures. After purification, highly pure amplicons will be ready for direct use in sequencing.

1. Prepare working solutions as indicated in the kit's instructions manual.
2. Follow the protocol for purification of DNA from a solution or an enzymatic reaction:
   1. *Sample capture:*
      a. Add 500 μL Capture buffer type 3 to up to 100 μL of sample. If sample volume is greater than 100 μL, divide the sample and purify using more than one GFX Micro-Spin column.
      b. Mix thoroughly. If the sample contains DNA fragments greater than 5 kbp, do not vortex as this may cause shearing of the DNA.
      c. Check that the Capture buffer type 3-sample mix is yellow or pale orange in color. If the color of the binding mixture is dark pink or red, add a small volume (~10 μL) of 3 M sodium acetate pH 5.0 and mix. Ensure that the binding mixture is a yellow or pale orange color before loading onto the GFX Microspin column.
      d. For each purification that is to be performed, place one GFX MicroSpin column into one Collection tube.
   2. *Sample Binding:*
      a. Centrifuge the Capture buffer type 3-sample mix briefly to collect the liquid at the bottom of the tube.
      b. Load the Capture buffer type 3-sample mix onto the assembled GFX MicroSpin column and Collection tube.
      c. Spin the assembled column and Collection tube at 16,000 × g for 30 s.
      d. Discard the flow-through by emptying the Collection tube. Place the GFX MicroSpin column back inside the Collection tube.
   3. *Wash and Dry:*
      a. Add 500 μL Wash buffer type 1 to the GFX MicroSpin column.
      b. Spin the assembled column and Collection tube at 16,000 × g for 30 s.

If purity is paramount (e.g., if the sample is to be used in a blunt-ended ligation), repeat the Wash & Dry step a and perform step b twice. After the first spin, discard the flow-through, place the GFX MicroSpin column back inside the Collection tube and centrifuge at 16,000 × g for an additional 30 s. This extra wash step may reduce the yield by 4%.

c. Discard the Collection tube and transfer the GFX MicroSpin column to a fresh DNasefree 1.5 mL microcentrifuge tube (supplied by user).

4. Elution of purified barcoded amplicons:
   a. Add 10–50 μL Elution buffer type 4 OR type 6 to the center of the membrane in the assembled GFX MicroSpin column and sample Collection tube.
   b. Incubate the assembled GFX MicroSpin column and sample Collection tube at room temperature for 1 min.
   c. Spin the assembled column and sample Collection tube at 16,000 × g for 1 min to recover the purified DNA.
5. Store at −20 °C until you begin preparing the Illumina sequencing library.
6. When ready for sequencing, quantify and determine the quality of the extracted DNA using Qubit® Fluorometer and the high sensitivity dsDNA assay kit.
7. Assign samples to be included in the same pool for sequencing.
8. Prepare equimolar pooling of indexed libraries, irrespective of pre-pooling library concentrations; combine samples in equimolar concentrations to be included in the same pool for sequencing. You can use The KAPA Library Quantification Kit, which enables accurate equimolar pooling.
9. Send the pools of libraries to an outsourced NGS service, who will assign each pool to the lane of the instrument for sequencing on a MiSeq instrument (Illumina). Unless one performs metagenomic analysis (16S sequencing) regularly, outsourcing NGS services can be the best cost-effective alternative, which allow access to expertise, reducing risk of errors and accelerating the research timeline.
10. Bioinformatic processing of data derived from 16S rDNA amplicon sequencing:

This is typically carried out using software QIIME software [48], Mothur, and UPARSE. Briefly, QIIME joins paired-end forward and reverse Illumina reads into contigs, extracts barcodes, and demultiplexes reads. Primers are then removed using the software program

Mothur [49] or similar. Using UPARSE [50], chimeras are removed and reads are clustered at 97% identity into operational taxonomic units (OTUs) using default settings. An OTU abundance table is generated within the UPARSE pipeline by mapping reads to representative sequences for each OTU. Using QIIME, a Biom file is created from the OTU table and reads are rarefied and singletons removed. A phylogenetic tree is constructed from representative sequences for each OTU, and is aligned using PYNAST [48] or similar software and filtered using default settings. Alpha diversity metrics (Shannon's, Simpson's, Simpson's reciprocal diversity index, Simpson's evenness) are calculated and Beta diversity analysis is conducted using generalized UniFrac (GUniFrac) [51] and principal coordinate analyses (PCoA). Samples are classified taxonomically with uclust within QIIME using taxonomic assignments from an updated reference database. Metagenome functional content can be predicted or inferred from 16S rRNA marker genes from a normalized OTU table and a reference genome database using the software PICRUSt.

3.1.2 Identification of Microbiota-Derived Metabolites: Analysis of SCFAs

SCFAs are the primary postbiotics of interest in HE. SCFAs are gut microbiota-derived bioactive products produced by the bacterial fermentation of non-digestible polysaccharides, and participate in maintaining gut barrier integrity and the host's immune response, both of which are features strongly linked to HE. Moreover, advanced liver disease and HE are associated with reduced intestinal SCFA levels. Most published studies have shown that SCFA-producing species, such as *Anaerostipes caccae, Bacteroides eggerthii,* and Clostridial species, are depleted in patients with HE. Additionally, evidence suggests that the modulation of SCFA levels is one of the main potential mechanisms by which microbiome-based strategies may impact in HE prevention or treatment.

Below we provide a detailed protocol to quantify the levels of the main SCFAs in stool samples.

Fecal SCFA Quantification

1. Remove from the refrigerator the previously prepared stock solutions of internal standard and fatty acid mix.
2. Weigh 0.2 gram of the fecal sample in an eppendorf tube.
3. Add 1 mL of Mili-Q water to each thawed sample and vortex well for about 3 min to homogenize the sample.

4. Adjust the pH of the suspension to pH 2–3 by adding 5 M HCl; leave the suspension to stand for 10 min at room temperature and centrifuge the samples at maximum speed for 5 min.
5. Filter (0.2 μm blue filters (Corning) and 2 mL syringes) the supernatant into clean fresh eppendorf tubes.
6. Add 30 μL of working internal standard (10 mM) to duplicate eppendorf tubes, and add to each 270 μL of filtered sample supernatant or standard.
7. Vortex to homogenize briefly and spin at maximum speed for 3 min.
8. Transfer the clear supernatant (do not transfer the small yellow pellet) to 250 μL inserts (Agilent) placed in amber glass 2-mL GC vials (Agilent) sealed with silicone/PTFE screw caps (Agilent).
9. Analyze the samples and standards by gas chromatography flame ionization detection (GC-FID) using a Varian 3800 GC system fitted with a Zebron ZB-FFAP column (30 mL × 0.32 mm ID × 0.25 μm df; Phenomenex) and a flame ionization detector with a CP-8400 auto-sampler.

Preparation of the 10 mM stock internal standard (IS, 2-ethylbutyric acid solution):

1. Wash a 5 mL volumetric flask with formic acid and leave to dry.
2. Add about 3 mL of formic acid and then 63 μL of 2-ETBA to the flask.
3. Fill the flask slowly to 5 mL with formic acid using a pipette.

Preparation of the 100 mM stock fatty acid mix

1. Wash a 5 mL volumetric flask with Milli-Q water and leave to dry.
2. Add about 3 mL of Milli-Q water.
3. Next add (always in this order): 45.7 μL of butyric acid, 46.5 μL of iso-butyric acid, 37.4 μL of propionic acid and 28.6 μL of acetic acid.
4. Bring the total volume in the flask to 5 mL with Mili-Q water.
5. Transfer to a glass bottle that has been pre-rinsed with Milli-Q water and store both the IS and fatty acid (FA) stocks in the fridge.
6. Before making the standards, ***prepare a 20 mM working solution of FA mix*** from the 100 mM FA mix stock solution as follow: 1000 μL of 100 mM FA mix stock +4000 μL of Mili-Q water.
7. To make the standards (Table 3).

Table 3
Preparation of fatty acids stansdards

mM	20 mM FA mix (μL)	10 mM IS (μL)	Mili-Q water (μL)
0.5	7.5	30	262.5
1.0	15	30	255
2.0	30	30	240
4.0	60	30	210
8.0	120	30	150
10.00	150	30	120

3.2 Animal Models to Study Causality and the Relationship Between the Gut Microbiota and Hepatic Encephalopathy

3.2.1 Fecal Microbiota Transplant

By conducting FMT, we can experimentally investigate the causal role of the gut microbiota on a host phenotype. This procedure involves using animals lacking the gut microbiota (recipients) that are colonized with the whole gut microbiota community provided by a donor.

To study the impact of the gut microbiota on the HE phenotype, we provide a two-step protocol for performing FMT in gut microbiota-depleted animals using an antibiotic cocktail: (a) depletion of the gut microbiota in recipient animals using antibiotics, and (b) transfer of the entire gut microbiota collected from a donor to the recipient animal. In the context of HE, we provide an approach in which the human gut microbiota from healthy individuals (donors) is transferred to antibiotic-treated rats with HE (recipients) to question whether a healthy gut microbiota has any beneficial effects on the HE phenotype. Turning the question on its head, we also provide an approach where the gut microbiota from a patient with HE (donor) is transferred to antibiotic-treated healthy rats (recipients), allowing us to determine whether the gut microbiota from a patient with HE can induce the typical symptoms of the pathology in the rat (Fig. 2).

(a) Depletion of the gut microbiota in rats using antibiotics

The administration of antibiotics to experimental animals (typically rodents) reduces the number of microbes in the gut. This facilitates the establishment of a new population of microbes from a donor by reducing competition [52]. Various combinations of antibiotics, along with different methods of administration, have shown varying levels of effectiveness [53]. Frequently, antibiotics are added to drinking water [54], but this can lead to variations in delivered doses. Also, antibiotics can be administrated via intramuscular injection [55] and by oral gastric gavage [56, 57]. In some cases, a combination of drinking water and oral gastric-gavage is used [58].

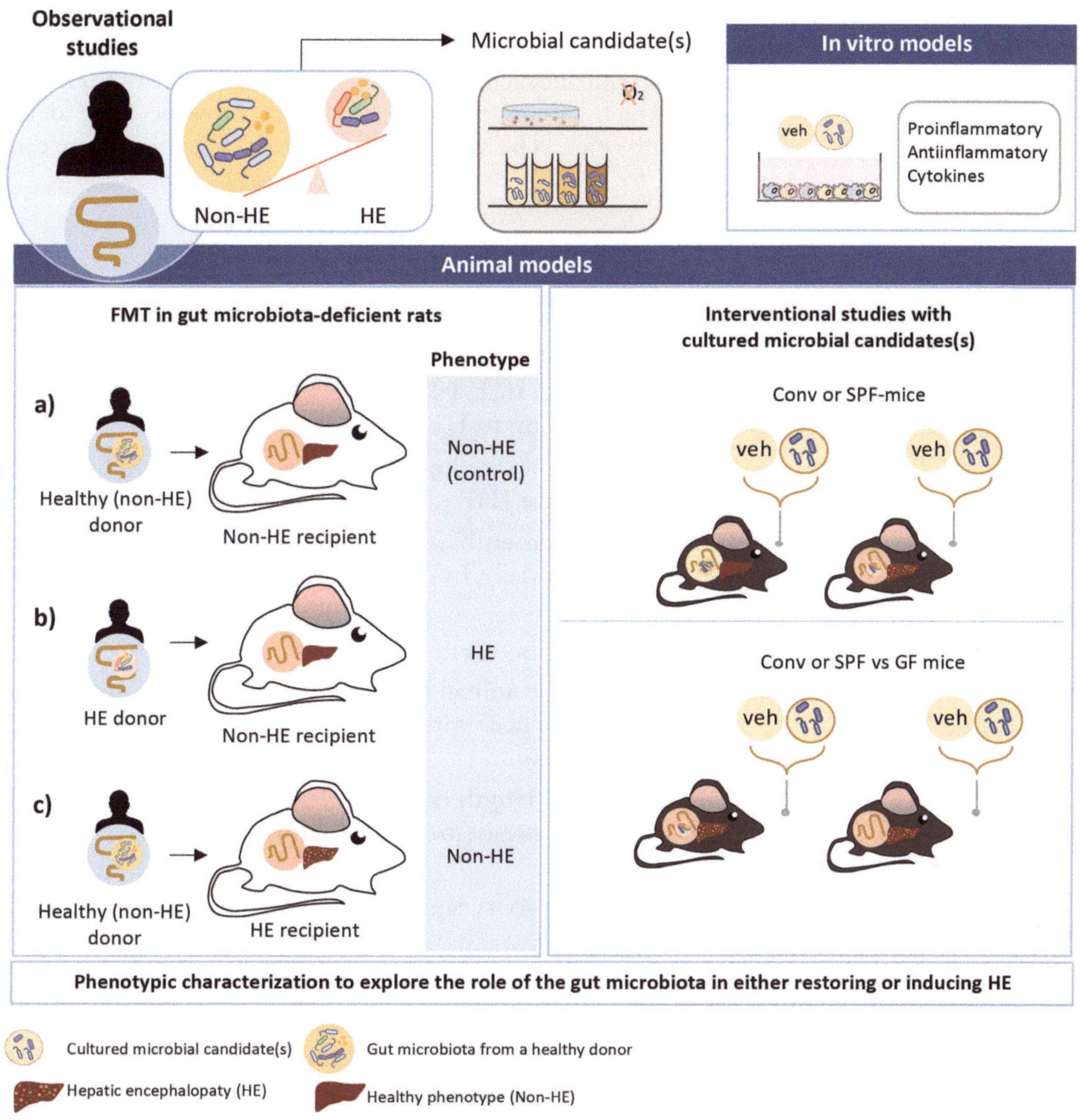

Fig. 2 Schematic representation of the experimental procedures to study the relationship between the gut microbiota and hepatic encephalopathy. To identify potential probiotics for hepatic encephalopathy (HE), researchers often select gut microbial taxa that are enriched in healthy individuals (non-HE) compared with patients with HE. These microbial candidates are then cultured in vitro under optimal anaerobic growth conditions. The mode of actions and effects of these microbial candidates are typically investigated through in vitro assays using cultured immune cells, and in animal models including conventional, specific pathogen-free (SPF) and/or germ-free (GF) mice (rats may also be used) with or without HE. Additionally, the transfer of gut microbiota from a donor (human) to a gut microbiota-deficient rat (fecal microbiota transplant, FMT) is conducted to explore whether the gut microbiota could restore the HE phenotype [transfer of the gut microbiota from non-HE donor to HE recipient (**b**)] or conversely, to decipher whether the whole gut microbiota community contributes to the HE phenotype [transfer of the gut microbiota from HE donor to non-HE recipient (**c**)]

We provide a protocol to deplete the gut microbiota in rats by administering a once-daily cocktail of ampicillin, metronidazole, vancomycin, ciprofloxacin HCl, and imipenem by oral-gavage [59, 60], which ensures that all animals receive the same daily dose of antibiotics. The period of antibiotic administration can vary from 7 to 28 days, depending on the study.

Protocol

1. Randomly divide rats into three different groups:
   (a) Healthy rats that will receive FMT from a healthy donor (control group).
   (b) Healthy rats that will receive FMT from an HE donor (healthy group to HE).
   (c) Rats that will develop HE and will receive FMT from a healthy donor (HE group to healthy).
2. All rats receive the antibiotic cocktail for the set period of time (7–28 days) once daily, by oral gavage following the indications below:
   - Fill a 1 mL syringe with antibiotics or vehicle (0.2 mL).
   - Immobilize the animal by firmly grasping the loose skin on the back of the neck with the hand that is not involved in the gavage procedure.
   - Insert the full length of the oral gavage into the mouth of the rat without resistance and administer the dose contained in the syringe.
   - Return the rat to its cage and check that it shows no signs of injury or pain.
3. Monitor the body weight once weekly.
4. Renew the bedding every other day (for all the groups) to minimize the chances of microbiota reestablishment in the antibiotic-treated animals.
5. After antibiotic treatment a washout must be performed by administrating deionised water for 3 days to all groups via oral gavage.
6. Verify the elimination of the gut microbiota in the cecal content of randomly selected sacrificed rats through 16 s rDNA sequencing, as described in Subheading 2.1.1.

Clearing the bowel using laxative-based techniques such as polyethylene glycol (PEG) is an alternative method to eliminate the existing gut microbiota before introducing the donor gut microbiota. Unlike antibiotics, laxatives induce the temporary reduction of the gut bacterial ecosystem in both humans and mice [61, 62]. Additionally, bowel cleansing preserves the indigenous microbiota after treatment when compared with the impact of antibiotics [62].

(b) Transferring the gut microbiota from a donor to a recipient animal.

FMT is an efficient strategy to modulate the gut microbiota and involves transferring processed stool from a donor to a recipient. Although some inter-individual variability has been observed, both pre-clinical and clinical studies have shown beneficial effects or some clinical success of FMT from a healthy donor to a HE animal model or patient in the treatment of HE [46, 63–66].

Protocol

Donor Stool Preparation

Source for FMT: frozen feces, fresh cecal content, or fresh feces can be used to perform FMT, as they have been shown to have similar microbial viability [67]. Freezing feces allows the storage of multiple samples for future use; however, the number of freeze-thaw cycles can affect microbial populations [68]. Although we have focused on human-to-rodent FMT, the procedure can also be performed rodent-to-rodent. We will describe both options in this protocol.

Storage, preservation, and processing of stool: storage conditions are also important. Short-term storage (one month) of fecal material at −20 °C showed an enteric colonization ability similar to that of samples frozen at −80 °C. For long-term storage (from 6 months up to 2 years), it is recommended to store the samples at −80 °C [69]. Cryoprotection prior to freezing is essential to maintain bacterial cell viability and integrity; 10% glycerol is recommended for the preservation of frozen fecal samples [70], but is not suitable for lyophilization. We use reduced PBS containing 15% (v/v) glycerol. Maltodextrin and trehalose are viable alternatives to glycerol and dimethyl sulfoxide as cryoprotectants. It is very important to ensure rapid handling to minimize exposure of the sample to oxygen, to ensure survival of anaerobic bacteria. It has been reported that approximately 62% of anaerobic bacteria remained alive when exposed to oxygen for less than 2 min; however, viability decreased to 55.5% after 90 min of exposure to oxygen [71]. According to another study, the viability of anaerobic microbes was approximately 50% when human stool was exposed to oxygen for 4–5 min. Viability dropped sharply to 0.1% after 40 min, and no cells remained alive after 2 h of oxygen exposure [72]. For human studies, stool samples are collected at home and immediately stored at −20 °C. When the sample arrives at the hospital, it is stored at −80 °C until processed.

Protocol for stool preparation from rodent donors (with or without HE):

1. Individually house each rat in cages without sawdust, and collect feces by gently massaging the lower abdomen to facilitate expulsion.
2. Using sterile forceps, carefully collect the feces and place in sterile 2 mL eppendorf tubes, which are then promptly placed on ice and rapidly processed.
3. Place 2 g of total fecal material into a sterile 15 mL centrifuge tube with 10 mL sterile PBS supplemented with 0.05% (w/v) cysteine and 15% (v/v) glycerol.
4. Homogenize the sample to produce a fecal slurry by vigorous vortex shaking.
5. Centrifuge at 2000 rpm for 5 min at 4 °C and collect the supernatant.
6. Store the supernatant at −80 °C in aliquots of 850 μL until further use.

Protocol for stool preparation from human donors (with or without HE):

1. Stool samples from healthy individuals and patients with HE are collected at home and immediately stored at −20 °C. Samples are then transported to the hospital in cooled plastic bags, avoiding sample thawing, and are stored at −80 °C until processing.
2. We then follow steps 3–6 from the above rodent donor protocol.

Dosage strategy and delivery routes for FMT: the optimal dose for FMT can vary widely depending on the disease model used and therefore needs to be optimized for each model. Administration ranges from a single dose to twice weekly for weeks [73, 74]. For rodents, the suggested maximum volume is 2% of body weight. Determining the ideal amount of fecal infusion is challenging, with suggested doses ranging from 100 μL [75] to 400 μL [76]. Oral gastric gavage is a commonly used method for FMT studies. Rectal enema can also be used but is more often used in humans than in animals, for example, to attenuate the progression of necrotizing enterocolitis [77]. In this protocol, we will only describe the oral gavage administration for rats.

Protocol for FMT in the study of HE:

1. After the antibiotic treatment, a washout period of 3 days must be performed by administering deionised water to all groups (control group, healthy to HE group, and HE to healthy group) by oral gavage [3].

2. To perform the FMT, administer 400 μL of the stored supernatant to each gut microbiota-depleted rat by oral gavage for 3 consecutive days (*see* Subheading 3.2.1, antibiotic cocktail administration for prior gut microbiota depletion).
3. To maintain the characteristics of the donor microbiota in the recipient animal, additional booster inoculations must be given twice weekly after the FMT throughout the entire study.

Analysis of recipient gut microbiota engraftment following FMT: alpha and beta diversity analysis in donors allows the assessment of gut microbiota engraftment. The Bayesian source tracker algorithm, which uses an OTU-based approach, can be utilized to monitor the degree of donor engraftment [78].

3.2.2 Pre-Clinical Trials to Identify Potential Probiotics for HE

NGS techniques have enabled the identification of associations between gut microbiota composition and disease in observational human studies [79]. These associations suggest that restoring the abundance of affected microbial taxa in a disease to a healthy state could improve health by restoring the phenotype. Historically, the identification of probiotics for a disease has been based on selecting gut microbial taxa enriched in healthy individuals compared with patients [80] (Fig. 2).

Developing novel probiotics for a disease also requires optimizing growth conditions, which is frequently challenging since bacterial candidates can be strict anaerobes [81]. Microbial growth can be enhanced through cross-feeding strategies, where the target microorganism is grown synergistically with other microorganism, and through strategies that increase microbial oxygen tolerance by creating oxidized conditions during consecutive subculture steps [82].

More recent approaches to probiotic development include the phenotypic strategy, which involves characterizing the effects of microbial candidates on the phenotype and modes of action in in vitro, ex vivo and animal models [80].

In vitro assays are often the first step in probiotic development due to their experimental simplicity and low cost compared with ex vivo and animal models. In the context of HE, in vitro assays for testing the probiotic properties of microorganisms may include the use of hepatocytes or intestinal cell lines. Focusing on the gut, exploring the protective effects of the microbial candidate against intestinal inflammation appears to be an appropriate approach, given that many HE symptoms are caused by an exacerbated proinflammatory tone in the intestine [19–21] (Fig. 2). In vitro assays should be complemented by animal experiments, using mouse or rat models, to assess the ability of the target microorganism to enhance health and confirm its probiotic properties in a specific disease. Depending on the complexity of the gut microbiota, different animal models are available for experimental design, including conventionally raised animals, SPF animals, or GF mice (Fig. 2).

(a) Animals with a complex (non-controlled) gut microbiota: conventionally raised and specific pathogen-free animals

The most widely used animal models to investigate effects of a target microorganism on host phenotype are interventions in conventional animals, which are raised without barrier conditions and have an uncontrolled gut microbiota. However, their use can lead to confounding results due to their unknown pathogen status. Therefore, the use of SPF animals is increasing gradually [83]. To maintain SPF status, standardized protocols specific to each facility are required, which include periodic pathogen monitoring and operational methods [84]. Nevertheless, the use of SPF animals as a health standard for animal experimentation does not guarantee reproducibility, as animal facilities often differ in the list of pathogens allowed to enter [84].

Importantly, the gut microbiota of rodent models is highly dependent on environmental factors, which can limit experimental reproducibility [85]. Therefore, researchers should take measures to eliminate environmental variables that affect the composition of the gut microbiota when designing an interventional experiment in animal models.

The following list includes relevant aspects that aid in ensuring experimental reproducibility using conventional or SPF animals:

1. The diet is an important source of variability when considering the composition of the gut microbiota. It is highly recommended to know the composition of the diet as well as the sterilization method (autoclaving or irradiation) and consistently use the same diet in different experiments.
2. It is also recommended to standardize housing conditions, as the type of cage, bedding and housing density influence gut microbiota [85]. As additional measures to eliminate housing-dependent variables, gut microbiota can be homogenized by mixing bedding (sawdust) containing feces from different cages, and then equally distributing it in each cage. This procedure should be performed once or twice one week before starting the experiment.
3. The composition of the gut microbiota also varies depending on the commercial vendor or repository providing the animals. When conducting experiments on conventionally raised animals, it is advisable to perform a microbiological analysis when purchasing animals, especially if the experiment involves the use of different strains or the use of animals provided by different vendors and/or repositories. The transfer of animals

from one facility to another can also impact the composition of the gut microbiota [85]. This can be mitigated by acclimatizing the animals prior to the experiment.

Experimental Design

Taking into consideration the aforementioned measures, below we provide a basic guide for designing an interventional experiment in animals to explore the effect of microbial candidates, one microorganism or a consortium of microorganisms, in restoring the phenotype of HE.

Protocol

1. Induce or not (controls) HE in animals by a previously selected method [3].
2. Before starting the intervention, collect fresh fecal samples and immediately freeze in liquid nitrogen and store at −80 °C until gut microbiota analysis.
3. Begin the intervention by oral gavage following the indications described in Subheading 3.2.1 (antibiotic cocktail administration).
4. Repeat the process for as many days as the microorganism is expected to induce an effect, frequently between 1 and 12 weeks.
5. Perform the necessary tests for the phenotypic characterization of HE, including behavioral characterization and ammonia blood levels [86–88].
6. Collect fecal samples as indicated in point 2 at the end of the intervention.
7. Sacrifice the animals and collect tissues of interest (intestinal content, small and large intestine, liver and brain) for further biochemical, molecular, and immunological analyses [89, 90].
8. Analyze the gut microbiota in the intestinal content and/or feces using the methods described in Subheading 2.1.1. (optional).

(b) Gut Microbiota-Deficient Animal Models

Here we provide relevant materials and methodological steps to consider when designing an intervention in GF mice in the context of HE. Of note, these experiments are usually costly and time consuming due to the maintenance of the gnotobiotic conditions and routine microbiological controls. Thus, it is recommended to validate the hypothesis beforehand with simpler approaches (i.e., in vitro assays or interventional experiments with conventional or SPF animals).

Gnotobiotic animals are generated from GF animals colonized with a single microorganism (monoxenic animals) or with a consortium of known microorganisms.

Monocolonization demonstrates the causal association of a single gut microorganism with a defined phenotype. Colonization with a consortium of bacterial strains is often performed to explore the impact of microorganism-microorganism interactions, such as cross-feeding interactions, on the phenotype [91]. Furthermore, this approach allows for standardization of the gut microbiota composition of rodent models, which is highly dependent on environmental factors, making it difficult to replicate experiments across different animal facilities [92]. There are two available models for standardizing the mouse gut microbiota, the Altered Schaedler Flora (ASF) model and the Oligo-Mouse-MI^{12} (Oligo-MM12) model, which consist of 8 or 12 bacterial strains, respectively, and are stably transmissible from generation to generation [93, 94].

Protocol

1. Induce HE in animals by a previously selected method [3].
2. Proceed with the colonization of the microbial candidate(s) by oral gavage as described in Subheading 3.2.1 (antibiotic cocktail administration).
3. Repeat the process for as many days as necessary for the microbial candidate(s) to colonize the intestine of the animal.
4. Determine the microbiological status to check colonization.
5. Perform the necessary tests for the phenotypic characterization of HE, including behavioral characterization and ammonia blood measure [86–88].
9. Sacrifice the animals and collect tissues of interest (intestinal content, small and large intestine, liver and brain) for further biochemical, molecular, and immunological analyses [90, 95].

Notes: all the steps described above must be performed under gnotobiotic conditions. Optimal conditions for colonization, such as the number of doses to be administered or the age of GF mice, must be previously determined.

4 Notes

4.1 Analysis of the Gut Microbiota Composition and Function

4.1.1 Identification of Microbial Taxa

(a) DNA extraction

It is important to wear gloves and avoid all skin contact with reagents and materials throughout the protocol. Furthermore, for fecal samples the same method should always be used to isolate DNA from a Microbial Community Standard. Specifically, we use the ZymoBIOMICS Microbial Community Standard, a commercially available standard mimicking the human gut microbiome for metagenomics studies. This microbial standard is a well-defined, well-characterized mock community consisting of Gram-negative and Gram-positive bacteria and yeasts of different sizes and cell wall composition. It is used as a defined input to assess the performance of metagenomic workflows by functioning as a positive control for sample processing. We also always include several negative (environmental) controls in each extraction and library preparation batch to identify potential contamination, bias in DNA extraction, and to optimize all steps performed during and after DNA extraction.

Of note, the extracted DNA can also be used for whole-genome shotgun metagenomic sequencing, which allows the analysis not only of the bacterial fraction of the gut microbiota using the 16S RNA amplicon sequencing approach (the protocol we describe in detail here), but also of all other microbial fractions of the gut microbiome. In whole-genome shotgun sequencing, the DNA is amplified using random primers, subcloned, and then sequenced to produce a DNA library representative of the microbial populations. The vast number of reads generated by whole-genome shotgun metagenomic sequencing can be analyzed to identify the species represented by the sequences and also the functional capacity of the community. In this case, after whole-genome shotgun metagenomic sequencing, quality trimming, removal of over-represented sequences, and read mapping to the human reference genome and removal can be performed using KneadData (v0.5.1). Taxonomic profiling of the total microbiome reads can be performed using MetaPhlAn2.0, and the abundance of microbial metabolic pathways determined using HUMAnN2 (http://huttenhower.sph.harvard.edu/humann2), which maps reads to a customized database of functionally annotated pan-genomes.

(b) Purification, library preparation, amplicon sequencing, and bioinformatic processing notes

As a general best practice, all PCR reagents and reactions should be kept on ice until PCR; the high proofreading activity of the enzyme will rapidly degrade primers at room temperature (this is generally true for all high-fidelity polymerases). Also, use high-quality DNA and work in a PCR-clean environment to avoid contamination. Normalize the amount of DNA used per PCR well. Generating similar amounts of 16S rRNA gene amplification per well will help to balance the amplicon pool for library preparation. We often use the barcoded primers S-D-Bact-0341-b-S-17 (CCTACGGGNGGCWGCAG) and S-D-Bact-0785-a-A-21 (GACTACHVGGGTATCTAATCC), which target and allow the detection of a wide range of bacteria. However, other barcoded primers can be used. It is important to emphasize that the choice of primers used to target the microbiome is critical to the results. In this sense, if you wish to compare the results with other previously published results or expected results from another research group, you must use the same primers to target the 16S rRNA gene, in addition to the same pipeline parameters for bioinformatics analyses, including software, reference databases, and alignment protocols. In this regard, it is important to note the enormous amount of data generated from microbiome research studies over the past decades and the high variability of published results and conclusions from these studies. This highlights the need to establish common standards for sampling and metadata harmonization to ensure data re-use for integrative and interdisciplinary microbiome analysis. Of note, to specifically investigate which species are present in a sample, the full-length bacterial 16S gene has to be amplified and the PCR product should show an amplicon size of ~1500 bp. In this case, you will need to use the Agilent Bioanalyzer System to check the size and quantity of the PCR product and send it for sequencing. After this, the sequences need to be BLASTed against the NCBI website to determine the species (cut-off value for the Query coverage is commonly set at 90% and samples with a Query coverage lower than 90% are typically not considered identified). Additionally, you can perform a multiplex PCR, combining primer sets that have approximately the same annealing temperature, to assess the presence of multiple species in the stool samples. It is important to note that even with these types of sequencing techniques, it may not be possible to determine the bacterial species and you will also need to sequence genes other that 16S rRNA to discriminate the target species.

4.1.2 Identification of Microbiota-Derived Metabolites: Analysis of SCFAs

HCl acidifies the samples, ensuring the protonation of SCFAs before analysis, which prevents the occurrence of split peaks in chromatograms [96]. Helium is employed as the carrier gas at an initial flow rate of 1.3 mL/min. The initial oven temperature should be set at 50 °C and maintained for 30 s, then raised to 140 °C at 10 °C/min and held for 30 s. The temperature should finally be increased to 240 °C at 20 °C/min and held for 5 min. The temperatures of the detector and the injection port should be 300 °C and 240 °C, respectively. A split-less injection of 0.2 μL is performed for each sample or standard using a 10 μL syringe (Agilent) installed on a CP-8400 auto-sampler (Varian). The SCFAs are identified on chromatograms by their specific retention times under the above conditions. Peak integration is carried out using Varian Star Chromatography Workstation version 6.0 software. Vials containing 1800 μL of water are run between each sample duplicates as blanks to prevent potential carryover and standards are included in each run to maintain the calibration. Using the chromatograms, the peak area is plotted against the concentration of the SCFA to obtain the calibration graph for each SCFA.

4.2 Animal Models to Study Causality and the Relationships Between the Gut Microbiota and Hepatic Encephalopathy Notes

4.2.1 Fecal Microbiota Transplant Notes

The timing of stool sampling is critical, as findings suggest high variability in gut microbiota abundance based on circadian rhythms. Some studies recommend collecting fecal samples from rodents either in the early morning, specifically between 7:00 AM and 11:00 AM [97], or in the afternoon, preferably between 3:00 PM and 5:00 PM, to minimize potential influences of circadian rhythms [98]. If the stool is not handled properly, it may result in a reduction of the viability of many anaerobic bacteria, potentially impacting the effectiveness of FMT. Therefore, the samples must be handled quickly. It is important to try to collect, transport, and process the feces under anaerobic conditions to avoid loss of anaerobic microbiota. Animal weight loss at the start of the antibiotic treatment is a common side effect, often related to the unpalatable taste of the antibiotic cocktail. All groups must receive fresh bedding every other day to minimize the chance of microbiota re-establishment in the antibiotic-treated animals. The age of the recipient is also important for the uptake of donor bacteria. Successful engraftment of donor microbiota has been found to be more pronounced in 3-week-old rodents than in 8-week-old rodents. FMT tends to be more effective when the recipient microbiota has a lower richness [67]. The gut microbiota is typically less diverse during the weaning period, likely due to the dietary transition from maternal milk to solid food. Despite the common use of oral gavage for FMT and faster fecal transfer, challenges associated with this technique include unintentional tracheal administration, bronchopneumonia, esophageal injury, gastric rupture, and weight loss [99–101].

The use of germ-free (GF) animals, often mice, is a highly recommended alternative to gut microbiota-depleted animals for FMT. GF animals completely lack microorganisms, and therefore need to be housed under the strictest biological barrier in animal facilities [102] that allow gnotobiotic conditions. Isolators are used for long-term maintenance of GF animals, while phenotypic characterization experiments are performed in simpler gnotobiotic systems [103]. Importantly, experienced staff following strict operating procedures are needed to maintain the sterility and the microbiological status of GF animals. Operating procedures for sterilization and handling are specific to each gnotobiotic facility [104].

5 Conclusions and Future Perspectives

The gut microbiome has recently emerged as an important potential modifiable contributor to HE. However, the precise gut microbiota changes causally associated with HE are uncertain and much remains to be understood about the potential functional consequences of HE-associated microbial changes on the disease. In relation to this, it is still unclear to what extent and how exactly gut microbiota alterations can be modified to provide benefits in the prevention and/or treatment of the disorder.

In this chapter, we have provided details of the most pertinent techniques, methods, and recommendations currently available, which may be crucial for the identification of new tailored microbiome-based strategies that could help in the prevention or treatment of HE. We would like to emphasize the importance of methodological harmonization between studies and on the need to carry out studies that could allow causality to be established in microbiome research. Furthermore, we highlight the interest for further studies based on multi-omics integrated analysis of microbiota-gut-brain axis parameters to elucidate the disease mechanisms related to gut microbiota in the context of HE. Finally, we also highlight the need for the development of precision models based on specific individual characteristics and metadata, including the microbiome profile, which would allow the classification of subjects with HE according to their predicted clinical response to potential microbiome-based interventions.

References

1. D'Amico G et al (1986) Survival and prognostic indicators in compensated and decompensated cirrhosis. Dig Dis Sci 31(5): 468–475
2. Ding A, Lee A, Callender M, Loughrey M, Quah SP, Dinsmore WW (2010) Hepatic encephalopathy as an unusual late complication of transjugular intrahepatic portosystemic shunt insertion for non-cirrhotic portal hypertension caused by nodular regenerative hyperplasia in an HIV-positive patient on highly active antiretroviral therapy. Int J STD AIDS 21(1):71–72
3. DeMorrow S et al (2021) 2021 ISHEN guidelines on animal models of hepatic encephalopathy. Liver Int 41(7):1474–1488
4. Butterworth RF (2003) Hepatic encephalopathy. Alcohol Res Health 27(3):240–246
5. Cornberg MTF, Karlsen TH (2010) European Association for the Study of the liver. Clinical practice guidelines of the European Association for the study of the liver—advancing methodology but preserving practicability. J Hepatol 70(1):5–7
6. Quigley EM, Stanton C, Murphy EF (2013) The gut microbiota and the liver, pathophysiological and clinical implications. J Hepatol 58(5):1020–1027
7. Sturgeon JA et al (2015) Physical and psychological correlates of fatigue and physical function: a collaborative health outcomes information registry (CHOIR) study. J Pain 16(3):291–8 e1
8. Guarner F, Malagelada JR (2003) Gut flora in health and disease. Lancet 361(9356): 512–519
9. Gill SR et al (2006) Metagenomic analysis of the human distal gut microbiome. Science 312(5778):1355–1359
10. Bäckhed F et al (2005) Host-bacterial mutualism in the human intestine. Science 307(5717):1915–1920
11. Topping DL, Clifton PM (2001) Short-chain fatty acids and human colonic function: roles of resistant starch and nonstarch polysaccharides. Physiol Rev 81(3):1031–1064
12. Cenit MC, Sanz Y, Codoner-Franch P (2017) Influence of gut microbiota on neuropsychiatric disorders. World J Gastroenterol 23(30): 5486–5498
13. Singh RK et al (2017) Influence of diet on the gut microbiome and implications for human health. J Transl Med 15(1):73
14. Fukui H, Wiest R (2016) Changes of intestinal functions in liver cirrhosis. Inflamm Intest Dis 1(1):24–40
15. Rhee SH, Pothoulakis C, Mayer EA (2009) Principles and clinical implications of the brain-gut-enteric microbiota axis. Nat Rev Gastroenterol Hepatol 6(5):306–314

16. Agusti A et al (2018) Bifidobacterium pseudocatenulatum CECT 7765 ameliorates neuroendocrine alterations associated with an exaggerated stress response and anhedonia in obese mice. Mol Neurobiol 55(6): 5337–5352
17. Wiest R, Albillos A, Trauner M, Bajaj JS, Jalan R (2017) Targeting the gut-liver axis in liver disease. J Hepatol 67(5):1084–1103
18. Mancini A et al (2018) Gut: liver: brain axis: the microbial challenge in the hepatic encephalopathy. Food Funct 9(3):1373–1388
19. Azhari H, Swain MG (2018) Role of peripheral inflammation in hepatic encephalopathy. J Clin Exp Hepatol 8(3):281–285
20. Manzhalii E, Virchenko O, Falalyeyeva T, Moiseienko V, Nykula T, Kondratiuk V, Savchuk O, Beregova T, Stremmel W (2019) Hepatic encephalopathy aggravated by systemic inflammation. Dig Dis 37(6):509–517
21. Jaffe A, Lim JK, Jakab SS (2020) Pathophysiology of hepatic encephalopathy. Clin Liver Dis 24(2):175–188
22. Arab JP, Martin-Mateos RM, Shah VH (2018) Gut-liver axis, cirrhosis and portal hypertension: the chicken and the egg. Hepatol Int 12(Suppl 1):24–33
23. Aguirre Valadez JM, Rivera-Espinosa L, Méndez-Guerrero O, Chávez-Pacheco JL, García Juárez I, Torre A (2016) Intestinal permeability in a patient with liver cirrhosis. Ther Clin Risk Manag 12:1729–1748
24. Milosevic I, Vujovic A, Barac A, Djelic M, Korac M, Radovanovic Spurnic A, Gmizic I, Stevanovic O, Djordjevic V, Lekic N, Russo E, Amedei A (2019) Gut-liver axis, gut microbiota, and its modulation in the management of liver diseases: a review of the literature. Int J Mol Sci 20(2):395
25. Trebicka J et al (2011) Endotoxin and tumor necrosis factor-receptor levels in portal and hepatic vein of patients with alcoholic liver cirrhosis receiving elective transjugular intrahepatic portosystemic shunt. Eur J Gastroenterol Hepatol 23(12):1218–1225
26. Liu J, Yantao X, Jiang B (2021) Novel insights into pathogenesis and therapeutic strategies of hepatic encephalopathy, from the gut microbiota perspective. Front Cell Infect Microbiol 11:586427
27. Fiuza C et al (2000) In vivo neutrophil dysfunction in cirrhotic patients with advanced liver disease. J Infect Dis 182(2):526–533
28. Katz S et al (1991) Liver bacterial clearance following hepatic artery ligation and portacaval shunt. J Surg Res 51(3):267–270
29. Sipeki N et al (2014) Immune dysfunction in cirrhosis. World J Gastroenterol 20(10): 2564–2577
30. Betrapally NS, Gillevet PM, Bajaj JS (2017) Gut microbiome and liver disease. Transl Res 179:49–59
31. Guerit JM et al (2009) Neurophysiological investigations of hepatic encephalopathy: ISHEN practice guidelines. Liver Int 29(6): 789–796
32. Gimenez-Garzó C, Urios A, Agustí A, González-López O, Escudero-García D, Escudero-Sanchis A, Serra MA, Giner-Durán R, Montoliu C, Felipo V (2015) Is cognitive impairment in cirrhotic patients due to increased peroxynitrite and oxidative stress? Antioxid Redox Signal 22(10): 871–877
33. Montagnese S, Amodio P, Morgan MY (2004) Methods for diagnosing hepatic encephalopathy in patients with cirrhosis: a multidimensional approach. Metab Brain Dis 19(3–4):281–312
34. Sung CM et al (2019) Predicting clinical outcomes of cirrhosis patients with hepatic encephalopathy from the fecal microbiome. Cell Mol Gastroenterol Hepatol 8(2): 301–318 e2
35. Zhang Z, Zhai H, Geng J, Yu R, Ren H, Fan H, Shi P (2013) Large-scale survey of gut microbiota associated with MHE via 16S rRNA-based pyrosequencing. Am J Gastroenterol 108:1601–1611
36. Bajaj JS, Hylemon PB, Ridlon JM, Heuman DM, Daita K, White MB, Monteith P, Noble NA, Sikaroodi M, Gillevet PM (2012a) Colonic mucosal microbiome differs from stool microbiome in cirrhosis and hepatic encephalopathy and is linked to cognition and inflammation. Am J Physiol Gastrointest Liver Physiol 303:G675–G685
37. Bajaj JS et al (2012b) Linkage of gut microbiome with cognition in hepatic encephalopathy. Am J Physiol Gastrointest Liver Physiol 302(1):G168–G175
38. Won SM, OK, Gupta H, Ganesan R, Sharma SP, Jeong JJ, Yoon SJ, Jeong MK, Min BH, Hyun JY, Park HJ, Eom JA, Lee SB, Cha MG, Kwon GH, Choi MR, Kim DJ, Suk KT (2022) The link between gut microbiota and hepatic encephalopathy. Int J Mol Sci 23(16): 8999
39. Berry SE et al (2020) Human postprandial responses to food and potential for precision nutrition. Nat Med 26(6):964–973

40. Wu Y, Perng W, Peterson KE (2020) Precision nutrition and childhood obesity: a scoping review. Meta 10(6)
41. Demetrowitsch TJ et al (2020) Precision nutrition in chronic inflammation. Front Immunol 11:587895
42. Bajaj JS (2010) Review article: the modern management of hepatic encephalopathy. Aliment Pharmacol Ther 31(5):537–547
43. Malaguarnera M et al (2010) Bifidobacterium combined with fructo-oligosaccharide versus lactulose in the treatment of patients with hepatic encephalopathy. Eur J Gastroenterol Hepatol 22(2):199–206
44. Bajaj JS et al (2014) Randomised clinical trial: lactobacillus GG modulates gut microbiome, metabolome and endotoxemia in patients with cirrhosis. Aliment Pharmacol Ther 39(10):1113–1125
45. Xia X et al (2018) Role of probiotics in the treatment of minimal hepatic encephalopathy in patients with HBV-induced liver cirrhosis. J Int Med Res 46(9):3596–3604
46. Kao D et al (2016) Fecal microbiota transplantation in the management of hepatic encephalopathy. Hepatology 63(1):339–340
47. Bloom PP et al (2022) Fecal microbiota transplant improves cognition in hepatic encephalopathy and its effect varies by donor and recipient. Hepatol Commun 6(8): 2079–2089
48. Caporaso JG et al (2010) QIIME allows analysis of high-throughput community sequencing data. Nat Methods 7(5):335–336
49. Schloss PD et al (2009) Introducing mothur: open-source, platform-independent, community-supported software for describing and comparing microbial communities. Appl Environ Microbiol 75(23):7537–7541
50. Edgar RC (2013) UPARSE: highly accurate OTU sequences from microbial amplicon reads. Nat Methods 10(10):996–998
51. Berry D et al (2012) Phylotype-level 16S rRNA analysis reveals new bacterial indicators of health state in acute murine colitis. ISME J 6(11):2091–2106
52. Sarrabayrouse G et al (2020) Mucosal microbial load in Crohn's disease: a potential predictor of response to faecal microbiota transplantation. EBioMedicine 51:102611
53. Bokoliya SC et al (2021) Procedures for fecal microbiota transplantation in murine microbiome studies. Front Cell Infect Microbiol 11:711055
54. O'Connor KM et al (2019) Manipulation of gut microbiota blunts the ventilatory response to hypercapnia in adult rats. EBioMedicine 44:618–638
55. Wu YY et al (2014) Toll-like receptor stimulation induces nondefensin protein expression and reverses antibiotic-induced gut defense impairment. Infect Immun 82(5):1994–2005
56. Kelly CJ, ZL, Campbell EL, Saeedi B, Scholz CC, Bayless AJ, Wilson KE, Glover LE, Kominsky DJ, Magnuson A, Weir TL, Ehrentraut SF, Pickel C, Kuhn KA, Lanis JM, Nguyen V, Taylor CT, Colgan SP (2015) Crosstalk between microbiota-derived short-chain fatty acids and intestinal epithelial HIF augments tissue barrier function. Cell Host Microbe 17(5):662–671
57. Grasa L et al (2015) Antibiotic-induced depletion of murine microbiota induces mild inflammation and changes in toll-like receptor patterns and intestinal motility. Microb Ecol 70(3):835–848
58. Stefka AT et al (2014) Commensal bacteria protect against food allergen sensitization. Proc Natl Acad Sci USA 111(36): 13145–13150
59. Kelly JR et al (2016) Transferring the blues: depression-associated gut microbiota induces neurobehavioural changes in the rat. J Psychiatr Res 82:109–118
60. Hoban AE et al (2017) Corrigendum to "Behavioural and neurochemical consequences of chronic gut microbiota depletion during adulthood in the rat" [Neuroscience 339 (2016) 463–477]. Neuroscience 344: 418
61. Jalanka J et al (2015) Effects of bowel cleansing on the intestinal microbiota. Gut 64(10): 1562–1568
62. Ji SK et al (2017) Preparing the gut with antibiotics enhances gut microbiota reprogramming efficiency by promoting xenomicrobiota colonization. Front Microbiol 8: 1208
63. Wang WW et al (2017) Fecal microbiota transplantation prevents hepatic encephalopathy in rats with carbon tetrachloride-induced acute hepatic dysfunction. World J Gastroenterol 23(38):6983–6994
64. Kang DJ, Hylemon PB, Bajaj JS (2015) Fecal transplant to mitigate hyperammonemia and hepatic encephalopathy in animal models. Ann Hepatol 14(5):762–763
65. Bajaj JS et al (2017) Fecal microbiota transplant from a rational stool donor improves hepatic encephalopathy: a randomized clinical trial. Hepatology 66(6):1727–1738
66. Bajaj JS et al (2019) Fecal microbial transplant capsules are safe in hepatic encephalopathy: A

phase 1, randomized, placebo-controlled trial. Hepatology 70(5):1690–1703
67. Ericsson AC, PA, Turner G, Dorfmeyer RA, Franklin CL (2017) Variable colonization after reciprocal fecal microbiota transfer between mice with low and high richness microbiota. Front Microbiol 8:196
68. Sergeant MJ et al (2012) High-throughput sequencing of 16S rRNA gene amplicons: effects of extraction procedure, primer length and annealing temperature. PLoS One 7(5): e38094
69. Shaw AG et al (2016) Latitude in sample handling and storage for infant faecal microbiota studies: the elephant in the room? Microbiome 4(1):40
70. Cammarota G, Ianiro G, Gasbarrini A (2014) Fecal microbiota transplantation for the treatment of Clostridium difficile infection: a systematic review. J Clin Gastroenterol 48(8): 693–702
71. Bellali S, LJ, Raoult D, Bou Khalil J (2019) Among live and dead bacteria, the optimization of sample collection and processing remains essential in recovering gut microbiota components. Front Microbiol 10:1606
72. Brusa T, Canzi E, Pacini N, Zanchi R, Ferrari A (1989) Oxygen tolerance of anaerobic bacteria isolated from human feces. Curr Microbiol 19(1):39–43
73. Hintze KJ, CJ, Rompato G, Benninghoff AD, Ward RE, Broadbent J, Lefevre M (2014) Broad scope method for creating humanized animal models for animal health and disease research through antibiotic treatment and human fecal transfer. Gut Microbes 5(2): 183–191
74. Wrzosek L, CD, Borentain P, Spatz M, Puchois V, Hugot C, Ferrere G, Mayeur C, Perlemuter G, Cassard AM (2018) Transplantation of human microbiota into conventional mice durably reshapes the gut microbiota. Sci Rep 8(1):6854
75. Sivan A et al (2015) Commensal Bifidobacterium promotes antitumor immunity and facilitates anti-PD-L1 efficacy. Science 350(6264):1084–1089
76. Lin DM et al (2019) Transplanting fecal virus-like particles reduces high-fat diet-induced small intestinal bacterial overgrowth in mice. Front Cell Infect Microbiol 9:348
77. Liu J et al (2020) Fecal microbiota transplantation by enema reduces intestinal injury in experimental necrotizing enterocolitis. J Pediatr Surg 55(6):1094–1098
78. Staley C, KT, Beura LK, Hamilton MJ, Weingarden AR, Bobr A, Kang J, Masopust D, Sadowsky MJ, Khoruts A (2017) Stable engraftment of human microbiota into mice with a single oral gavage following antibiotic conditioning. Microbiome 5(1):87
79. Clemente JC et al (2012) The impact of the gut microbiota on human health: an integrative view. Cell 148(6):1258–1270
80. Veiga P et al (2020) Moving from probiotics to precision probiotics. Nat Microbiol 5(7): 878–880
81. O'Toole PW, Marchesi JR, Hill C (2017) Next-generation probiotics: the spectrum from probiotics to live biotherapeutics. Nat Microbiol 2:17057
82. Khan MT et al (2023) Synergy and oxygen adaptation for development of next-generation probiotics. Nature 620(7973): 381–385
83. Convenor M, Berard M, Feinstein R, Gallagher A, Illgen-Wilcke B, Pritchett-Corning K, Raspa M (2014) FELASA recommendations for the health monitoring of mouse, rat, hamster, Guinea pig and rabbit colonies in breeding and experimental units. Lab Anim 48(3):178–192
84. Luchins K, Langan G (2018) Chapter 33 managing animal colony health. In: Weichbrod RH, Thompson GA, Norton JN (eds) Management of animal care and use programs in research, education, and testing. Boca Raton (FL)
85. Franklin CL, Ericsson AC (2017) Microbiota and reproducibility of rodent models. Lab Anim (NY) 46(4):114–122
86. Agusti A, CO, Rodrigo R, Llansola M, Hernández-Rabaza V, Felipo V (2011) p38 MAP kinase is a therapeutic target for hepatic encephalopathy in rats with portacaval shunts. Gut 60(11):1572–1579
87. Hernandez-Rabaza V et al (2016) Neuroinflammation increases GABAergic tone and impairs cognitive and motor function in hyperammonemia by increasing GAT-3 membrane expression. Reversal by sulforaphane by promoting M2 polarization of microglia. J Neuroinflammation 13(1):83
88. Agusti A, H-RV, Balzano T, Taoro-Gonzalez-L, Ibañez-Grau A, Cabrera-Pastor A, Fustero S, Llansola M, Montoliu C, Felipo V (2017) Sildenafil reduces neuroinflammation in cerebellum, restores GABAergic tone, and improves motor in-coordination in rats with hepatic encephalopathy. CNS Neurosci Ther 23(5):386–394
89. Dadsetan S et al (2016) Reducing peripheral inflammation with infliximab reduces neuroinflammation and improves cognition in

rats with hepatic encephalopathy. Front Mol Neurosci 9:106

90. Hernandez-Rabaza V et al (2015) Sildenafil reduces neuroinflammation and restores spatial learning in rats with hepatic encephalopathy: underlying mechanisms. J Neuroinflammation 12:195
91. Olson CA et al (2018) The gut microbiota mediates the anti-seizure effects of the ketogenic diet. Cell 173(7):1728–1741 e13
92. Macpherson AJ, McCoy KD (2015) Standardised animal models of host microbial mutualism. Mucosal Immunol 8(3):476–486
93. Wymore Brand M et al (2015) The altered Schaedler Flora: continued applications of a defined murine microbial community. ILAR J 56(2):169–178
94. Brugiroux S et al (2016) Genome-guided design of a defined mouse microbiota that confers colonization resistance against salmonella enterica serovar typhimurium. Nat Microbiol 2:16215
95. Dadsetan S et al (2016b) Infliximab reduces peripheral inflammation, neuroinflammation, and extracellular GABA in the cerebellum and improves learning and motor coordination in rats with hepatic encephalopathy. J Neuroinflammation 13(1):245
96. Zhao G, Nyman M, Åke Jönsson J (2006) Rapid determination of short-chain fatty acids in colonic contents and faeces of humans and rats by acidified water-extraction and direct-injection gas chromatography. Biomed Chromatogr 20:674–682. https://doi.org/10.1002/bmc.580
97. Ericsson AC et al (2018) The influence of caging, bedding, and diet on the composition of the microbiota in different regions of the mouse gut. Sci Rep 8(1):4065
98. Wang S et al (2018) Gut microbiota mediates the anti-obesity effect of calorie restriction in mice. Sci Rep 8(1):13037
99. de Meijer VE et al (2010) Repetitive orogastric gavage affects the phenotype of diet-induced obese mice. Physiol Behav 100(4):387–393
100. Arantes-Rodrigues R et al (2012) The effects of repeated oral gavage on the health of male CD-1 mice. Lab Anim (NY) 41(5):129–134
101. Kinder JM et al (2014) Long-term repeated daily use of intragastric gavage hinders induction of oral tolerance to ovalbumin in mice. Comp Med 64(5):369–376
102. A.B.M.A.B (2018) The gut microbiome in health and disease. Chapter: Gnotobiology. Vol. First Online: 28 July 2018
103. Hecht G et al (2014) A simple cage-autonomous method for the maintenance of the barrier status of germ-free mice during experimentation. Lab Anim 48(4):292–297
104. Boyd C, Eaton KA, Schoeb TR, Swennes AG, Bickerton H, Parsons BJ, Sinclair S, Vowles C (2017) Gnotobiotics. Chapter 3—Procedures for Rodents

Chapter 11

Specialized Established Techniques for Assessing Multiorgan Damages at Different Levels of the Gut-Liver-Brain Axis in Experimental Animal Models of Hepatic Encephalopathy

Ali Sepehrinezhad and Ali Shahbazi

Abstract

Hepatic encephalopathy (HE) is a brain disorder caused by liver failure or chronic liver diseases, and its pathophysiology is currently being debated. The majority of the clues acquired about the pathophysiology of HE come from various preclinical and animal studies. HE is a striking example of a disorder that affects the vital organs of the gut-liver-brain axis. This chapter introduces animal models of HE and describes some of the most applied techniques for investigating pathological alterations in the key implicated organs (brain, liver, and intestines) in these models. Researchers with a background in cirrhosis, liver failure, cerebral edema, or HE may skip this chapter.

Key words Hepatic encephalopathy, Gut-liver-brain axis, Blood-brain barrier permeabilization assay, Electron microscopy, Liver function tests, Intestinal permeability, Immunohistochemistry

1 Introduction

Hepatic encephalopathy (HE) is identified as a spectrum of neurological complications that arise from advanced liver diseases and/or acute liver failure or portosystemic shunt. HE has a poor prognosis and is still considered a life-threatening condition in patients with liver disease especially those with acute liver failure and coma [1, 2]. Furthermore, the most recently accessible medicines simply manage the disease rather than treat it, and the only available treatment is liver transplantation, which, due to the difficulties of identifying donor tissue, makes this last treatment difficult to obtain [3, 4]. The disease's pathogenesis is extremely complex, and studies have emphasized the significance of ammonia, systemic inflammation, and oxidative stress in its pathophysiology. Despite substantial research, the disease's mechanisms remain poorly

Tiziano Balzano (ed.), *Experimental and Clinical Methods in Hepatic Encephalopathy Research*, Neuromethods, vol. 219,
https://doi.org/10.1007/978-1-0716-4354-9_11,

understood [5, 6]. HE is a good example of a condition that affects the gut-liver-brain axis on multiple levels [7]. As the literature discussed, cirrhosis induces gut dysbiosis, increases gut-derived ammonia products, disrupts intestinal barriers, and promotes bacterial translocation. These altered metabolites reach the injured liver, which is unable to detoxify them, and so enter the systemic circulation. In liver tissue, these metabolites activated resident macrophages, satellite cells, and Kupffer cells, causing inflammation. The portal-systemic shunt allows these altered metabolites to access the brain parenchyma via a permeable BBB, a barrier whose intact structure is vital to preserving. In the brain tissue, these neurotoxic agents trigger glial cells in particular microglia and astrocytes to induce neuroinflammation and neuronal injuries [8–10]. A large portion of the insights now available on the pathogenesis of the disease is derived from extensive pre-clinical research conducted on HE-affected animal models. Therefore, understanding common and particular methods for examining all levels of the gut-liver-brain axis in animal models of HE is critical in pre-clinical research. This chapter describes in detail some of the most essential methods for investigating the pathological alterations of this critical axis in animal models of HE.

1.1 A Brief Introduction of Animal Models for Study of HE

The primary focus in animal models of HE is liver injury, which can be acute or chronic. The primary key factor of the disease, hyperammonemia, should be able to be induced in the most effective animal model of HE. Naturally, these animal models should also be able to induce systemic inflammation, hepatic inflammation, oxidative stress, and neuroinflammation because the pathophysiology of HE is multifactorial [11]. There are a number of methods to induce an animal model of HE, including by administering chemical compounds like thioacetamide, azoxymethane, acetaminophen, galactosamine, and carbon tetrachloride, as well as surgical techniques like hepatic artery ligation, portacaval anastomosis, and bile duct ligation. A particular kind of HE is induced by these models [12] (Table 1).

1.2 Assessment of Neurological Score and the Stage of HE

Patients with HE may experience a wide range of neurological and psychiatric symptoms such as bradykinesias, slurred speech, reduced alertness, confusion, personality changes, shortened attention span, ataxia, asterixis, nystagmus, decreased short-term memory, depressive, and anxiety. According to West Haven Criteria, HE is classified into five grades from neuropsychological alterations without perceptible mental changes and lack of awareness and euphoria in minimal HE and grad I to coma stage in grade IV. Assessment of the grade of HE is a key indicator for determining the induction of animal models of HE. A series of clinical examinations are commonly used to determine the grade of HE (Table 2; Fig. 1).

Table 1
Methods for induction of different types of HE

Species	Method	Reference
Type A		
Mice	Azoxymethane- A single intraperitoneal injection (100 mg/kg)	[33]
Mice	Thioacetamide-three consecutive days- intraperitoneal injections (600 mg/kg)	[36]
Rats	Thioacetamide- three consecutive days- intraperitoneal injections (200–300 mg/kg)	[90–92]
Mice	Acetaminophen- A single intraperitoneal injection (300 mg/kg)	[93, 94]
Rats	Acetaminophen- A single intraperitoneal injection (500–800 mg/kg)	[95, 96]
Pigs	Acetaminophen- 250 mg/kg through intestinal catheter +1000–3000 mg every hour for maintenance	[97, 98]
Rats	Galactosamine- three consecutive days- intraperitoneal injections (1.5 g/kg)	[99]
Pigs	Galactosamine- A bolus infusion (0.75–1.5 g/kg)	[100, 101]
Type B		
Rats	Portacaval anastomosis- laparotomy, end-to-side anastomosis portal vein with inferior vena cava	[102–104]
Pigs	Portacaval anastomosis- laparotomy, end-to-side anastomosis portal vein with inferior vena cava	[7]
Type C		
Mice	Carbon tetrachloride- intraperitoneal injection- twice a week for two months- (50% solution of CCL_4 in olive oil; 1 mL/kg)	[105]
Rats	Carbon tetrachloride-intraperitoneal injection- three times per week for two months- (0.4 g/kg in mineral oil)	[106]
Rats	Thioacetamide- intraperitoneal injection- twice per week for three months- (200 mg/kg)	[107]
Mice	Bile duct ligation- laparotomy, common bile duct ligate using silk sutures one at the liver hilum, one near the duodenum, and resect the bile duct between two sutures.	[108, 109]
Rats	Bile duct ligation- laparotomy, common bile duct ligate using silk sutures one at the liver hilum, one near the duodenum, and resect the bile duct between two sutures.	[47, 110–112]

2 Material and Methods

Under physiological conditions, the brain microenvironment is protected from neurotoxic circulating agents by the blood-brain barrier (BBB), a highly selective and semi-permeable barrier [13]. In HE, the BBB breaks down, which causes the brain's water regulation mechanisms to become disrupted. Additionally, a

Table 2
Two samples of behavioral checklists for determining the stage of encephalopathy and clinical evaluations in animal models of acute HE

Sample 1					
Stage of encephalopathy	**Definition**		**Day1**	**Day2**	**Day3**
Stage I	Lethargy (generalized reduction in spontaneous activity)		□□	□□	□□
Stage II	Mild ataxia		□□	□□	□□
Stage III	Lack of spontaneous movement but with intact righting reflex		□□	□□	□□
Stage IV	Loss of righting reflex but intact pain reflex (measured by reaction to tail pinch)		□□	□□	□□
Stage V	Coma (Unresponsive/No eyelash reflex)		□□	□□	□□
Sample 2					
Parameters	**Neurological score**		**Day1**	**Day2**	**Day3**
Pinna reflex	Intact □ ②	Absent □ ⓪			
Corneal reflex	Intact □ ②	Absent □ ⓪			
Tail flexion	Intact □ ②	Absent □ ⓪			
Escape response	Intact □ ②	Absent □ ⓪			
Righting reflex	Intact □ ②	Absent □ ⓪			
Ataxia	Intact □ ②	Absent □ ⓪			
Total neurological score (Min score = 0 Max score = 12)					

All clinical examinations and neurological scores should be done by an investigator blind to the groups

2.1 Blood-Brain Barrier Permeabilization Assay

2.1.1 Background

variety of harmful compounds and metabolites (such as lipopolysaccharide, pro-inflammatory cytokines, ammonia, and aberrant gut-derived metabolites) enter the brain parenchyma, causing hyperammonemic conditions, and neuroinflammation [14–16]. One typical technique for determining the permeability of the BBB is the measurement of extravasation of Evans blue dye. This technique is based on the dye's molecular weight and its capacity to bind to plasma proteins, particularly albumin, which can do so reversibly [17]. The intact BBB prevents molecules with a molecular weight greater than 180 Da from passing through. Evans Blue (960 Da molecular weight) binds to the albumin macromolecule to generate a massive protein complex (68,500 Da). This complex will be detected in regions of the brain where the blood-brain barrier has compromised [17–19].

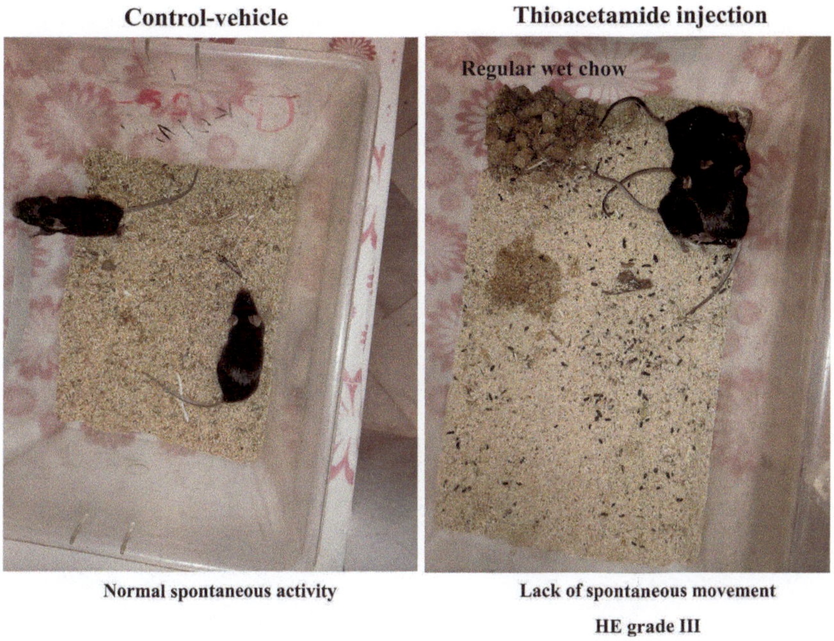

Fig. 1 Grade of encephalopathy progressed to stage III 72 h after injection of thioacetamide in mice

2.1.2 Method

1. Prepare Evans blue 2% solution in normal saline (37 °C) (*see* **Note 1**).
2. Weigh the animals to determine the appropriate anesthetic and injection dye dosage.
3. Sedate the animal with 10% ketamine (80 mg/kg) and 2% xylazine (10 mg/kg) intraperitoneally.
4. Make an oblique incision in the groin to expose the femoral vein in rats (*see* **Note 2**).
5. Inject the dye solution corresponding to the animal's weight gently using a 32-gauge needle (4 mL/kg).
6. Immediately after injecting the dye, set a timer for 30 min and allow Evans blue to circulate throughout the body. The tips of the nose and ears will become blue as soon as the dye is injected, and the skin will follow suit a few minutes later (Fig. 2a).
7. Put the animal back in its home cage and use a rectal thermometer to check its body temperature.
8. Using small surgical devices, open the thoracic cavity to expose the heart.
9. Perfuse animals transcardially with ice-cold PBS (50 mL for mice and 200 mL for rats). The right atrium is pinched and a blunted needle connected to the manual perfusion system or perfusion pump is inserted into the left ventricle (maintain

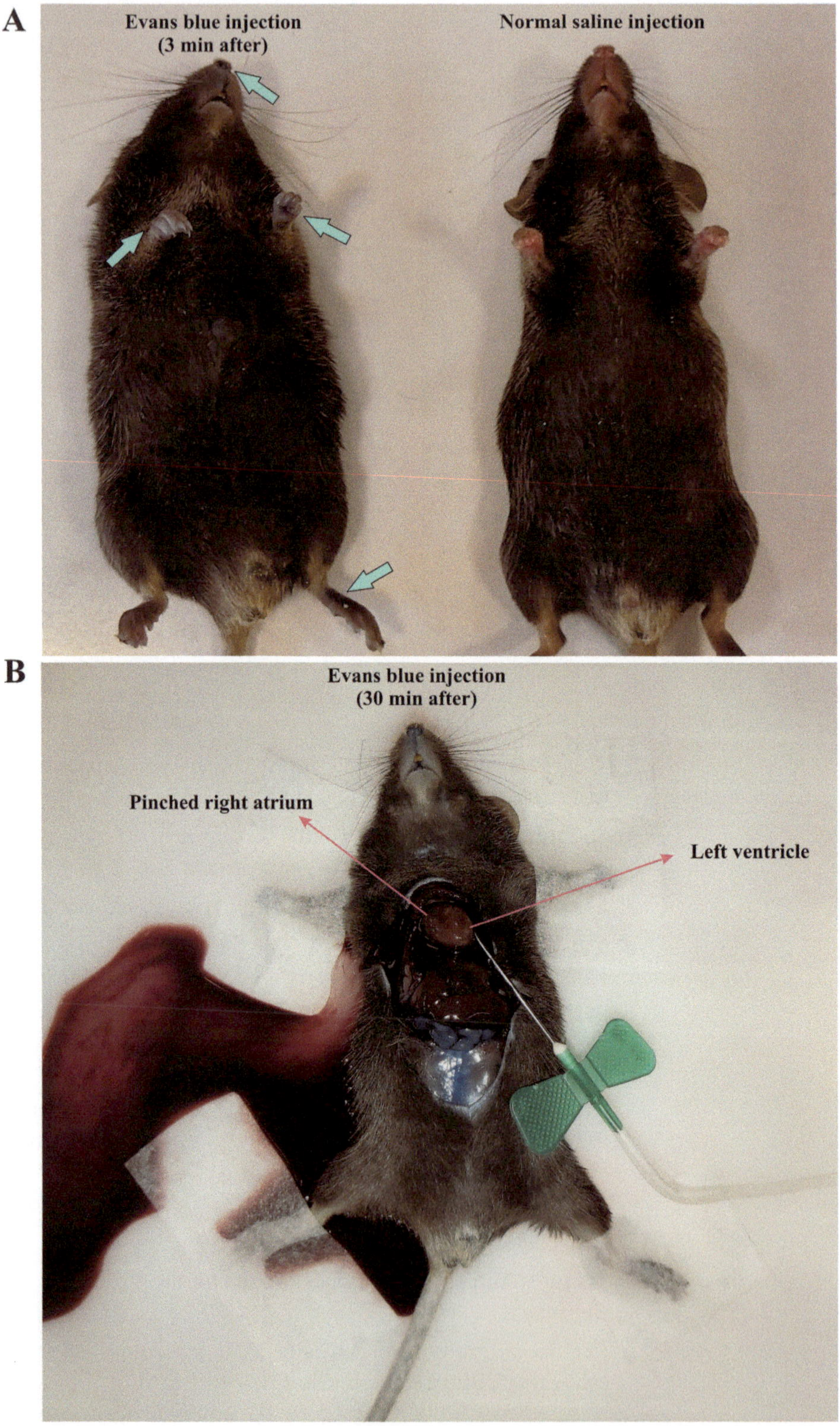

Fig. 2 Animal skin color changes following Evans blue injection. (**a**) Animal's nose, hands, and feet (skin) altered in color 3 min after retro-orbital injection of dye. (**b**) Transcardially perfusion of a C57BL/6 mouse 30 min after injection of Evans blue

liquid flow with 110 mmHg pressure) (Fig. 2b). The flow continues until the liquid flowing from the right atrium is completely clear.

10. Decapitate the animal, dissect the regions of interest (i.e., cerebral cortex, hippocampus, striatum), and store tissues at −80 ° C for further analysis.
11. Weight samples and manually homogenize 50 mg of pieces in 250 μL of ice-cold 50% trichloroacetic acid for 10 min (this step should be done on ice).
12. Centrifuge homogenized samples at 9000 rpm for 20 min (4 °C).
13. Plate supernatants duplicate or triplicate in a 96-well plate and measure fluorescence intensity at 620–630 nm.
14. The concentration of Evans Blue in the desired samples should be calculated using the standard concentration of Evans blue prepared in ice-cold 50% trichloroacetic acid solution (Fig. 3).

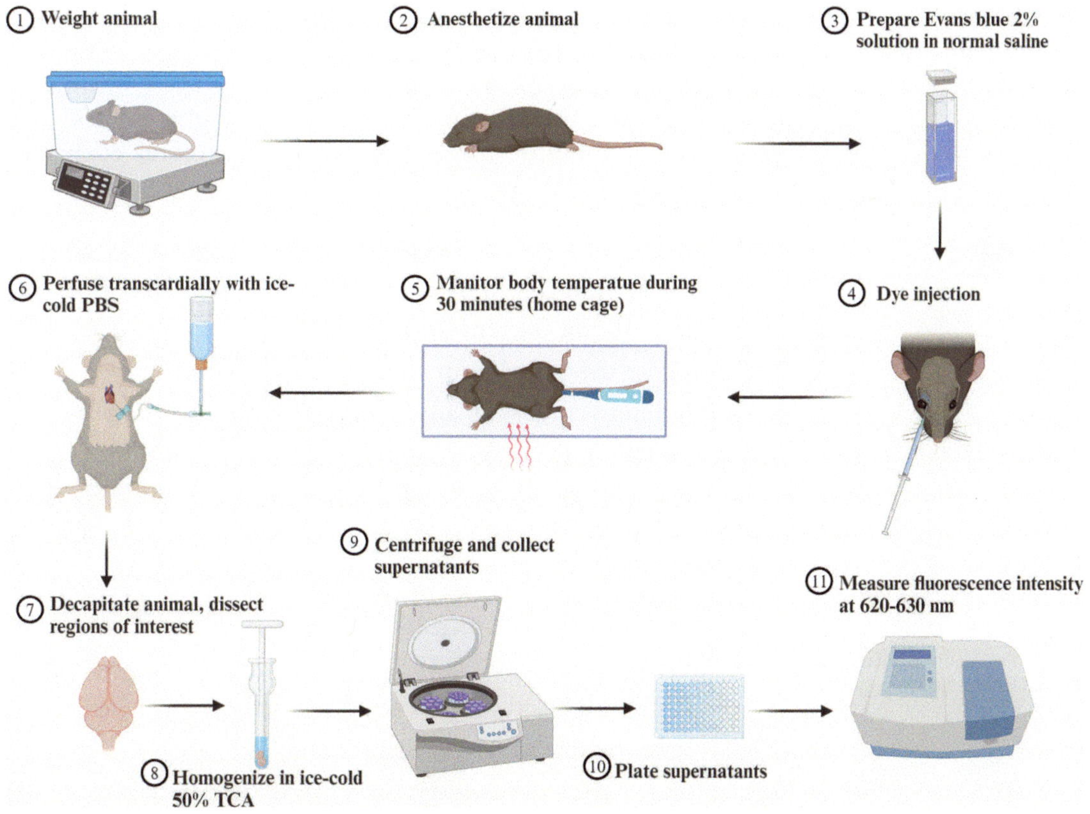

Fig. 3 A schematic overview of the protocol for assessing BBB permeabilization in mouse models of HE. (Created with BioRender.com)

2.2 Assessment of Brain Edema

2.2.1 Background

Normal blood supply to the brain tissue requires equilibrium and precise regulation of three critical components that are localized in the fixed space of the cranial cavity. These components include brain matter, blood, and cerebrospinal fluid. Cerebral edema can be caused by any factor that increases the amount of one of these three components in the skull's confined space [20]. Cerebral edema is a fatal condition that severely interrupts the blood supply to brain tissue, causing irreversible damage to neuronal cells. Cerebral edema can be induced by a breakdown in the blood-brain barrier, in which case a huge and unmanageable level of plasma solutes and proteins enter the extravascular space, and subsequently fluid flows into this area, increasing the interstitial space (vasogenic edema). Another type of edema, cytotoxic edema, affects a number of the central nervous system's major supporting cells, astrocytes [21–23]. In HE, hyperammonemia raised ammonia levels in brain tissue, where astrocytes may detoxify ammonia into glutamine using the glutamine synthetase enzyme. An excessive amount of ammonia in brain tissue reduces the ability of astrocytes to detoxify, and high levels of glutamine accumulate inside the cell, increasing intracellular osmotic pressure and allowing fluid to flow into the cell, resulting in astrocyte swelling [24–26]. Some animal models and clinical studies of HE have identified both types of edema, which are linked to brain dysfunctions and the emergence of neurological symptoms [27–35]. Among other protocols, measuring the brain water content using the wet/dry weight method is a typical way to monitor the development of cerebral edema in animal models of HE [34, 36–38] (Fig. 4).

2.2.2 Method

1. Weight the animals to establish the right dosage for anesthesia.
2. Sedate the animal with 10% ketamine (80 mg/kg) and 2% xylazine (10 mg/kg) intraperitoneally.
3. Decapitate the animal immediately after sedation.
4. Gently remove the skull and meninges with small forceps.
5. Dissect 10 mg (a) of cortical samples (more than three pieces) into a pre-weighed tube and determine its weight using an accurate digital scale (b = weight of wet cortical samples + tube).
6. Place the tube with the door open in a pre-heated oven (120 °C) for 24 h.
7. Weight the tube using an accurate digital scale and record it (c = weight of dried cortical samples + tube).
8. Calculate the dry weight of samples (d) as b − c.
9. Calculate the percentage of brain water content as: $x = [(a - d)/a] \times 100$.

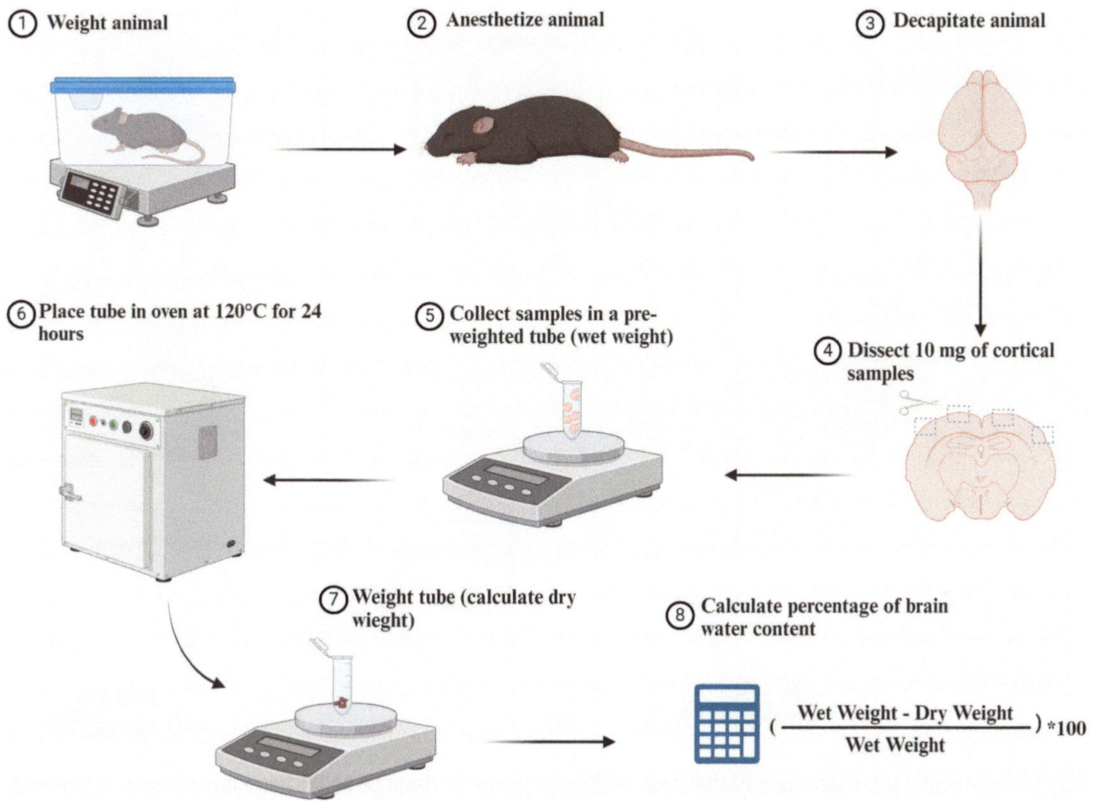

Fig. 4 A schematic summary of the steps for determining brain water content as an indicator of cerebral edema in mouse models of HE. (Created with BioRender.com)

2.3 Measurement of Intracranial Pressure

2.3.1 Background

As was previously indicated cerebral edema is one of the most significant pathological findings in the brain that appears in patients with HE and its experimental models, particularly in the acute form of the disease [29, 39, 40]. Because the brain is located in the fixed cranial space, the increase in the volume of brain interstitial fluid (cerebral edema) leads to disruption of blood flow autoregulation mechanisms and an increase in intracranial pressure, also known as intracranial hypertension [41–43]. Increased intracranial pressure has the potential to fatally alter cerebral structures, such as midline shift and herniation, and could potentially lead to the onset of neurological manifestations in HE [44–46]. Cannulation of the cisterna magna in rats under anesthesia is one of the primary techniques for both acute and long-term intracranial pressure monitoring [47]. Cannulation of the cisterna magna is a direct access route to cerebrospinal fluid (CSF) that causes no injury to the skull or brain tissue and allows for continuous monitoring of intracranial pressure (Fig. 5).

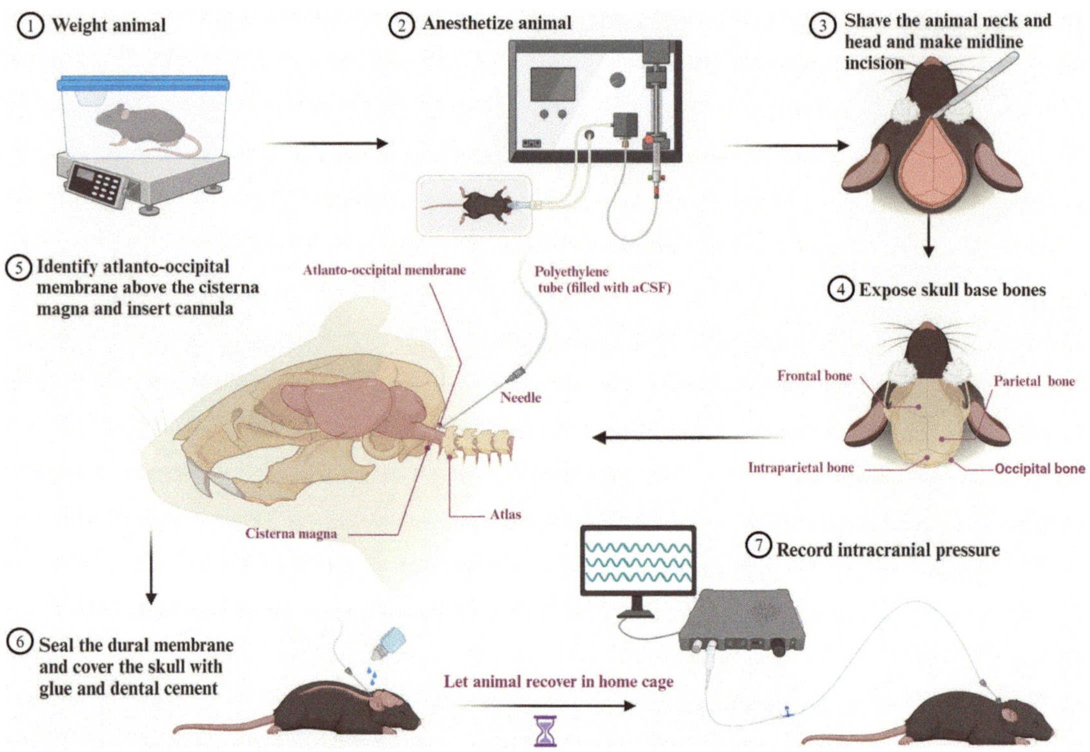

Fig. 5 A schematic representation of the step-by-step protocol for measuring intracranial pressure in a mouse model of HE. (Created with BioRender.com)

2.3.2 Method

1. Weight animals to determine the appropriate anesthetic and injection dye dosage.
2. Sedate the animal with 10% ketamine (80 mg/kg) and 2% xylazine (10 mg/kg) intraperitoneally (*see* **Note 3**).
3. Check toe pinch reflexes, place the animal on a heating pad and a stereotaxic frame (flexing its head to a 50° angle) and carefully monitor its body temperature using a rectal digital thermometer.
4. Apply eye ointment on the animal's eye and close the eye with hypoallergenic paper glue.
5. Shave the animal (head and neck), then use 70% ethanol and povidone-iodine to sanitize the skin.
6. Make a 1 cm midline incision between the skull base and the occipital crest. Using forceps, the overlying skin and muscles are pushed aside to expose the atlanto-occipital membrane and dura mater (placed above the cisterna magna) (*see* **Note 4**).
7. Attach a 30G dental blunted needle (mouse) or 23G blunted needle (rat) to a polyethylene tube (PE10) that is already filled with fresh artificial CSF. The artificial CSF contains 126 mM NaCl, 26 mM NaHCO3, 10 mM glucose, 2.5 mM KCl, 2 mM MgSO4, 2 mM CaCl2 and 1.25 mM NaH2PO4 (pH = 7.4).

8. Connect the other end of the PE10 cannula to a pressure transducer.
9. Gently, insert the cannula (needle) to a depth of a maximum of 2 mm into the cisterna magna (45° relative to the animal head).
10. Immediately, seal the dural membrane with cyanoacrylate glue. Cover the skull with dental cement and cyanoacrylate glue.
11. Allow the animal to recover for 20 min and then record intracranial pressure.
12. In cases with chronic monitoring of intracranial pressure, inject lidocaine (1 mg per mouse) and bupivacaine (0.25 mg per mouse) at the site of injury subcutaneously (for induction of analgesia) as well as inject carprofen (5 mg/kg) subcutaneously.
13. Put the animal back in its home cage singly and use a heating pad to control its body temperature.

2.4 Ultra Structural Analysis (Electron Microscopy)

Electron microscopy (EM) uses electrons as a source of illumination to produce high-quality images of the structural characteristics of tissues at the level of individual cells, intracellular organelles, and specific macromolecules. Two types of electron microscopy are available to researchers. Scanning electron microscopy (SEM) produces high-resolution three-dimensional images from the cell surface, whereas transmission electron microscopy (TEM) generates images from intracellular compartments at incredibly small length scales [48, 49]. In neuroscience research, TEM is widely used to investigate the pathological changes of cerebral microvasculature, the structure of tight junction proteins, and the phenotype of glial cells as well as neurons [50]. In animal models of HE, changes in the structure of the BBB, cell swelling, and organelle changes in neurons and glial cells can be directly depicted using TEM [36, 51, 52].

In the brain, the differentiation of distinct types of nervous system cells (neuron, astrocyte, oligodendrocyte, microglia, pericyte, and endothelial cell) in TEM images is critical, which necessitates a careful study of the images utilizing previous experiences. In electron microscope images, each cell in the CNS has distinct properties that set it apart from others [50, 53–55] (*see* Table 3; Fig. 6).

2.5 In Vitro Evaluation of HE with Ammonia-Exposed Astrocytes

2.5.1 Background

Astrocytes are the most affected cell types in hyperammonemic conditions such as HE. These cells, the most abundant glial cells in the brain parenchyma, are responsible for several essential metabolic processes that support the integrity of the BBB and preserve neuronal health [56]. When the concentration of ammonia in the CNS rises, astrocytes are the only cells in the brain that can produce the enzyme glutamine synthetase, which detoxifies neurotoxic

Table 3
Distinct properties of brain cell types in TEM micrograph

Cells	Description of properties
Endothelial cells	Slender, crescent-shaped, heterochromatic nuclei; in close contact with blood, linked to neighboring endothelial cells via tight junction proteins, with little cytosol surrounding the nuclei.
Pericytes	Embrace the endothelial cells (ring-formed cells) and reside in the basement membrane from the parenchymal side, crescent-shaped and slender nuclei with little cytosol.
Astrocytes	Irregular nucleus; clear cytosol rich in the number of organelles such as mitochondria, rough endoplasmic reticulum, Golgi complex, intermediate filaments and lysosomes; processes embrace brain capillaries as astrocyte end-feet.
Microglia	Pleomorphic, small and heterochromatic nucleus, irregular cytoplasm rich in lysosomes and extended rough endoplasmic reticulum (activated microglia).
Neurons	Round euchromatic nucleus; cytoplasm enriched with round mitochondria, lysosomes, Golgi complex, free ribosomes, neurofilaments and microtubules; extensive synapses with neighboring cells.
Oligodendrocytes	Round heterochromatin nucleus; short rough endoplasmic reticulum, small mitochondria; close contact to neurons.

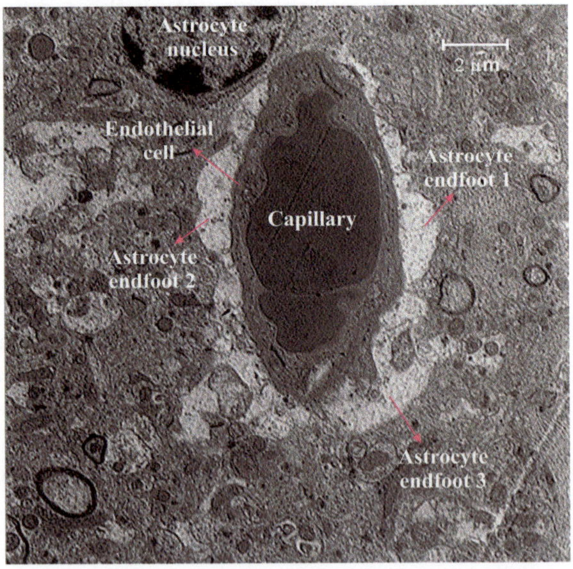

Fig. 6 Transmission electron micrograph from the cerebral cortex of a C57BL/6 mouse

ammonia levels. This enzyme catalyzes ammonia and glutamate to glutamine, lowering the concentration of ammonia in the CNS. However, if liver damage is not controlled, excessive levels of ammonia enter the brain, exceeding astrocytes' ability to detoxify them. In this condition, glutamine levels rise in astrocytes, raising osmotic pressure and causing cell swelling. Aquaporin 4 water channels facilitate flow through astrocytes [24, 57, 58]. Cerebral edema is caused by cell swelling, which can lead to fatal complications such as intracranial hypertension and brain herniation in HE [29, 59]. Given the importance of astrocytes in the pathophysiology of HE, HE can be mimicked by cultivating these cells and subjecting them to different concentrations of ammonia (5 mM, 10 mM, and 20 mM) [60–62]. Physical isolation by shaking is one of the standard methods for isolating astrocytes from the cerebral cortex of mouse or rat pups. Nevertheless, extra steps are required to preserve the high purity of these cells and reduce contamination with other types of glial cells and neurons [63, 64]. We go into further depth about these stages below (Fig. 7).

2.5.2 Method

1. Hold newborn mouse or rat pups (P1–P3) gently, remove any wood chips or dust from their skin, and use 70% ethanol to sterilize their head and neck (*see* **Note 5**).

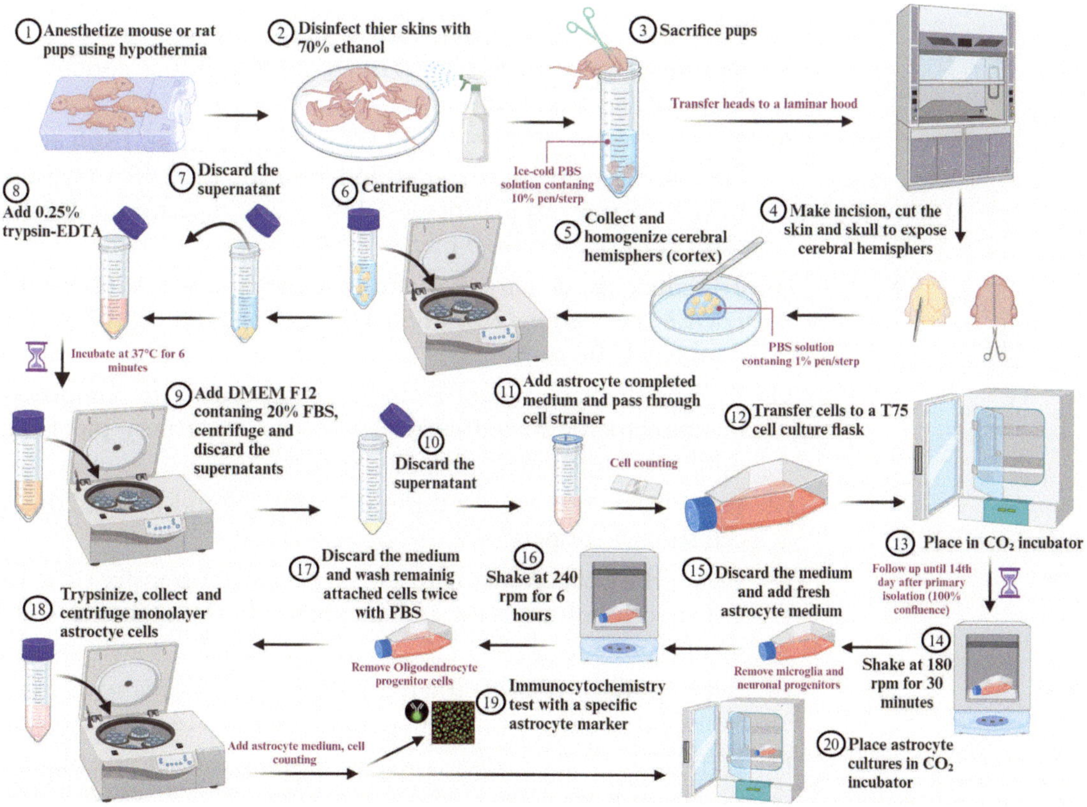

Fig. 7 A schematic illustrating the sequential process for isolating and purifying astrocytes from mouse pups' cerebral pieces, which are the primary affected cells in HE. (Created with BioRender.com)

2. Sacrifice the animal with a sterile scissor and place the head in an ice-cold PBS solution containing 10% penicillin and streptomycin (*see* **Note 6**).
3. To expose the skull, make a longitudinal incision in the midline from the base of the head to the animal's nose with fine, sterile scissors.
4. Make a longitudinal incision from caudal to rostral in the skull's thin bone, which is flexible like a membrane (*see* **Note 7**).
5. To expose the cerebral cortex, carefully remove the cerebellum and meninges with a stereomicroscope and fine forceps (*see* **Note 8**).
6. Using forceps, remove the cortices from the two hemispheres and place them in a petri dish containing 100 microliters of ice-cold PBS and 1% penicillin-streptomycin. Repeat these steps for the other three pups immediately and collect the cortical pieces in a petri dish.
7. Using a sterile scalpel, cut the brain fragments into the smallest pieces possible in less than 3 min (mechanical digestion).
8. Pour the softened tissues into a sterile 50 mL Falcon after diluting them with 5–10 mL of sterile PBS buffer containing 1% penicillin-streptomycin.
9. To pellet cortical pieces, centrifuge for 5 min at 1500 rpm. Discard the supernatant and add 600 μL of 0.25% trypsin-EDTA to the residual tissues. Enzymatic digestion will take place at 37 °C for 5–10 min.
10. To reduce the activity of the trypsin enzyme, add 1800 mL of DMEM containing 20% FBS to the Falcon, and centrifuge for 5 min at 1500 rpm.
11. Discard the supernatant and add 20 mL astrocyte completed medium to the mixed cortical cells and vigorously pipetting until turns into single cells (*see* **Note 9**).
12. Pass mixed cortical cells through a 100 μm and then a 70 μm cell strainer, respectively.
13. Count the isolated cells with a hemocytometer transfer them to a T75 cell culture flask and incubate at 37 °C in the CO_2 incubator.
14. Replace the media of isolated mixed cortical cells with fresh astrocyte medium every 2 days until days 10–14.
15. If the cells achieve 100% confluence, split (passage) the flask.
16. On the 14th day following the initial isolation of mixed cortical cells, when the cells have reached 100% confluence, place the flasks in the shaker incubator and shake for 30 min at 180 rpm to detach microglia and neural progenitor cells from cultures.

17. Discard the supernatants to remove microglia and neural progenitor cells.
18. Add fresh astrocyte media to the remaining cells and return them to the shaker incubator. Shake for 6 h at 240 rpm to detach the oligodendrocyte precursor cells.
19. After this time has passed, physically shake the flasks for 1 min to ensure that there is as little contamination with oligodendrocyte precursor cells as feasible.
20. Rinse the remaining attached cells in the flask twice with PBS containing 1% penicillin-streptomycin.
21. Incubate monolayer astrocytes with 2 mL trypsin-EDTA for 5 min at 37 °C.
22. Add 6 mL DMEM containing 20% FBS to the flask, then centrifuge the detached astrocytes for 5 min at 1500 rpm.
23. Discard the supernatant and add 20 mL astrocyte completed medium to purified cells and plate them to new T75 tissue culture flasks (*see* **Note 10**).
24. Replace the medium 2 days after the purification process with new fresh astrocyte medium containing DMEM low glucose, 10% FBS, 1% penicillin-streptomycin, and 1% amphotericin B (astrocyte maturation medium).
25. Mature astrocytes are ready for exposure to ammonia and other interventions about 26–27 days after initial isolation.

2.6 Liver Function Indicators

An animal model of HE should be able to accurately mimic liver injury found in patients. In clinical studies, several tests are used to confirm liver damage, including non-invasive methods such as measuring bilirubin, albumin, aspartate transferase (AST), alanine aminotransferase (ALT), alkaline phosphatase (ALP), and gamma-glutamyl transferase (GGT) [65–67]. Also, ultrasound can identify a variety of pathological alterations in the liver parenchyma, including hepatocyte swelling, inflammation, steatotic changes, and fibrosis [68–70]. Blood ammonia levels are also a good indicator of hepatocyte injury [71, 72]. These tests can also be performed in animal models of liver disease [73–75]. However, in animal models, due to complete access to the entire liver tissue, it is possible to directly assess the liver histological alterations using particular staining methods (Table 4). In histological examinations of liver tissue, several standard scoring methods are typically employed to determine the severity of tissue damage in at least five randomized regions of a tissue segment [76–78] (Table 5).

Table 4
Direct and indirect methods for determining liver injury in animal models of liver disease and HE

Methods	Description	Reference
Liver function tests	Blood markers used to assess liver injury include AST, ALT, ALP, GGT, albumin, prothrombin time, and bilirubin.	[113–117]
Blood ammonia	After taking the samples, immediately transfer them to an ice-chilled tube. Analysis must be completed within 25 min or samples stored at −20 °C for 48 h.	[113, 114, 118–120]
Hematoxylin and eosin	Infiltration of inflammatory cells, centrilobular hepatocyte necrosis, fibrosis, sinusoidal dilatation, sinusoidal congestion, hepatocyte vacuolization and nuclear pyknosis were all identified.	[28, 121–123]
Masson's trichrome staining	Detection of collagen deposition around portal tracts.	[28, 124, 125]
Picrosirius red staining	Visualization of collagen I and III fibers, detection of fibrosis in the liver parenchyma.	[121, 122, 126]
Periodic acid–Schiff (PAS) staining	Visualization of glycogen deposits in liver parenchyma. Glycogen (red to purple); Nuclei (blue)	[36, 127]
Liver weight	Indicator of inflammation and fibrosis.	[119]

2.7 Hematoxylin and Eosin, and PAS Staining (Paraffin-Embedded Tissue)

2.7.1 Methods

Deparaffinization and Rehydration of Tissue Sections

1. Prepare paraffin-embedded tissue sections (5–6 μm).
2. Place the slides in a xylene bath for 10 min (three times, each with fresh xylene).
3. Shake off excess liquid on the slide and place it in 100% ethanol for 6 min.
4. Shake off excess liquid and place the slide in 90% ethanol for 5 min.
5. Place slides in 70% ethanol for 4 min.
6. Place slides in distilled water for 2 min.
7. Hematoxylin and Eosin.
   - A. Place slides in Hematoxylin solution for 5–8 min at room temperature.
   - B. Rinse slides with tap water for 10 min.
   - C. Place slides in distilled water for 2 min.
   - D. Place slides in alcoholic acid (1% hydrochloric acid in 70% ethanol) for 20 s.
   - E. Place slides in distilled water for 2 min.
   - F. Stain sections with Eosin for 2–3 min.
   - G. Place slides in distilled water for 2 min.

Table 5
Scoring systems for assessing the severity of acute and chronic liver injury in animal models

Grading of acute liver injury	Zone 1	Zone 2	Zone 3	Zone 4	Zone 5	Zone 6
0, Intact parenchyma						
1, Hepatocyte vacuolation Focal nuclear pyknosis						
2, Cytoplasmic hypereosinophilia and severe nuclear pyknosis						
3, Hemorrhage, hepatocyte necrosis and infiltration of polymorphonuclear cells						
Total						
Ishak scoring system (chronic liver injury)	**Zone 1**	**Zone 2**	**Zone 3**	**Zone 4**	**Zone 5**	**Zone 6**
0, No fibrosis						
1, Fibrosis in some portal regions with or without fibrous septa						
2, Fibrosis of most portal areas, with or without fibrous septa						
3, Fibrosis of most portal areas, some portal-to-portal bridging						
4, Fibrosis portal areas with severe portal-to-portal bridging and portal-to-central bridging						
5, Severe portal to portal bridging and portal to central bridging, occasional nodules as incomplete cirrhosis						
6, Cirrhosis						
Total						

8. PAS staining.
   A. Incubate sections with 1% Periodic acid for 8–10 min at room temperature.
   B. Place slides in distilled water for 2 min.
   C. Incubate slides with Schiff reagent for 15 min.
   D. Shake off excess liquid on the slide and incubate it with Sodium metabisulphite solution for 2 min (repeat step 10 three times with fresh agent).
   E. Rinse slides with tap water for 10 min.
   F. Place slides in distilled water for 2 min.
   G. Incubate sections with Mayer hemalum solution for 8–10 min.

H. Rinse slides with tap water for 10 min.

I. Place slides in a distilled water bath for 2 min.

Dehydration

9. Place slides in 70% ethanol for 2 min.
10. Shake off excess liquid and place the slide in 90% ethanol for 3 min.
11. Shake off excess liquid and place slides in 100% ethanol for 5 min.
12. Place the slides in a xylene bath for 7 min (three times, each time using fresh xylene).
13. Mount slides for further analyses under a light microscope.

2.8 Intestinal Barrier Permeabilization Assay

2.8.1 Background

The intestinal barrier, which is mostly made up of the mucus layer, epithelial barrier, and immunological barrier, plays a crucial role in nutrient absorption and prevents the transfer of numerous unregulated luminal contents into the circulation [79]. The intestinal epithelial cells are the primary components of the intestinal barrier that severely regulate paracellular pathways and selective uptake of ions and nutrients through their tight junction proteins and transporters, respectively [80]. There is evidence that disruption of the intestinal barrier is related to bacterial translocation, endotoxemia, severity of liver disease, and the onset of HE [14, 81–84]. In animal investigations, the integrity of the intestinal barrier can be directly evaluated by visualizing the structure of its tight junction proteins by electron microscopy and immunohistochemistry, and indirectly through dye-based approaches such as Evans blue [82, 85, 86] (Figs. 8 and 9).

2.8.2 Methods

Protocol 1

1. Weighing animals to establish proper anesthesia dosage.
2. Sedate the animal with 10% ketamine and 2% xylazine intraperitoneally.
3. Shave the abdomen and disinfect it with 70% ethanol and povidone-iodine.
4. Make a midline incision to expose the large intestine.
5. A thin catheter should be gently inserted via the anus into the large intestine and placed in the colon's distal section.
6. Ligate the proximal and distal parts of the colon with a silk surgical suture.
7. Inject 0.5 mL of 1.5% Evans blue (in PBS) into the colon through the inserted catheter.

8. Close the abdomen with three single sutures to prevent viscera from drying out, and allow the dye to remain in the closed colon for 120 min [59] (*see* **Note 11**).
9. Sacrifice the animal under deep anesthesia and remove the closed loop of the colon from the abdominal cavity.
10. Open the dissected colon and rinse it three times with acetylcysteine (6 mmol/L) to remove any unabsorbed dyes in the mucus layer.
11. Dry dissected colon on filter paper at 37 °C for 24 h.
12. Weight colon, homogenate and incubate with 3 mL of dimethylformamide per 0.5 g sample at 60 °C for 24 h.
13. Centrifugation at 5000 rpm for 15 min at 25 °C.
14. Collect supernatant and plate in triplicate into a 96-well plate.
15. Use a spectrophotometer at a wavelength of 620 nm to calculate the concentration of Evans blue in the intestinal sample based on a prepared standard curve of Evans blue in formamide (Fig. 8).

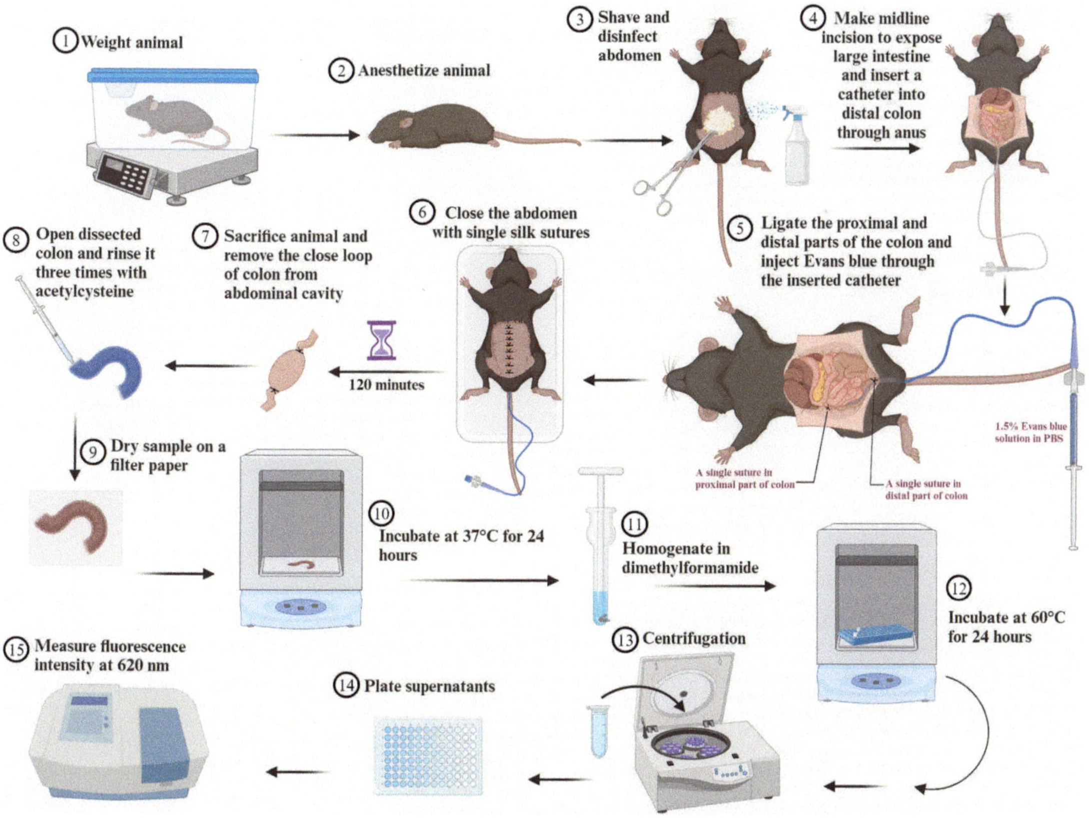

Fig. 8 An outline of the detailed procedure for determining the level of Evans blue in the mouse large intestine as an indicator of intestinal barrier permeabilization. (Created with BioRender.com)

Protocol 2

1. Weight animals to establish proper anesthesia dosage.
2. Sedate the animal with 10% ketamine and 2% xylazine intraperitoneally.
3. Disinfect the animal's tail with 70% ethanol, then warm it with a lamp or 45 °C water to make the vessels easier to access.
4. Inject 50 mg/kg of Evans blue (in PBS) into the lateral vein of the tail.
5. Allow the dye to circulate for 180 min (*see* **Note 12**).
6. Sacrifice the animal under deep anesthesia and remove the colon from the abdominal cavity.
7. Open the dissected colon and rinse it twice with PBS.
8. Weight colon, homogenate and incubate with 3 mL of dimethylformamide per 0.5 g sample at 60 °C for 24 h.
9. Centrifugation at 5000 rpm for 15 min at 25 °C.
10. Collect supernatant and plate in triplicate into a 96-well plate.
11. Use a spectrophotometer at a wavelength of 620 nm to calculate the concentration of Evans blue in the intestinal sample based on a prepared standard curve of Evans blue in formamide (Fig. 9).

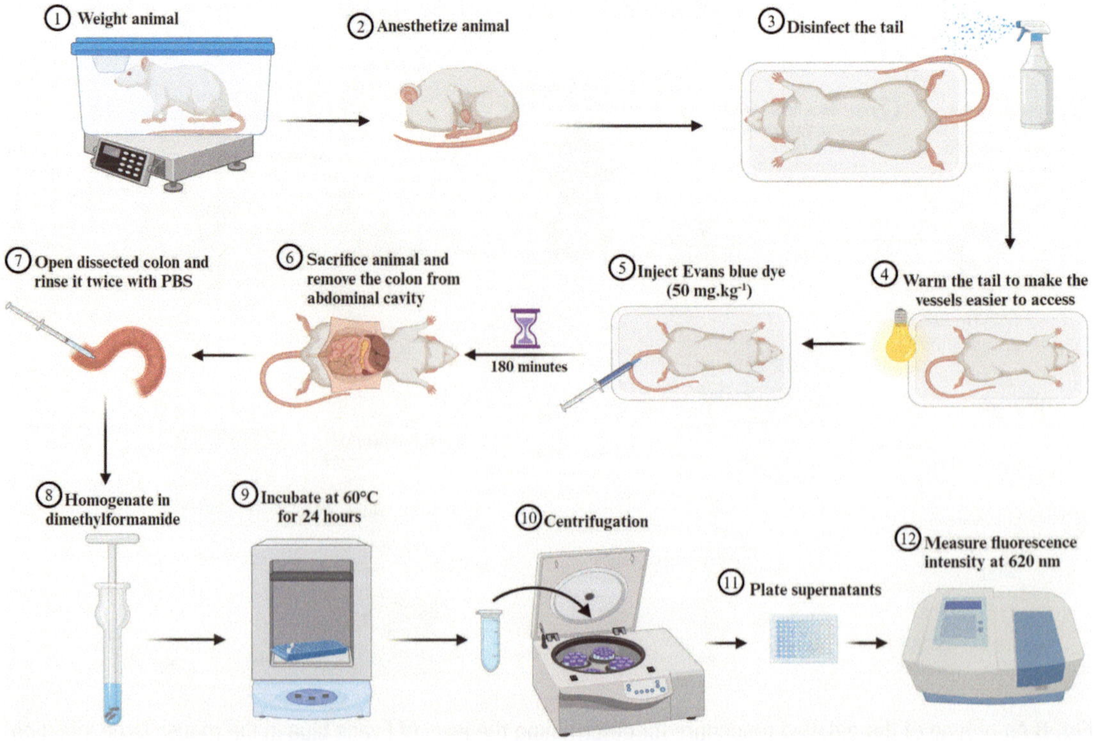

Fig. 9 The detailed procedure for determining the level of Evans blue in the large intestine as an indicator of intestinal barrier permeabilization in rats is shown schematically

2.9 Immunohistochemistry: An Immune-Based Staining Approach for Detecting Injuries to the Gut, Liver, and Brain

2.9.1 Background

Immunohistochemistry (IHC) is a strong immunostaining technique that uses particular antibodies to identify specific antigens in organs of interest, which can be visualized using either fluorescent or chromogenic immunostaining [87]. This technique can be used on a variety of tissues, including biopsies, animal tissues, and autopsy samples [88, 89]. Fixation, antigen retrieval, blocking, antibody labeling, and visualization are the general phases of IHC, and each one will be taught in depth. IHC is a commonly utilized approach to examine cellular and molecular pathological alterations at multiple levels of the gut, liver, and brain axis in animal models of HE (Table 6).

2.9.2 Methods (Paraffin-Embedded Tissue)

Deparaffinization and Rehydration

1. Prepare paraffin-embedded tissue sections (5 μm).
2. Place the slides in a xylene bath for 10 min (three times, each with fresh xylene).
3. Shake off excess liquid on the slide and place it in 100% ethanol for 6 min.
4. Shake off excess liquid and place the slide in 90% ethanol for 5 min.
5. Place slides in 70% ethanol for 4 min.
6. Rinse slides in the tap water bath for 3 min.
7. Place slides in distilled water for 2 min.
8. Place slides in a PBS bath for 10 min at room temperature (*see* **Note 13**).

Antigen Retrieval, Blocking, and Inactivation of Endogenous Peroxidase

9. Place the slide in pre-heated PBS (100 °C; pH ~ 7.4) on a hotplate for 25 min (unmasking of antigen) (*see* **Note 14**).
10. Place the slide in a PBS bath for 3 min at room temperature.
11. Cover sections with 1% bovine serum albumin (BSA) and 2% triton X100 solution for 30 min at room temperature (*see* **Note 15**).
12. Rinse the slide with PBS for 7 min (twice, each with fresh PBS).
13. To decrease the endogenous peroxidase activity, place the slide in 3% hydrogen peroxide in a methanol solution for 15 min at room temperature (protect it from light).
14. Rinse the slide with PBS for 7 min (twice, each with fresh PBS).
15. Incubate sections with normal goat serum (blocking agent) for 45 min at room temperature.
16. Rinse the slide with PBS for 7 min (twice, each with fresh PBS).

Table 6
Main markers for detecting pathological alterations at different levels of the gut-live-brain axis in HE animal models using the IHC approach

Organ	Marker/antibody	Target cells/description	Reference
Brain	GFAP	Swollen (hypertrophied) and reactive astrocytes	[28, 36, 108, 128, 129]
	AQP4	Perivascular astrocytes (end-feet)	[28]
	Kir4.1	Spatial potassium (K+) buffering in astrocytes	[28]
	GLT-1 (EAAT2)	Astrocyte glutamate transporter	[130, 131]
	IBA1	Microglia activation (reactive)	[108, 129, 132–135]
	CD11b/c (OX-42)	Microglia activation	[136]
	CD11b	Microglia activation	[137]
	CD68	Microglial activation (classical marker)	[138]
	NeuN	Neurons	[132, 134]
	Claudin-5, claudin 4, occludin, zona occluden-1 (ZO-1)	Endothelial cells (tight junction proteins; BBB integrity)	[38, 139–142]
Liver	CD68	Macrophages/Kupffer cells	[123, 143]
	RM-4	Kupffer cells	[144]
	anti-Collagen IV	Collagen fiber deposition	[145]
	FOXO	Apoptosis	[114]
	CD68 (ED1), CD163 (ED2)	Resident macrophages and recruited blood monocytes	[146]
	GFAP	Myofibroblasts and hepatic stellate cells	[147]
	Desmin, alpha-smooth muscle actin	Hepatic stellate cells	[148, 149]
	BrdU	Cell proliferation	[147]
	CD11b, CD3	Inflammatory infiltrated immune cells	[150]
	VCAM-1, ICAM-1	Cell adhesion molecules (infiltration of inflammatory cells in the perivascular area)	[150]
	ki-67	Proliferative cells	[28]
	Vimentin, alpha-smooth muscle actin	Fibrosis markers	[147, 151]
Intestine	Claudin-4, E-cadherin, occludin, ZO-1	Epithelial cells (tight junction proteins; intestinal barrier permeabilization)	[152–155]
	PCNA	Proliferative cells	[155]
	TNF-α, NF-κBp65	Inflammation	[156]

Primary Antibody

17. Shake off excess liquid on slides and gently dry all liquids around the section with tissue paper.
18. Put the slide on a flat surface and incubate tissue sections with the desired primary antibody (*see* **Note 16**).
19. Place the slides at 4 °C overnight (*see* **Note 17**).

Secondary Antibody, Visualization, and Dehydration

20. Rinse the primary antibody in PBS containing 0.1% tween 20 for 7 min (two times, each with a fresh solution).
21. Incubate the slide with the appropriate secondary antibody for 150 min at room temperature.
22. Rinse the secondary antibody with PBS containing 0.1% tween 20 for 7 min (two times, each with a fresh solution).
23. Incubate slide with 3,3′-Diaminobenzidine (DAB; chromogen) for 6 min at room temperature (dark brown reaction should be seen).
24. Rinse slides in the tap water bath for 5 min.
25. Place slides in the distilled water bath for 3 min.
26. Counterstaining with hematoxylin for 5 min.
27. Rinse slides in the tap water bath for 5 min.
28. Place slides in the distilled water bath for 3 min.
29. Place slides in 70% ethanol for 2 min.
30. Shake off excess liquid and place slide in 90% ethanol for 3 min.
31. Shake off excess liquid and place slides in 100% ethanol for 5 min.
32. Place the slides in a xylene bath for 7 min (three times, each time using fresh xylene).
33. Mount slides for further analyses under a light microscope.

3 Conclusion

In this chapter, we introduced the best animal models for inducting different types of HE. In addition, we went over in-depth, step-by-step explanations of the key practical specialized methods for assessing multiorgan damages (i.e., brain, liver, and intestines) in HE animal models. The strategies provided in this chapter are among the most essential protocols, covering a wide range of practical methods that HE field researchers may encounter.

4 Notes

1. Because Evans blue is light-sensitive, it is best to prepare the solution and inject it into the animals with the least amount of ambient light possible.
2. Since the femoral vein is difficult to access in mice, retro-orbital injection (through retro-orbital sinus) can be utilized instead.

3. If it is required to monitor the animal's intracranial pressure for an extended period of time, it is preferable to utilize anesthetic gas (an isoflurane chamber; 5% isoflurane in 1 L/min oxygen for induction and 2.5% isoflurane in the air for maintenance).
4. If the animals must be monitored for an extended period, all surgical devices should be sterile and sterile gloves should be used.
5. This process involves preparing brain tissues from four pups for a T75 tissue culture flask. Use the hypothermia approach (keep mouse or rat pups on the cooling platform) to induce anesthesia in mouse or rat pups.
6. Sterilize all surgical devices in an oven at 183 °C for at least 30 min.
7. When cutting the skull, take care not to injure the cerebral cortex.
8. This is a crucial step to prevent contamination with fibroblasts and other non-neural cells.
9. Astrocyte completed medium contains basic DMEM medium (low glucose), 15% fetal bovine serum (FBS), 1% penicillin-streptomycin, and 1% amphotericin B (astrocyte proliferative medium).
10. Immunocytochemistry using a specific antibody such as GFAP, S100β, EAAT2 and AQP4 for the detection of astrocytes is mandatory for validation and determination of percentage of purification.
11. During this period, monitor the animal's body temperature, examine its pain reflexes, and apply ketamine if necessary to maintain anesthesia.
12. During this period, monitor the animal's body temperature, examine its pain reflexes, and apply ketamine if necessary to maintain anesthesia.
13. Do not let the sections dry out in each step.
14. Antigen retrieval is required to expose all epitopes of the target antigen in the samples, and it can be done using a variety of ways, including enzymatic antigen retrieval, chemical antigen retrieval, and heat-induced antigen retrieval.
15. Triton should only be used for nuclear or cytoplasmic antigens, not membrane antigens.
16. To prepare the primary antibody at the manufacturer's specified concentration, use an antibody buffer solution containing 1% BSA in PBS.
17. Keep the slides in a water bath to prevent the tissue pieces' surfaces from drying out.

References

1. Häussinger D, Dhiman RK, Felipo V, Görg B, Jalan R, Kircheis G et al (2022) Hepatic encephalopathy. Nat Rev Dis Primers 8(1):43
2. Nardelli S, Gioia S, Faccioli J, Riggio O, Ridola L (2023) Hepatic encephalopathy–recent advances in treatment and diagnosis. Expert Rev Gastroenterol Hepatol 17(3): 225–235
3. Ridola L, Faccioli J, Nardelli S, Gioia S, Riggio O (2020) Hepatic encephalopathy: diagnosis and management. J Transl Int Med 8(4):210–219
4. Montagnese S, Rautou P-E, Romero-Gómez M, Larsen FS, Shawcross DL, Thabut D et al (2022) EASL clinical practice guidelines on the management of hepatic encephalopathy. J Hepatol 77(3):807–824
5. Rose CF, Amodio P, Bajaj JS, Dhiman RK, Montagnese S, Taylor-Robinson SD et al (2020) Hepatic encephalopathy: novel insights into classification, pathophysiology and therapy. J Hepatol 73(6):1526–1547
6. Won S-M, Oh KK, Gupta H, Ganesan R, Sharma SP, Jeong J-J et al (2022) The link between gut microbiota and hepatic encephalopathy. Int J Mol Sci 23(16):8999
7. Luo M, Xin R-J, Hu F-R, Yao L, Hu S-J, Bai F-H (2023) Role of gut microbiota in the pathogenesis and therapeutics of minimal hepatic encephalopathy via the gut-liver-brain axis. World J Gastroenterol 29(1):144
8. Sepehrinezhad A, Shahbazi A. The pathophysiology of hepatic encephalopathy at the level of gut-liver-brain axis: the role of resident innate immune cells. 2024
9. Bajaj JS, Pimentel M (2022) Is hepatic encephalopathy the first true disorder of gut-brain interaction? Off J Am College Gastroenterol:10.14309
10. Zhu R, Liu L, Zhang G, Dong J, Ren Z, Li Z (2023) The pathogenesis of gut microbiota in hepatic encephalopathy by the gut–liver–brain axis. Biosci Rep 43(6)
11. Lima LCD, Miranda AS, Ferreira RN, Rachid MA, Simões ESAC (2019) Hepatic encephalopathy: lessons from preclinical studies. World J Hepatol 11(2):173–185. Epub 2019/03/02. https://doi.org/10.4254/wjh.v11.i2.173. PubMed PMID: 30820267; PubMed Central PMCID: PMCPMC6393717
12. DeMorrow S, Cudalbu C, Davies N, Jayakumar AR, Rose CF (2021) ISHEN guidelines on animal models of hepatic encephalopathy. Liver Int 41(7):1474–1488. Epub 2021/04/27. https://doi.org/10.1111/liv.14911. PubMed PMID: 33900013; PubMed Central PMCID: PMCPMC9812338
13. Wu D, Chen Q, Chen X, Han F, Chen Z, Wang Y (2023) The blood–brain barrier: structure, regulation, and drug delivery. Signal Transduct Target Ther 8(1):217
14. Shahbazi A, Sepehrinezhad A, Vahdani E, Jamali R, Ghasempour M, Massoudian S et al (2023) Gut dysbiosis and blood-brain barrier alteration in hepatic encephalopathy: from gut to brain. Biomedicines 11(5):1272. PubMed PMID: https://doi.org/10.3390/biomedicines11051272
15. Dhanda S, Sandhir R (2018) Blood-brain barrier permeability is exacerbated in experimental model of hepatic encephalopathy via MMP-9 activation and downregulation of tight junction proteins. Mol Neurobiol 55: 3642–3659
16. Claeys W, Van Hoecke L, Lefere S, Geerts A, Verhelst X, Van Vlierberghe H et al (2021) The neurogliovascular unit in hepatic encephalopathy. JHEP Rep 3(5):100352
17. Waithe OY, Peng X, Childs EW, Tharakan B (2023) Measurement of blood-brain barrier hyperpermeability using Evans blue extravasation assay. In: Vascular hyperpermeability: methods and protocols. Springer, pp 177–184
18. Goldim MPS, Della Giustina A, Petronilho F (2019) Using Evans blue dye to determine blood-brain barrier integrity in rodents. Curr Protoc Immunol 126(1):e83. Epub 2019/09/05. https://doi.org/10.1002/cpim.83
19. Peko L, Katz S, Gattegno R, Ilovitsh T (2024) Protocol to assess extravasation of fluorescent molecules in mice after ultrasound-mediated blood-brain barrier opening. STAR Protocols 5(1):102770
20. Jha SK (2003) Cerebral edema and its management. Medical J Armed Forces India 59(4):326–331. Epub 2003/10/01. https://doi.org/10.1016/s0377-1237(03)80147-8. PubMed PMID: 27407555; PubMed Central PMCID: PMCPMC4923559
21. Nag S, Manias JL, Stewart DJ (2009) Pathology and new players in the pathogenesis of brain edema. Acta Neuropathol 118:197–217
22. Adeva MM, Souto G, Donapetry C, Portals M, Rodriguez A, Lamas D (2012)

Brain edema in diseases of different etiology. Neurochem Int 61(2):166–174

23. Badaut J, Plesnila N (2017) Brain edema: from molecular mechanisms to clinical practice. Academic
24. Sepehrinezhad A, Zarifkar A, Namvar G, Shahbazi A, Williams R (2020) Astrocyte swelling in hepatic encephalopathy: molecular perspective of cytotoxic edema. Metab Brain Dis 35(4):559–578. https://doi.org/10.1007/s11011-020-00549-8
25. Carlos P-M, Aldo T-D (2017) Astrocyte pathophysiology in liver disease. In: Maria Teresa G, Luca Colucci DA (eds) Astrocyte. IntechOpen, Rijeka
26. Rama Rao KV, Jayakumar AR, Tong X, Alvarez VM, Norenberg MD (2010) Marked potentiation of cell swelling by cytokines in ammonia-sensitized cultured astrocytes. J Neuroinflammation 7(1):66. https://doi.org/10.1186/1742-2094-7-66
27. Pierzchala K, Hadjihambi A, Mosso J, Jalan R, Rose CF, Cudalbu C (2024) Lessons on brain edema in HE: from cellular to animal models and clinical studies. Metab Brain Dis 39(3): 403–437. https://doi.org/10.1007/s11011-023-01269-5
28. Elsherbini DMA, Ghoneim FM, El-Mancy EM, Ebrahim HA, El-Sherbiny M, El-Shafey M et al (2022) Astrocytes profiling in acute hepatic encephalopathy: possible enrolling of glial fibrillary acidic protein, tumor necrosis factor-alpha, inwardly rectifying potassium channel (Kir 4.1) and aquaporin-4 in rat cerebral cortex. Front Cell Neurosci 16:896172. Epub 2022/09/06. https://doi.org/10.3389/fncel.2022.896172. PubMed PMID: 36060277; PubMed Central PMCID: PMCPMC9428715
29. Cudalbu C, Taylor-Robinson SD (2019) Brain edema in chronic hepatic encephalopathy. J Clin Exp Hepatol 9(3):362–382. Epub 2019/07/31. https://doi.org/10.1016/j.jceh.2019.02.003. PubMed PMID: 31360029; PubMed Central PMCID: PMCPMC6637228
30. Lu K (2023) Cellular pathogenesis of hepatic encephalopathy: an update. Biomolecules 13(2):396. PubMed PMID: https://doi.org/10.3390/biom13020396
31. Jaeger V, DeMorrow S, McMillin M (2019) The direct contribution of astrocytes and microglia to the pathogenesis of hepatic encephalopathy. J Clin Transl Hepatol 7(4): 352–361. Epub 2020/01/10. https://doi.org/10.14218/jcth.2019.00025. PubMed PMID: 31915605; PubMed Central PMCID: PMCPMC6943208
32. Kato M, Hughes RD, Keays RT, Williams R (1992) Electron microscopic study of brain capillaries in cerebral edema from fulminant hepatic failure. Hepatology (Baltimore, Md) 15(6):1060–1066. Epub 1992/06/01. https://doi.org/10.1002/hep.1840150615
33. Grant S, McMillin M, Frampton G, Petrescu AD, Williams E, Jaeger V et al (2018) Direct comparison of the thioacetamide and azoxymethane models of type A hepatic encephalopathy in mice. Gene Exp 18(3):171–185. Epub 2018/06/14. https://doi.org/10.3727/105221618x15287315176503. PubMed PMID: 29895352; PubMed Central PMCID: PMCPMC6190119
34. Jia W, Liu J, Hu R, Hu A, Tang W, Li L et al (2020) Xiaochaihutang improves the cortical astrocyte edema in thioacetamide-induced rat acute hepatic encephalopathy by activating NRF2 pathway. Front Pharmacol 11:11. https://doi.org/10.3389/fphar.2020.00382
35. Abdelaziz RR, Elkashef WF, Said E (2015) Tranilast reduces serum IL-6 and IL-13 and protects against thioacetamide-induced acute liver injury and hepatic encephalopathy. Environ Toxicol Pharmacol 40(1):259–267. https://doi.org/10.1016/j.etap.2015.06.019
36. Sepehrinezhad A, Shahbazi A, Joghataei MT, Larsen FS, Sahab NS (2023) Inhibition of autotaxin alleviates pathological features of hepatic encephalopathy at the level of gut–liver–brain axis: an experimental and bioinformatic study. Cell Death Dis 14(8):490. https://doi.org/10.1038/s41419-023-06022-5
37. Hajipour S, Farbood Y, Dianat M, Nesari A, Sarkaki A (2023) Effect of Berberine against cognitive deficits in rat model of Thioacetamide-induced liver cirrhosis and hepatic encephalopathy (behavioral, biochemical, molecular and histological evaluations). Brain Sci 13(6):944. PubMed PMID: https://doi.org/10.3390/brainsci13060944
38. Abo El gheit RE, Atef MM, Badawi GA, Elwan WM, Alshenawy HA, Emam MN (2020) Role of serine protease inhibitor, ulinastatin, in rat model of hepatic encephalopathy: aquaporin 4 molecular targeting and therapeutic implication. J Physiol Biochem 76(4):573–586. https://doi.org/10.1007/s13105-020-00762-0
39. Rabinstein AA (2010) Treatment of brain edema in acute liver failure. Curr Treat Options Neurol 12(2):129–141. Epub 2010/09/16. https://doi.org/10.1007/s11940-010-0062-0

40. Blei AT, Traber PG (1989) Brain edema in experimental fulminant hepatic failure. In: Butterworth RF, Layrargues GP (eds) Hepatic encephalopathy: pathophysiology and treatment. Humana Press, Totowa, pp 231–244
41. Tadevosyan A, Kornbluth J (2021) Brain herniation and intracranial hypertension. Neurol Clin 39(2):293–318. Epub 2021/04/27. https://doi.org/10.1016/j.ncl.2021.02.005
42. O'Reilly MW, Westgate CS, Hornby C, Botfield H, Taylor AE, Markey K et al (2019) A unique androgen excess signature in idiopathic intracranial hypertension is linked to cerebrospinal fluid dynamics. JCI Insight 4(6) Epub 2019/02/13. https://doi.org/10.1172/jci.insight.125348. PubMed PMID: 30753168; PubMed Central PMCID: PMCPMC6483000
43. Mondragon J, Klovenski V (2024) Pseudotumor Cerebri. StatPearls. Treasure Island (FL) ineligible companies. Disclosure: Victoria Klovenski declares no relevant financial relationships with ineligible companies.: StatPearls Publishing Copyright © 2024, StatPearls Publishing LLC
44. Bernal W, Hall C, Karvellas CJ, Auzinger G, Sizer E, Wendon J (2007) Arterial ammonia and clinical risk factors for encephalopathy and intracranial hypertension in acute liver failure. Hepatology (Baltimore, Md) 46(6): 1844–1852
45. Mohsenin V (2013) Assessment and management of cerebral edema and intracranial hypertension in acute liver failure. J Crit Care 28(5):783–791
46. Kamat P, Kunde S, Vos M, Vats A, Gupta N, Heffron T et al (2012) Invasive intracranial pressure monitoring is a useful adjunct in the management of severe hepatic encephalopathy associated with pediatric acute liver failure. Pediatr Crit Care Med 13(1):e33–ee8
47. Hadjihambi A, Harrison IF, Costas-Rodríguez M, Vanhaecke F, Arias N, Gallego-Durán R et al (2019) Impaired brain glymphatic flow in experimental hepatic encephalopathy. J Hepatol 70(1):40–49. Epub 2018/09/12. https://doi.org/10.1016/j.jhep.2018.08.021. PubMed PMID: 30201461; PubMed Central PMCID: PMCPMC7613052
48. Inkson BJ (2016) Scanning electron microscopy (SEM) and transmission electron microscopy (TEM) for materials characterization. In: Materials characterization using nondestructive evaluation (NDE) methods. Elsevier, pp 17–43
49. Tang C, Yang Z (2017) Transmission electron microscopy (TEM). In: Membrane characterization. Elsevier, pp 145–159
50. Sobaniec-Lotowska ME (2003) Ultrastructure of astrocytes in the cortex of the hippocampal gyrus and in the neocortex of the temporal lobe in experimental valproate encephalopathy and after valproate withdrawal. Int J Exp Pathol 84(3):115–126
51. Dhanda S, Sunkaria A, Halder A, Sandhir R (2018) Mitochondrial dysfunctions contribute to energy deficits in rodent model of hepatic encephalopathy. Metab Brain Dis 33(1):209–223. https://doi.org/10.1007/s11011-017-0136-8
52. Bai Y, Wang Y, Yang Y (2018) Hepatic encephalopathy changes mitochondrial dynamics and autophagy in the substantia nigra. Metab Brain Dis 33(5):1669–1678. https://doi.org/10.1007/s11011-018-0275-6
53. Nahirney PC, Tremblay M-E (2021) Brain ultrastructure: putting the pieces together. Front Cell Dev Biol 9. https://doi.org/10.3389/fcell.2021.629503
54. Wechsler W, Meller K (1967) Electron microscopy of neuronal and glial differentiation in the developing brain of the chick. In: Bernhard CG, Schadé JP (eds) Progress in brain research. Elsevier, 26, pp 93–144
55. García-Cabezas MÁ, John YJ, Barbas H, Zikopoulos B (2016) Distinction of neurons, glia and endothelial cells in the cerebral cortex: an algorithm based on cytological features. Front Neuroanat 10. https://doi.org/10.3389/fnana.2016.00107
56. Zhang YM, Qi YB, Gao YN, Chen WG, Zhou T, Zang Y et al (2023) Astrocyte metabolism and signaling pathways in the CNS. Front Neurosci 17:1217451. Epub 2023/09/21. https://doi.org/10.3389/fnins.2023.1217451. PubMed PMID: 37732313; PubMed Central PMCID: PMCPMC10507181
57. Drews L, Zimmermann M, Westhoff P, Brilhaus D, Poss RE, Bergmann L et al (2020) Ammonia inhibits energy metabolism in astrocytes in a rapid and glutamate dehydrogenase 2-dependent manner. Dis Models Mech 13(10) Epub 2020/09/13. https://doi.org/10.1242/dmm.047134. PubMed PMID: 32917661; PubMed Central PMCID: PMCPMC7657470
58. Hadjihambi A, Rose C, Jalan R (2014) Novel insights into ammonia mediated neurotoxicity pointing to potential new therapeutic

strategies. Hepatology (Baltimore, Md) 60. https://doi.org/10.1002/hep.27282

59. Kumar G, Taneja A, Kandiah PA (2017) Brain and the liver: cerebral edema, hepatic encephalopathy and beyond: hepatic. Crit Care 7:83–103. https://doi.org/10.1007/978-3-319-66432-3_8
60. Ohara K, Aoyama M, Fujita M, Sobue K, Asai K (2009) Prolonged exposure to ammonia increases extracellular glutamate in cultured rat astrocytes. Neurosci Lett 462(2):109–112
61. Rangroo Thrane V, Thrane AS, Wang F, Cotrina ML, Smith NA, Chen M et al (2013) Ammonia triggers neuronal disinhibition and seizures by impairing astrocyte potassium buffering. Nat Med 19(12):1643–1648
62. Rao KR, Panickar K, Jayakumar A, Norenberg M (2005) Astrocytes protect neurons from ammonia toxicity. Neurochem Res 30:1311–1318
63. Schildge S, Bohrer C, Beck K, Schachtrup C (2013) Isolation and culture of mouse cortical astrocytes. J Vis Exp (71) Epub 2013/02/06. https://doi.org/10.3791/50079. PubMed PMID: 23380713; PubMed Central PMCID: PMCPMC3582677
64. Güler BE, Krzysko J, Wolfrum U (2021) Isolation and culturing of primary mouse astrocytes for the analysis of focal adhesion dynamics. STAR Protocols 2(4):100954. https://doi.org/10.1016/j.xpro.2021.100954
65. Labenz C, Nagel M, Kremer WM, Hilscher M, Schilling CA, Toenges G et al (2020) Association between diabetes mellitus and hepatic encephalopathy in patients with cirrhosis. Aliment Pharmacol Ther 52(3):527–536
66. Gairing SJ, Mangini C, Zarantonello L, Gioia S, Nielsen EJ, Danneberg S et al (2022) Prevalence of minimal hepatic encephalopathy in patients with liver cirrhosis: a multicenter study. Off J Am College Gastroenterol:10.14309
67. Long L, Li H, Deng G, Wang X, Lu S, Li B et al (2021) Impact of hepatic encephalopathy on clinical characteristics and adverse outcomes in prospective and multicenter cohorts of patients with acute-on-chronic liver diseases. Front Med 8:709884
68. Kh AF (2024) Ultrasound diagnostics for liver cirrhosis. Web of Semantics: J Interdiscip Sci 2(6):347–351
69. Khan S, Butt NI, Tareen MQK, Sami H, Ashfaq F, Mahmood K (2022) Hyponatremia in liver cirrhosis patients presenting with hepatic encephalopathy. Prof Med J 29(07):970–974
70. Li W, Duan Y, Liu Z, Lu X, She J, Qing J et al (2022) Clinical value of hemodynamic changes in diagnosis of hepatic encephalopathy after transjugular intrahepatic portosystemic shunt. Scand J Gastroenterol 1-6:1
71. Thanapirom K, Treeprasertsuk S, Choudhury A, Verma N, Dhiman RK, Al Mahtab M et al (2024) Ammonia is associated with liver-related complications and predicts mortality in acute-on-chronic liver failure patients. Sci Rep 14(1):5796. https://doi.org/10.1038/s41598-024-56401-x
72. Shawcross DL, Wright GA, Stadlbauer V, Hodges SJ, Davies NA, Wheeler-Jones C et al (2008) Ammonia impairs neutrophil phagocytic function in liver disease. Hepatology (Baltimore, Md) 48(4):1202–1212
73. de David C, Rodrigues G, Bona S, Meurer L, Gonzalez-Gallego J, Tunon MJ et al (2011) Role of quercetin in preventing thioacetamide-induced liver injury in rats. Toxicol Pathol 39(6):949–957
74. Shareef SH, Juma AS, Agha DN, Alzahrani AR, Ibrahim IAA, Abdulla MA (2023) Hepatoprotective effect of alpinetin on thioacetamide-induced liver fibrosis in sprague dawley rat. Appl Sci 13(9):5243
75. Moghadamnia D, Mokhtari M, Khatamsaz S (2017) Protective effects of omega supplement on induced hepatic mal-function by Thioacetamide in male rats. Zahedan. J Res Med Sci 19(3)
76. Heo J, Jung D-K (2011) Bioimpedance changes in rats with ccl 4-induced liver fibrosis. J Sensor Sci Technol 20(2):71–76
77. El-Sisi AE, Zakaria S (2019) Potential antifibrotic effect of direct acting antiviral drugs on CCl4 induced hepatic fibrosis in rats. Egypt J Basic Clin Pharmacol 9
78. Chen TM, Subeq YM, Lee RP, Chiou TW, Hsu BG (2008) Single dose intravenous thioacetamide administration as a model of acute liver damage in rats. Int J Exp Pathol 89(4):223–231
79. Vancamelbeke M, Vermeire S (2017) The intestinal barrier: a fundamental role in health and disease. Expert Rev Gastroenterol Hepatol 11(9):821–834. Epub 2017/06/27. https://doi.org/10.1080/17474124.2017.1343143. PubMed PMID: 28650209; PubMed Central PMCID: PMCPMC6104804
80. Pellegrini C, Fornai M, D'Antongiovanni V, Antonioli L, Bernardini N, Derkinderen P (2023) The intestinal barrier in disorders of

the central nervous system. Lancet Gastroenterol Hepatol 8(1):66–80. https://doi.org/10.1016/S2468-1253(22)00241-2
81. Tsiaoussis GI, Assimakopoulos SF, Tsamandas AC, Triantos CK, Thomopoulos KC (2015) Intestinal barrier dysfunction in cirrhosis: Current concepts in pathophysiology and clinical implications. World J Hepatol 7(17):2058–2068. Epub 2015/08/25. https://doi.org/10.4254/wjh.v7.i17.2058. PubMed PMID: 26301048; PubMed Central PMCID: PMCPMC4539399
82. Pijls KE, Jonkers DMAE, Elamin EE, Masclee AAM, Koek GH (2013) Intestinal epithelial barrier function in liver cirrhosis: an extensive review of the literature. Liver Int 33(10): 1457–1469. https://doi.org/10.1111/liv.12271
83. Bajaj JS, Ridlon JM, Hylemon PB, Thacker LR, Heuman DM, Smith S et al (2012) Linkage of gut microbiome with cognition in hepatic encephalopathy. American journal of physiology-gastrointestinal and liver. Physiology 302(1):G168–GG75. https://doi.org/10.1152/ajpgi.00190.2011
84. Chopyk DM, Grakoui A (2020) Contribution of the intestinal microbiome and gut barrier to hepatic disorders. Gastroenterology 159(3):849–863. Epub 2020/06/23. https://doi.org/10.1053/j.gastro.2020.04.077. PubMed PMID: 32569766; PubMed Central PMCID: PMCPMC7502510
85. Wang WW, Zhang Y, Huang XB, You N, Zheng L, Li J (2017) Fecal microbiota transplantation prevents hepatic encephalopathy in rats with carbon tetrachloride-induced acute hepatic dysfunction. World J Gastroenterol 23(38):6983–6994. Epub 2017/11/04. https://doi.org/10.3748/wjg.v23.i38.6983. PubMed PMID: 29097871; PubMed Central PMCID: PMCPMC5658316
86. Schäfer TV, Ivnitsky JJ, Rejniuk VL (2019) Permeability of intestinal and blood—tissue barriers in rats for Evans blue dye under conditions of acute intoxication with cyclophosphamide. Bull Exp Biol Med 168(1):38–40. https://doi.org/10.1007/s10517-019-04640-8
87. Magnani B, Taylor CR (2023) Immunohistochemistry should be regulated as an assay. Arch Pathol Lab Med 147:1229–1231
88. Yigzaw D, Dagnaw GG (2024) Review of immunohistochemistry techniques: applications, current status, and future perspectives. In: Seminars in diagnostic pathology. Elsevier
89. Lin F, Prichard JW, Liu H, Wilkerson ML (2022) Handbook of practical immunohistochemistry: frequently asked questions. Springer Nature
90. El-Latif El-Ghazaly MA, Rashed ER, Shafey GM, Zaki HF, Attia AS (2020) Amelioration of thioacetamide-induced hepatic encephalopathy in rats by low-dose gamma irradiation. Environ Sci Pollut Res 27(1):334–343. https://doi.org/10.1007/s11356-019-06934-w
91. Farjam M, Dehdab P, Abbassnia F, Mehrabani D, Tanideh N, Pakbaz S et al (2012) Thioacetamide-induced acute hepatic encephalopathy in rat: behavioral, biochemical and histological changes. Iran Red Crescent Med J 14(3):164–170. Epub 2012/06/28. PubMed PMID: 22737573; PubMed Central PMCID: PMCPMC3372030
92. Sepehrinezhad A, Shahbazi A, Sahab Negah S, Joghataei MT, Larsen FS (2021) Drug-induced-acute liver failure: a critical appraisal of the thioacetamide model for the study of hepatic encephalopathy. Toxicol Rep 8:962–970. https://doi.org/10.1016/j.toxrep.2021.04.011
93. Huffman AM, Syed M, Rezq S, Anderson CD, Yanes Cardozo LL, Romero DG (2023) Loss of microRNA-21 protects against acetaminophen-induced hepatotoxicity in mice. Arch Toxicol 97(7):1907–1925. Epub 2023/05/14. https://doi.org/10.1007/s00204-023-03499-z. PubMed PMID: 37179516; PubMed Central PMCID: PMCPMC10919897
94. Catumbela CSG, Morales R (2024) Elderly mice with history of acetaminophen intoxication display worsened cognitive impairment and persistent elevation of astrocyte and microglia burden. Sci Rep 14(1):14205. https://doi.org/10.1038/s41598-024-65185-z
95. Saad MA, Rastanawi AA, El-Yamany MF (2018) Alogliptin abates memory injuries of hepatic encephalopathy induced by acute paracetamol intoxication via switching-off autophagy-related apoptosis. Life Sci 215: 11–21. https://doi.org/10.1016/j.lfs.2018.10.069
96. Posadas I, Santos P, Blanco A, Muñoz-Fernández M, Ceña V (2010) Acetaminophen induces apoptosis in rat cortical neurons. PLoS One 5(12):e15360. https://doi.org/10.1371/journal.pone.0015360
97. Thiel C, Lauber J, Klingert W, Klingert K, Morgalla MH, Beschorner R et al (2017) Pathophysiological central nervous system changes in a porcine model of acetaminophen-induced acute liver failure.

Toxicol Lett 281:119–126. https://doi.org/10.1016/j.toxlet.2017.09.018

98. Thiel C, Thiel K, Etspueler A, Morgalla MH, Rubitschek S, Schmid S et al (2011) A reproducible porcine model of acute liver failure induced by Intrajejunal acetaminophen administration. Eur Surg Res 46(3): 118–126. https://doi.org/10.1159/000323411
99. Yokoyama T, Banta S, Berthiaume F, Nagrath D, Tompkins RG, Yarmush ML (2005) Evolution of intrahepatic carbon, nitrogen, and energy metabolism in a D-galactosamine-induced rat liver failure model. Metab Eng 7(2):88–103. https://doi.org/10.1016/j.ymben.2004.09.003
100. Ho DW, Fan ST, To J, Woo YH, Zhang Z, Lau C et al (2002) Selective plasma filtration for treatment of fulminant hepatic failure induced by D-galactosamine in a pig model. Gut 50(6):869–876. Epub 2002/05/16. https://doi.org/10.1136/gut.50.6.869. PubMed PMID: 12010892; PubMed Central PMCID: PMCPMC1773226
101. DU W, Pan X, Chen J, Zhang Y, Chen Y, LI L. (2009) Establishment and further improvement of D-galactosamine induced acute hepatic failure in Duroc breeding pigs. Chinese J Infect Dis:212–216
102. Oria M, Raguer N, Chatauret N, Bartolí R, Odena G, Planas R et al (2006) Functional abnormalities of the motor tract in the rat after portocaval anastomosis and after carbon tetrachloride induction of cirrhosis. Metab Brain Dis 21(4):297–308. https://doi.org/10.1007/s11011-006-9036-z
103. Muñoz-Ortega M, Macías-Segura N, Ventura-Juárez J, Ávila-Blanco ME, Ponce-Damian LD, González-Blas D et al (2021) Recovery from liver failure and fibrosis in a rat Portacaval anastomosis model after Neurointermediate pituitary lobectomy. J Immunol Res 2021(1):5529784. https://doi.org/10.1155/2021/5529784
104. Coy DL, Srivastava A, Gottstein J, Butterworth RF, Blei AT (1991) Postoperative course after portacaval anastomosis in rats is determined by the portacaval pressure gradient. American journal of physiology-gastrointestinal and liver. Physiology 261(6): G1072–G10G8. https://doi.org/10.1152/ajpgi.1991.261.6.G1072
105. Zhang D, Jiang M, Miao D (2011) Transplanted human amniotic membrane-derived mesenchymal stem cells ameliorate carbon tetrachloride-induced liver cirrhosis in mouse. PLoS One 6(2):e16789. https://doi.org/10.1371/journal.pone.0016789
106. Galicia-Moreno M, Rodríguez-Rivera A, Reyes-Gordillo K, Segovia J, Shibayama M, Tsutsumi V et al (2009) N-acetylcysteine prevents carbon tetrachloride-induced liver cirrhosis: role of liver transforming growth factor-beta and oxidative stress. Eur J Gastroenterol Hepatol 21(8):908–914
107. Aydın AF, Küskü-Kiraz Z, Doğru-Abbasoğlu S, Güllüoğlu M, Uysal M, Koçak-Toker N (2010) Effect of carnosine against thioacetamide-induced liver cirrhosis in rat. Peptides 31(1):67–71. https://doi.org/10.1016/j.peptides.2009.11.028
108. Claeys W, Van Hoecke L, Geerts A, Van Vlierberghe H, Lefere S, Van Imschoot G et al (2022) A mouse model of hepatic encephalopathy: bile duct ligation induces brain ammonia overload, glial cell activation and neuroinflammation. Sci Rep 12(1): 17558. https://doi.org/10.1038/s41598-022-22423-6
109. Tag CG, Sauer-Lehnen S, Weiskirchen S, Borkham-Kamphorst E, Tolba RH, Tacke F et al (2015) Bile duct ligation in mice: induction of inflammatory liver injury and fibrosis by obstructive cholestasis. J Vis Exp (96) Epub 2015/03/06. https://doi.org/10.3791/52438. PubMed PMID: 25741630; PubMed Central PMCID: PMCPMC4354634
110. Braissant O, Rackayová V, Pierzchala K, Grosse J, McLin VA, Cudalbu C (2019) Longitudinal neurometabolic changes in the hippocampus of a rat model of chronic hepatic encephalopathy. J Hepatol 71(3):505–515. https://doi.org/10.1016/j.jhep.2019.05.022
111. Hsu SJ, Zhang C, Jeong J, Lee SI, McConnell M, Utsumi T et al (2021) Enhanced meningeal lymphatic drainage ameliorates neuroinflammation and hepatic encephalopathy in cirrhotic rats. Gastroenterology 160(4):1315-29.e13. Epub 2020/11/24. https://doi.org/10.1053/j.gastro.2020.11.036. PubMed PMID: 33227282; PubMed Central PMCID: PMCPMC7956141
112. Hadjihambi A, Harrison I, Rizzo G, Macnaughtan J, Beaton N, Davies N et al (2017) Brain glymphatic clearance is impaired in a rat model of cirrhosis. J Clin Exp Hepatol 7:S53
113. Essam RM, Saadawy MA, Gamal M, Abdelsalam RM, El-Sahar AE (2023) Lactoferrin averts neurological and behavioral impairments of thioacetamide-induced hepatic encephalopathy in rats via modulating HGMB1/TLR-4/MyD88/Nrf2 pathway.

Neuropharmacology 236:109575. https://doi.org/10.1016/j.neuropharm.2023.109575

114. Abo El-Magd NF, El-Kashef DH, El-Sherbiny M, Eraky SM (2023) Hepatoprotective and cognitive-enhancing effects of hesperidin against thioacetamide-induced hepatic encephalopathy in rats. Life Sci 313:121280. https://doi.org/10.1016/j.lfs.2022.121280
115. Eraky SM, El-Kashef DH, El-Sherbiny M, Abo El-Magd NF (2023) Naringenin mitigates thioacetamide-induced hepatic encephalopathy in rats: targeting the JNK/Bax/caspase-8 apoptotic pathway. Food Funct 14(2):1248–1258. https://doi.org/10.1039/D2FO03470K
116. Tamnanloo F, Chen X, Oliveira MM, Tremblay M, Rose CF (2024) Excessive intragastric alcohol administration exacerbates hepatic encephalopathy and provokes neuronal cell death in male rats with chronic liver disease. J Neurosci Res 102(5):e25337. https://doi.org/10.1002/jnr.25337
117. Sedik AA, Hussein DT, Fathy K, Mowaad NA (2024) Neuroprotective and cognitive enhancing effects of herbecetin against thioacetamide induced hepatic encephalopathy in rats via upregulation of AMPK and SIRT1 signaling pathways. Sci Rep 14(1):11396. https://doi.org/10.1038/s41598-024-61639-6
118. Tamnanloo F, Ochoa-Sanchez R, Oliveira MM, Lima C, Lépine M, Dubois K et al (2023) Multiple ammonia-induced episodes of hepatic encephalopathy provoke neuronal cell loss in bile-duct ligated rats. JHEP Rep 5(12):100904
119. Bayraktar M, Hacımüftüoğlu A, Okkay U, Koçak MN, Kösedağ M, Tekin E et al (2024) Use of bee venom in preventive medicine: an experimental hepatic encephalopathy study in rats. Vet Med Sci 10(1):e1336. https://doi.org/10.1002/vms3.1336
120. Ramakrishna K, Sinku S, Majumdar S, Singh N, Gajendra TA, Rani A et al (2023) Indole-3-carbinol ameliorated the thioacetamide-induced hepatic encephalopathy in rats. Toxicology 492:153542. https://doi.org/10.1016/j.tox.2023.153542
121. Dwivedi DK, Jena G, Kumar V (2020) Dimethyl fumarate protects thioacetamide-induced liver damage in rats: studies on Nrf2, NLRP3, and NF-κB. J Biochem Mol Toxicol 34(6):e22476. Epub 2020/02/16. https://doi.org/10.1002/jbt.22476
122. Wallace M, Hamesch K, Lunova M, Kim Y, Weiskirchen R, Strnad P et al (2015) Standard operating procedures in experimental liver research: thioacetamide model in mice and rats. Lab Anim 49(1_suppl):21–29
123. Wu K-C, Huang H-C, Chang T, Lee W-S, Chuang C-L, Hsin IF et al (2018) Effect of sirolimus on liver cirrhosis and hepatic encephalopathy of common bile duct-ligated rats. Eur J Pharmacol 824:133–139. https://doi.org/10.1016/j.ejphar.2018.02.016
124. El Awdan SA, Abdel Rahman RF, Ibrahim HM, Hegazy RR, El Marasy SA, Badawi M et al (2019) Regression of fibrosis by cilostazol in a rat model of thioacetamide-induced liver fibrosis: up regulation of hepatic cAMP, and modulation of inflammatory, oxidative stress and apoptotic biomarkers. PloS One 14(5):e0216301. Epub 2019/05/09. https://doi.org/10.1371/journal.pone.0216301. PubMed PMID: 31067255; PubMed Central PMCID: PMCPMC6505801
125. Murad H, Gazzaz Z, Ali S, Ibraheem M (2017) Candesartan, rather than losartan, improves motor dysfunction in thioacetamide-induced chronic liver failure in rats. Braz J Med Biol Res 50(11):e6665
126. Chobert M-N, Couchie D, Fourcot A, Zafrani E-S, Laperche Y, Mavier P et al (2012) Liver precursor cells increase hepatic fibrosis induced by chronic carbon tetrachloride intoxication in rats. Lab Investig 92(1):135–150. https://doi.org/10.1038/labinvest.2011.143
127. Jin WX, Wang B, Zhang YL, Dong R, Wang XB, Guo JM et al (2018) Effects of hepatic blood inflow on liver ultrastructure and regeneration after extensive liver resection in rats with cirrhosis. Exp Ther Med 16(3):2573–2583. Epub 2018/09/14. https://doi.org/10.3892/etm.2018.6467. PubMed PMID: 30210605; PubMed Central PMCID: PMCPMC6122590
128. Suárez I, Bodega G, Rubio M, Fernández B (2009) Induction of NOS and nitrotyrosine expression in the rat striatum following experimental hepatic encephalopathy. Metab Brain Dis 24(3):395–408. https://doi.org/10.1007/s11011-009-9154-5
129. Chen J-R, Wang B-N, Tseng G-F, Wang Y-J, Huang Y-S, Wang T-J (2014) Morphological changes of cortical pyramidal neurons in hepatic encephalopathy. BMC Neurosci 15(1):15. https://doi.org/10.1186/1471-2202-15-15
130. Morita T, Mizutani Y, Michimae Y, Sawada M, Sato K, Hikasa Y et al (2004) Severe involvement of cerebral neopallidum in a dog with hepatic encephalopathy. Vet

Pathol 41(4):442–445. https://doi.org/10.1354/vp.41-4-442

131. Suárez I, Bodega G, Fernández B (2000) Modulation of glutamate transporters (GLAST, GLT-1 and EAAC1) in the rat cerebellum following portocaval anastomosis. Brain Res 859(2):293–302. https://doi.org/10.1016/S0006-8993(00)01993-4
132. McMillin M, Frampton G, Thompson M, Galindo C, Standeford H, Whittington E et al (2014) Neuronal CCL2 is upregulated during hepatic encephalopathy and contributes to microglia activation and neurological decline. J Neuroinflammation 11(1):121. https://doi.org/10.1186/1742-2094-11-121
133. El-Mansoury B, Smimih K, El Khiat A, Draoui A, Aimrane A, Chatoui R et al (2024) Short working memory impairment associated with hippocampal microglia activation in chronic hepatic encephalopathy. Metabolites 14(4):193. https://doi.org/10.3390/metabo14040193
134. McMillin M, Grant S, Frampton G, Andry S, Brown A, DeMorrow S (2016) Fractalkine suppression during hepatic encephalopathy promotes neuroinflammation in mice. J Neuroinflammation 13(1):198. https://doi.org/10.1186/s12974-016-0674-8
135. Rodrigo R, Cauli O, Gomez-Pinedo U, Agusti A, Hernandez-Rabaza V, Garcia-Verdugo JM et al (2010) Hyperammonemia induces Neuroinflammation that contributes to cognitive impairment in rats with hepatic encephalopathy. Gastroenterology 139(2):675–684. https://doi.org/10.1053/j.gastro.2010.03.040
136. Jiang W, Desjardins P, Butterworth RF (2009) Cerebral inflammation contributes to encephalopathy and brain edema in acute liver failure: protective effect of minocycline. J Neurochem 109(2):485–493. https://doi.org/10.1111/j.1471-4159.2009.05981.x
137. Kong X, Luo S, Wu JR, Wu S, De Cecco CN, Schoepf UJ et al (2016) (18)F-DPA-714 PET imaging for detecting neuroinflammation in rats with chronic hepatic encephalopathy. Theranostics 6(8):1220–1231. Epub 2016/06/10. https://doi.org/10.7150/thno.15362. PubMed PMID: 27279913; PubMed Central PMCID: PMCPMC4893647
138. Sun X, Han R, Cheng T, Zheng Y, Xiao J, So K-F et al (2019) Corticosterone-mediated microglia activation affects dendritic spine plasticity and motor learning functions in minimal hepatic encephalopathy. Brain Behav Immun 82:178–187. https://doi.org/10.1016/j.bbi.2019.08.184
139. McMillin MA, Frampton GA, Seiwell AP, Patel NS, Jacobs AN, DeMorrow S (2015) TGFβ1 exacerbates blood–brain barrier permeability in a mouse model of hepatic encephalopathy via upregulation of MMP9 and downregulation of claudin-5. Lab Investig 95(8):903–913. https://doi.org/10.1038/labinvest.2015.70
140. Vairappan B, Sundhar M, Srinivas BH (2019) Resveratrol restores neuronal tight junction proteins through correction of ammonia and inflammation in CCl4-induced cirrhotic mice. Mol Neurobiol 56(7):4718–4729. https://doi.org/10.1007/s12035-018-1389-x
141. Sarada SKS, Titto M, Himadri P, Saumya S, Vijayalakshmi V (2015) Curcumin prophylaxis mitigates the incidence of hypobaric hypoxia-induced altered ion channels expression and impaired tight junction proteins integrity in rat brain. J Neuroinflammation 12(1):113. https://doi.org/10.1186/s12974-015-0326-4
142. Mavrakis AG, Havaki S, Marinos E, Chroni E, Konstantinou D (2012) Occludin dislocation in brain capillary endothelium of rats with bile duct ligation induced cholestasis. Neurosci Lett 528(2):180–184. https://doi.org/10.1016/j.neulet.2012.08.066
143. Liu C, Tao Q, Sun M, Wu JZ, Yang W, Jian P et al (2010) Kupffer cells are associated with apoptosis, inflammation and fibrotic effects in hepatic fibrosis in rats. Lab Investig 90(12):1805–1816. https://doi.org/10.1038/labinvest.2010.123
144. Tsujimoto T, Kawaratani H, Kitazawa T, Hirai T, Ohishi H, Kitade M et al (2008) Decreased phagocytic activity of Kupffer cells in a rat nonalcoholic steatohepatitis model. World J Gastroenterol 14(39):6036–6043. Epub 2008/10/22. https://doi.org/10.3748/wjg.14.6036. PubMed PMID: 18932283; PubMed Central PMCID: PMCPMC2760201
145. Faisul AG, Al-Saidya AM (2022) Development of an experimental hepatic encephalopathy in a rabbit model: biochemical and immunohistochemical study. Iraqi J Vet Sci 36:179–185
146. Orfila C, Lepert JC, Alric L, Carrera G, Beraud M, Vinel JP et al (1999) Expression of TNF-alpha and Immunohistochemical distribution of hepatic macrophage surface markers in carbon tetrachloride-induced chronic liver injury in rats. Histochem J 31(10):677–685. https://doi.org/10.1023/A:1003851821487
147. Tennakoon AH, Izawa T, Wijesundera KK, Murakami H, Katou-Ichikawa C, Tanaka M

et al (2015) Immunohistochemical characterization of glial fibrillary acidic protein (GFAP)-expressing cells in a rat liver cirrhosis model induced by repeated injections of thioacetamide (TAA). Exp Toxicol Pathol 67(1):53–63. https://doi.org/10.1016/j.etp.2014.09.008

148. Vilaseca M, García-Calderó H, Lafoz E, García-Irigoyen O, Avila MA, Reverter JC et al (2017) The anticoagulant rivaroxaban lowers portal hypertension in cirrhotic rats mainly by deactivating hepatic stellate cells. Hepatology (Baltimore, Md) 65(6):2031–2044. Epub 2017/02/01. https://doi.org/10.1002/hep.29084

149. Zimmermann A, Zimmermann H, Fellay M, Reichen J (1999) Cells with morphological and immunohistochemical features of hepatic stellate cells (Ito cells) form an extralittoral (extrasinusoidal) compartment in the cirrhotic rat liver. Histol Histopathol 14(3): 719–727. Epub 1999/07/30. https://doi.org/10.14670/hh-14.719

150. Tugues S, Fernandez-Varo G, Muñoz-Luque J, Ros J, Arroyo V, Rodés J et al (2007) Antiangiogenic treatment with sunitinib ameliorates inflammatory infiltrate, fibrosis, and portal pressure in cirrhotic rats. Hepatology (Baltimore, Md) 46(6): 1919–1926. Epub 2007/10/16. https://doi.org/10.1002/hep.21921

151. Di Pascoli M, Diví M, Rodríguez-Vilarrupla A, Rosado E, Gracia-Sancho J, Vilaseca M et al (2013) Resveratrol improves intrahepatic endothelial dysfunction and reduces hepatic fibrosis and portal pressure in cirrhotic rats. J Hepatol 58(5):904–910. https://doi.org/10.1016/j.jhep.2012.12.012

152. Gao J-H, Wen S-L, Tong H, Wang C-H, Yang W-J, Tang S-H et al (2016) Inhibition of cyclooxygenase-2 alleviates liver cirrhosis via improvement of the dysfunctional gut-liver axis in rats. American journal of physiology-gastrointestinal and liver. Physiology 310(11):G962–GG72. https://doi.org/10.1152/ajpgi.00428.2015

153. Zhou Z, Li K, Guo J, Wang Y, Wei Y, Duan J et al (2023) Green tea Catechin EGCG ameliorates Thioacetamide-induced hepatic encephalopathy in rats via modulation of the microbiota-gut-liver Axis. Mol Nutr Food Res 67(8):2200821. https://doi.org/10.1002/mnfr.202200821

154. Yang DH, Ye ZY, Xie YJ, He XJ, Xu WJ, Zhou WM (2012) Effect of salvianolate on intestinal epithelium tight junction protein zonula occludens protein 1 in cirrhotic rats. World J Gastroenterol 18(47):7040–7047. Epub 2013/01/17. https://doi.org/10.3748/wjg.v18.i47.7040. PubMed PMID: 23323006; PubMed Central PMCID: PMCPMC3531692

155. Wang PF, Yao DH, Hu YY, Li Y (2019) Vitamin D improves intestinal barrier function in cirrhosis rats by upregulating heme oxygenase-1 expression. Biomol Ther 27(2): 222–230. Epub 2018/09/04. https://doi.org/10.4062/biomolther.2018.052. PubMed PMID: 30173501; PubMed Central PMCID: PMCPMC6430230

156. Tsai H-C, Li T-H, Huang C-C, Huang S-F, Liu R-S, Yang Y-Y et al (2018) Beneficial effects of the peroxisome proliferator-activated receptor α/γ agonist Aleglitazar on progressive hepatic and splanchnic abnormalities in cirrhotic rats with portal hypertension. Am J Pathol 188(7):1608–1624. https://doi.org/10.1016/j.ajpath.2018.03.018

Chapter 12

Methods for Measuring Brain Mitochondrial Impairment and Oxidative Stress in Hepatic Encephalopathy

Reza Heidari, Mohammad Mehdi Ommati, and Hossein Niknahad

Abstract

Hepatic encephalopathy (HE) is a clinical situation that refers to neurological complications of liver failure and hyperammonemia. Ammonia (ammonium ion in biological fluids; NH_4^+) is the primary culprit of brain injury in HE. Although the precise mechanism of NH_4^+-induced neurotoxicity is far from clear, it has been well-established that this chemical can induce significant oxidative stress in the brain. At subcellular levels, mitochondria are critical targets for NH_4^+ cytotoxicity. Therefore, potential therapeutic approaches in HE targeted oxidative stress and mitochondrial impairment. Hence, developing convenient, reliable, and reproducible methods is crucial for assessing and monitoring therapeutic interventions in HE. In the current protocol, an animal model of acute HE is described. Then, techniques for evaluating oxidative stress biomarkers in the brain, isolating brain mitochondria, and assessing mitochondrial indices are outlined.

Key words Bioenergetics, Energy crisis, Hyperammonemia, Neurotoxicity

1 Introduction

Hepatic encephalopathy (HE) refers to the signs and symptoms of liver failure in the central nervous system (CNS) [1]. The liver is the main organ responsible for ammonia (ammonium ion in biological fluids; NH_4^+) detoxification through the urea cycle in hepatocytes' cytoplasm, and mitochondria [2, 3]. NH_4^+ is a well-known neurotoxic agent [2, 3]. Hence, liver failure, with any etiology, could lead to hyperammonemia and NH_4^+-induced neurotoxicity. However, other molecules such as bilirubin, cytotoxic bile acids, and manganese also could play a role in the pathogenesis of HE [4]. The role of mentioned molecules in the pathogenesis of HE depends on the nature/type of this complication [4]. For example, in cholestasis/cirrhosis-induced liver failure and HE, blood and brain levels of bile acids and manganese reach a neurotoxic level [4].

Tiziano Balzano (ed.), *Experimental and Clinical Methods in Hepatic Encephalopathy Research*, Neuromethods, vol. 219, https://doi.org/10.1007/978-1-0716-4354-9_12,

The significance of mitochondrial impairment and oxidative stress in the pathogenesis of brain injury during HE is considered in many investigations [5–8]. The elevated level of NH_4^+ in the CNS is associated with increased levels of reactive oxygen species (ROS), lipid peroxidation, impaired antioxidant systems capacity, and damage to vital proteins [4]. NH_4^+-induced oxidative stress could be related to the direct effect of this molecule on enzymatic and non-enzymatic antioxidant systems or, more importantly, to its role in enhancing ROS formation from secondary sources such as mitochondria [4].

Oxidative stress and mitochondrial impairment are mechanistically interrelated events (Fig. 1). Mitochondria are a significant source of intracellular ROS formation [9]. Mitochondrial ROS are known to act as critical determinants in cell function and participate in many complex networks of cell signaling [10]. Therefore, the conditions that promote the changes in mitochondria-mediated ROS formation (e.g., elevated brain level of NH_4^+) may participate in mitochondria-mediated cell death and organ injury [10]. On the other hand, it is well-known that excess ROS could inhibit normal mitochondrial function and energy metabolism [9] (Fig. 1). In the current protocol, biomarkers of oxidative stress in the brain tissue of an experimental HE model are described.

As mentioned, brain mitochondria are critical targets for NH_4^+-induced neurotoxicity [4]. NH_4^+-induced mitochondrial impairment could be mediated through several mechanisms. It has been reported that NH_4^+ can inhibit the tricarboxylic acid (TCA) cycle enzymes and decrease production of vital molecules such as NADH, which is needed for proper mitochondrial function and energy metabolism [4] (Fig. 1). High levels of NH_4^+ also can inhibit proper function of mitochondrial electron transport chain components and consequently suppress ATP production [4] (Fig. 1). Adverse effect of NH_4^+ on mitochondrial membrane potential also has been documented [4]. It has also been well-known that NH_4^+ could induce mitochondrial permeabilization and release of cell death mediators (e.g., cytochrome *c*) from this organelle [4, 11] (Fig. 1). It is noteworthy to mention here that not only NH_4^+ could induce mitochondrial impairment but also molecules such as manganese and bile acids could deteriorate mitochondrial function in some forms of HE (e.g., cirrhosis-associated HE) [12–14]. Regarding brain mitochondrial dysfunction in HE, detailed protocols for assessing the brain mitochondrial indices are described in an experimental HE model in this chapter.

HE could be induced in laboratory animals using various hepatotoxic agents. Different chemicals, such as carbon tetrachloride, thioacetamide, acetaminophen, azoxymethane, and galactosamine, have been applied to induce acute liver failure and HE [15, 16]. The length of treatment and dose of toxicant dictate the type of HE [15]. Thioacetamide (TAA) is a well-established

Figures

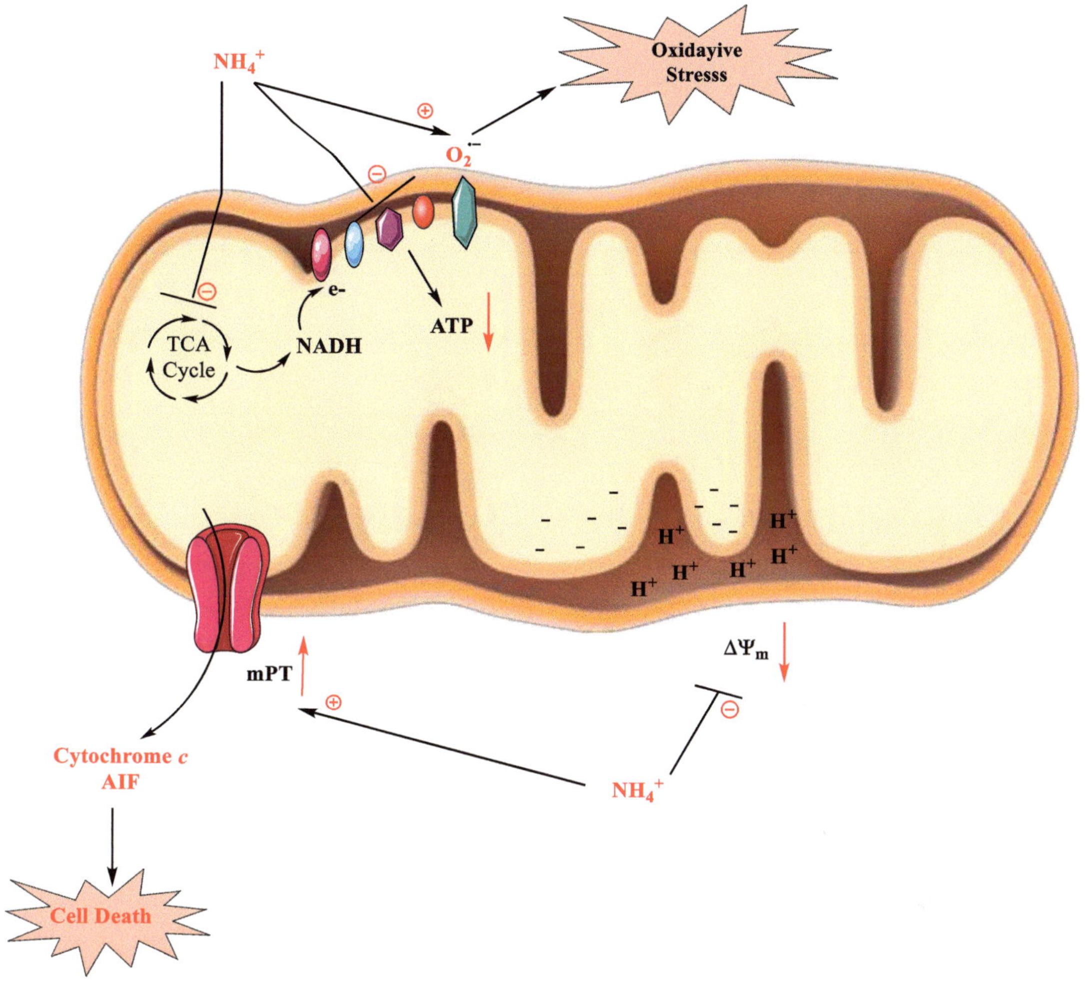

Fig. 1 Ammonia (NH_4^+)-induced mitochondrial impairment during hepatic encephalopathy could be mediated through several mechanisms. Induction of mitochondrial permeabilization and release of cell death mediators from this organelle, suppression of mitochondrial membrane potential ($\Delta\Psi$m), inhibition of mitochondrial electron transport chain components, enhancement of mitochondria-mediated reactive oxygen species (ROS) formation and oxidative stress, and the inhibition of tricarboxylic acid (TCA) cycle enzymes are reported mechanisms for NH_4^+-induced mitochondrial dysfunction in the CNS. mPT Mitochondria permeability transition, ATP adenosine triphosphate, AIF Apoptosis-inducing factor

hepatotoxic agent for inducing acute HE [15, 17] (Fig. 2). It has been mentioned that this model is relative to the clinical status of HE and complications such as brain edema are well-developed in TAA-treated animals [15]. The practicality and reproducibility of TAA-induced liver failure and HE are good in laboratory animals such as rats [15] (Fig. 2). Moreover, due to animal size, rats give sufficient samples (blood and tissue) to assess both mitochondrial indices and oxidative stress biomarkers (Fig. 2).

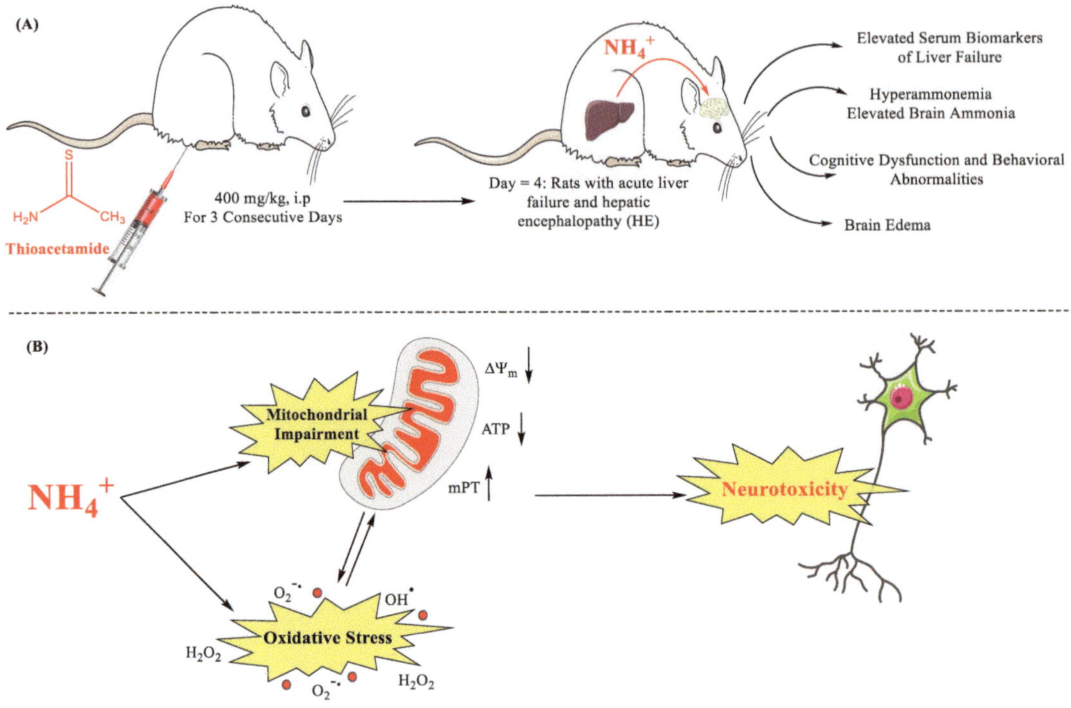

Fig. 2 A schematic representation of the animal model of acute liver failure and hepatic encephalopathy (**a**). Thioacetamide is frequently used as a hepatotoxicant to induce liver failure and hyperammonemia. Due to its large body size and sufficient sample supply, thioacetamide-induced liver injury in rats could be an appropriate model to investigate brain biomarkers of oxidative stress and mitochondrial function in hyperammonemia and hepatic encephalopathy (**a**). Oxidative stress and mitochondrial impairment are two interconnected events involved in the mechanism of ammonia (NH_4^+)-induced neurotoxicity (**b**). mPT Mitochondrial permeability transition, ATP adenosine triphosphate, $\Delta\Psi$ Mitochondrial membrane potential

2 Methods to Study Brain Mitochondrial Impairment and Oxidative Stress in HE

2.1 Materials

2.1.1 Materials for the Establishment of the In Vivo Model of Hepatic Encephalopathy

1. Thioacetamide.
2. Sterile dextrose solution containing sodium chloride and potassium chloride: Dissolve 5 g of D-glucose, 450 mg of NaCl, and 200 mg of KCl in 100 mL of distilled water in a volumetric flask and make up the volume to 100 mL. Sterile in an autoclave and use.
3. Syringe (1, 2.5, and 5 mL syringes).
4. Sterile isotonic NaCl solution (0.9% w: v; Normal saline): Dissolve 9 g of NaCl in 500 mL of distilled water in a graduated cylinder and make up the volume to 1000 mL. Sterile in an autoclave and use. Sterile commercially manufactured normal saline solutions could alternatively be utilized.

5. Thiopental for animal deep anesthesia. Thiopental should be used based on 100 mg/kg doses for intraperitoneal (i.p) injection.
6. Electric shaver.
7. Cotton gauze swabs.
8. Tissue paper.
9. Paraformaldehyde solution (10% v: v): Mix 27 mL of 37% paraformaldehyde solution with 73 mL of phosphate buffer saline (PBS).
10. Tubes for preparing blood serum.
11. Scissors.
12. Forceps.
13. Commercial kits for evaluating serum level of liver failure (alanine aminotransferase, aspartate aminotransferase, and lactate dehydrogenase).

2.1.2 Materials for Assessing Blood and Brain Ammonia Levels in the Animal Model of Hepatic Encephalopathy (HE)

1. Sodium nitroprusside 5% w: v solution: Dissolve 5 g of sodium nitroprusside in 90 mL of distilled water and bring the volume to 100 mL. Protect the solution from light.
2. Phenol 10% w: v: Dissolve 10 g of phenol in 80 mL of distilled water and add water to 100 mL (*see* **Note 1**).
3. Oxidizing agent (containing sodium hypochlorite dissolved in alkaline citrate buffer): First, to prepare the alkaline citrate buffer, add 800 mL of distilled water to a suitable container. Add 25.703 g of sodium citrate dihydrate and 2.421 g of citric acid to the solution. Adjust solution to final desired pH = 9. Then, add distilled water until the volume is 1 L. Add 1 mL of sodium hypochlorite to 4 mL of the alkaline citrate buffer to prepare the oxidizing agent (*see* **Note 2**).
4. Trichloroacetic acid 6% w: v: Dissolve 6 g of TCA in 90 mL of distilled water and bring the volume to 100 mL.
5. Ammonium chloride: use to prepare the standard curve.
6. Plastic tubes (10 mL).
7. Anticoagulant-coated (e.g., sodium citrate) tubes for blood plasma preparation.
8. Shaking incubator.
9. 96-Well plate.
10. Tissue homogenizer.
11. Centrifuge.
12. Spectrophotometer plate reader.

2.1.3 Materials for Evaluating Biomarkers of Oxidative Stress in the Brain Tissue of an Animal Model of Hepatic Encephalopathy

1. Tissue homogenization buffer (40 mM Tris–HCl buffer): Weigh 6.30 g of Tris–HCl and dissolve in 500 mL of distilled water. Adjust the pH to 7.4, and add water to increase the volume to 1 L. Store on ice or at 4 °C during use.
2. 2′,7′-Dichlorofluorescein diacetate (DCF-DA): Dissolve 48 mg of DCF-DA in 1 mL of dimethyl sulfoxide (DMSO) to obtain the stock solution (100 mM). Wrap the DCF-DA stock solution in aluminum foil and store it in the dark (−20 ° C). On the experiment day, add 10 μL of the 100 mM stock solution to 990 μl of 40 mM Tris–HCl buffer to obtain a DCF-DA working solution (1 mM) (*see* **Note 3**).
3. Dinitrophenylhydrazine (DNPH): Dissolve 19.8 mg of DNPH in 10 mL of 12 N HCl (*see* **Note 4**).
4. Trichloroacetic acid (TCA) 20% w: v solution: Add 20 g of TCA to 80 mL of distilled water in a volumetric flask and bring the volume to 100 mL.
5. Thiobarbituric acid reactive substances (TBARs) assay buffer: 10% w: v trichloroacetic acid (TCA), 0.375% w: v thiobarbituric acid (TBA), and 1 mL of hydrochloric acid (12 N) (pH = 2). Weigh 10 g of TCA and 0.375 g of TBA. Dissolve in 50 mL of distilled water (*see* **Note 5**). Add 1 mL of HCl (12 N) and bring the volume to 100 mL.
6. Ethanol: ethyl acetate (1 mL of 1:1 v: v) solution: Mix 100 mL of absolute ethanol with 100 mL of ethyl acetate (*see* **Note 6**).
7. Ferric reducing antioxidant power (FRAP) working solution: Mix 25 mL of 300 mM acetate buffer (pH = 2) with 2.5 mL of 20 mM iron chloride hexahydrate ($FeCl_3.6H_2O$) and 2.5 mL of 40 mM 2, 4, 6-Tris (2-pyridyl)-s-triazine; TPTZ (in HCl 6 N) to prepare FRAP working solution (*see* **Note 7**).
8. Tris–HCl buffer (40 mM) for glutathione content assay (pH = 8.9): Weigh 6.30 g of Tris–HCl and dissolve in 500 mL of distilled water in a volumetric flask. Add water to bring the volume up to 1 L. Adjust the pH to 8.9 (*see* **Note 8**).
9. 10 mM 5, 5′-Dithiobis (2-nitrobenzoic acid) (DTNB): Dissolve 19.8 mg of DTNB in 5 mL of methanol, wrap with aluminum foil, and keep in the dark. The DTNB solution should be prepared freshly on the day of the experiment.
10. Triton X-100.
11. Fluorimeter.
12. Coomassie Brilliant Blue.
13. 96-well plate.
14. n-Butanol,
15. Shaking incubator.
16. Bovine serum albumin.

17. Tissue homogenizer.
18. Centrifuge (4 °C).
19. Plastic tubes (10 mL).
20. n-Butanol.
21. Plate reader.

2.1.4 Materials for the Isolation of Mitochondria from the Brain in an Animal Model of Hepatic Encephalopathy

1. Extra pure water (e.g., Milli-Q® systems).
2. Mitochondria isolation buffer: 75 mM mannitol, 10 mM 3-(N-morpholino) propanesulfonic acid (MOPS), 175 mM sucrose, 1% w: v bovine serum albumin fraction V (BSA), and 1 mM ethylene glycol-bis (β-aminoethyl ether)-N, N, N′, N′--tetraacetic acid (EGTA). Weigh 59.9, 13.66, 2.09, 1 g, and 0.380 of sucrose, mannitol, MOPS, BSA, and EGTA, respectively. Dissolve the reagents in a volumetric flask in 800 mL of extra pure water. Adjust the pH to 7.2 and bring the volume to the final 1 L. Freeze (−20 °C) aliquots mitochondria isolation buffer solution to prevent fungal and bacterial infection. Thaw and use on the experiment day.
3. Mitochondria incubation buffer: The mitochondria incubation buffer in the current protocol consisted of 75 mM KCl, 150 mM sucrose, and 5 mM MOPS (pH = 7.2). The buffer type for preparing mitochondria suspension will depend on each experiment. To prepare 1 L of mitochondria incubation buffer: Weigh 51.34, 1.045 g, and 5.59 g of sucrose, MOPS, and KCl, respectively. Dissolve the reagents in 500 mL of extra pure water, adjust pH to 7.2, and bring the volume to the final 1 L.
4. Mitochondria incubation buffer for permeabilization and depolarization assay: 125 mM sucrose, 10 mM 4-(2-hydroxyethyl)-1-piperazineethanesulfonic acid (HEPES), 65 mM KCl, pH = 7.2. Weigh 42.78, 2.38, and 4.91 g of sucrose, HEPES, and KCl, respectively, and dissolve them in 500 mL of extra pure water in a volumetric flask. Adjust the pH to 7.2 and bring the volume to 1 L.
5. Bradford reagent: Weigh 100 mg of Coomassie Brilliant Blue and dissolve in 50 mL of absolute ethanol in a volumetric flask. Then, add 100 mL of phosphoric acid (85%) and bring the volume to 1000 mL by adding distilled water. Mix well and store in a dark glass bottle (*see* **Note 9**).
6. Plate reader.
7. Homogenizer.
8. Centrifuge (4 °C).
9. 50 mL tubes.
10. Microtube (2 mL).

2.1.5 Materials for the Assessment of the Brain Mitochondrial Indices in a Hepatic Encephalopathy Animal Model

1. Rhodamine 123: Weigh 38.1 mg of rhodamine and dissolve it in 1 mL of absolute ethanol to obtain 100 mM stock solution. The stock should be divided to aliquots of 250 μL samples, wrapped with aluminum foil, and stored in the dark (−20 °C). To prepare the rhodamine 123 working solution (on the day of the experiment), add 10 μL of the stock solution to 990 μL of incubation buffer to obtain 1 mM rhodamine. Add 10 μL of 1 mM rhodamine to each sample tube for each experiment (≈1 mL; Final concentration 10 μM). Tubes should be protected from light during the experiment.
2. Methyl tetrazolium solution (10 mg/mL solution): Weigh 100 mg of MTT and dissolve it in 10 mL of phosphate buffer saline (PBS; pH = 7.4) (*see* **Note 10**).
3. Plate reader (Capable of temperature setting, continuous shaking, and measuring absorbance in a kinetic mode).
4. 96-well plate.
5. Methyl tetrazolium (MTT).
6. ATP detection kit (Luminescence based).
7. Trichloroacetic acid (TCA; 0.5% w: v): Weigh 500 mg of TCA into a 50 mL volumetric flask and add 50 mL of extra pure water. Dissolve and bring the volume up to the final 100 mL.
8. Luminometer.
9. Calcium chloride stock solution (1 mM): Weigh 11 mg of $CaCl_2$ and dissolve it in 100 mL of extra pure water to prepare a 1 mM $CaCl_2$ stock solution.
10. Microcentrifuge (4 °C).
11. Microtube (2 mL).

3 Methods

3.1 Establishment of the In Vivo Model of Hepatic Encephalopathy (HE)

1. Weigh rats.
2. Inject 2.5 mL/kg of normal saline to control (vehicle-treated) animals.
3. Inject thioacetamide (400 mg/kg; dissolved in normal saline) to animals subjected to hepatic encephalopathy.
4. Repeat the thioacetamide dose for three consecutive days.
5. Inject KCl and NaCl-containing dextrose 5% solution by subcutaneous (*s.c*) route every day after thioacetamide administration (*see* **Note 11**).
6. On day = 4, thioacetamide-treated animals are hyperammonemic and show signs and symptoms of hepatic encephalopathy.

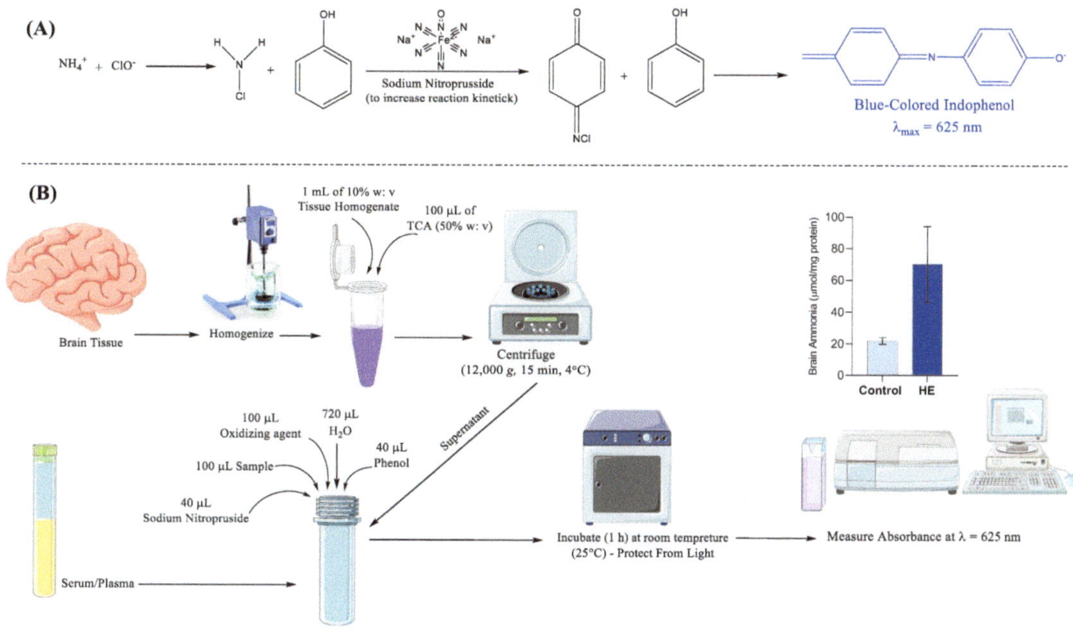

Fig. 3 The phenate-perchlorate reaction, also known as the Bertholet reaction, is a convenient procedure to assess ammonium levels in biological samples (**a**). Ammonium chloride could be used to prepare the standard curve. Brain tissue homogenization should be digested by trichloroacetic acid (6% w: v) before ammonia measurement (**b**). Samples such as urine, serum, and plasma could be directly used in the Bertholet reaction for ammonia detection (**b**). The homogenized brain tissue (10% w: v in 40 mM Tris–HCl buffer; pH = 7.4) could be used in other experiments explained for assessing biomarkers of oxidative stress in this protocol

7. On the day of the experiment (Day = 5), anesthetize rats (thiopental 100 mg/kg, i.p) and collect the liver tissue and blood samples for further assessments such as serum biomarkers of acute liver failure, hyperammonemia, and histopathological examinations (*see* **Note 12**). Excise the brain for assessing tissue ammonia, biomarkers of oxidative stress, and mitochondrial indices (Fig. 3).

3.2 Plasma and Brain Ammonia Level

1. Homogenize brain tissue (≈500 mg) in 5 mL of ice-cooled Tris–HCl buffer (40 mM, 4 °C, pH = 7.4) (Fig. 3).
2. Add 100 μL of 50% w: v trichloroacetic acid (TCA) to 1 mL of brain tissue homogenate.
3. Centrifuge (12,000 *g*, 15 min, 4 °C).
4. Mix 100 μL of the supernatant with 40 μL of sodium nitroprusside,100 μL of oxidizing agent, 720 μL of distilled water, and 40 μL of phenol (Fig. 3) (*see* **Note 13**).
5. Incubate (25 °C) samples in the dark for 1 h.
6. Measure the absorbance at λ = 625 nm (Fig. 3).

3.3 Biomarkers of Oxidative Stress in the Brain Tissue

3.3.1 Estimation of Brain Tissue Reactive Oxygen Species (ROS) Formation

1. Weigh the brain (≈500 mg) and homogenize in 5 mL of ice-cooled (4 °C) 40 mM Tris–HCl buffer (pH = 7.4) (Fig. 4).
2. Add 100 μl of well-mixed tissue homogenate to 900 μL of 40 mM Tris–HCl buffer (pH = 7.4) in a 2 mL microtube.
3. Add 10 μL of 1 mM DCF-DA (Final concentration of 10 μM) to the samples (Fig. 4).
4. Incubate the mixture in the dark (shaking incubator, 37 °C) for 15 min.
5. Add 100 μL of samples to a 96-well plate and measure the fluorescence intensity ($\lambda_{\text{excitation}} = 485$ and $\lambda_{\text{emission}} = 525$) (Fig. 4).

3.3.2 Brain Tissue Protein Carbonylation

1. Add 500 μL of brain tissue homogenate (described in Subheading 3.3.1) to a 2 mL microtube (Fig. 5).
2. Treat tissue samples with 100 μL of Triton X-100 and vortex.
3. Centrifuge (700 *g*, 10 min, 4 °C).
4. Collect the supernatant and add 500 μL of 10 mM DNPH solution.
5. Incubate samples in the dark (25 °C) for 1 h. Vortex every 15 min (Fig. 5).
6. Add 100 μL of TCA (20% w: v), mix well, and centrifuge (10,000 *g*, 10 min, 4 °C).

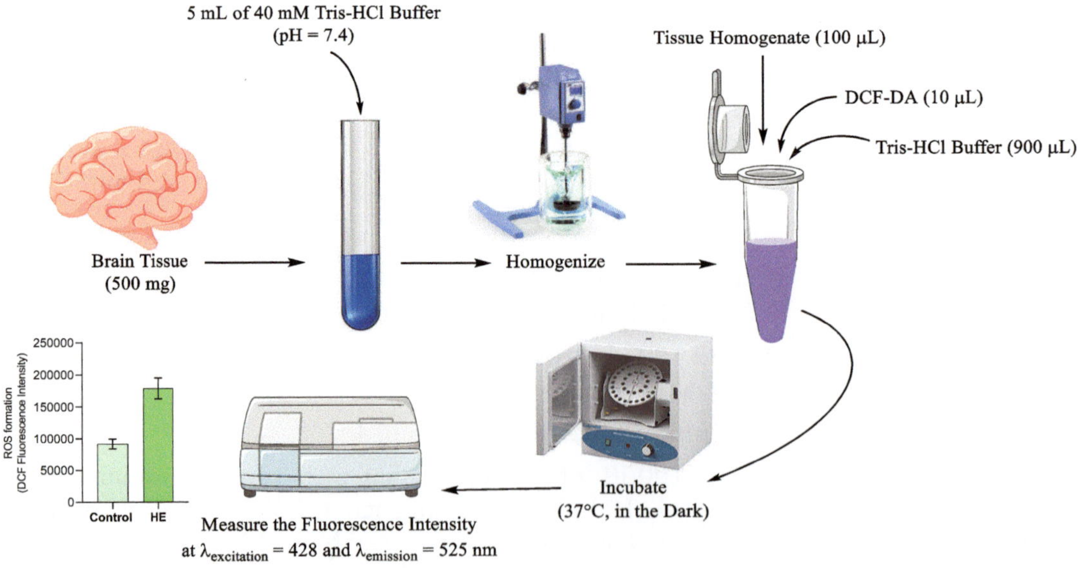

Fig. 4 Dichlorofluorescein diacetate (DCF-DA) is commonly used as a probe to detect reactive oxygen species (ROS) in biological samples. DCF solution should be freshly prepared on the experiment day and protected from light. DCF (non-fluorescent) is converted to fluorescent DCF upon interaction with ROS. Higher fluorescent intensity indicates a higher level of ROS

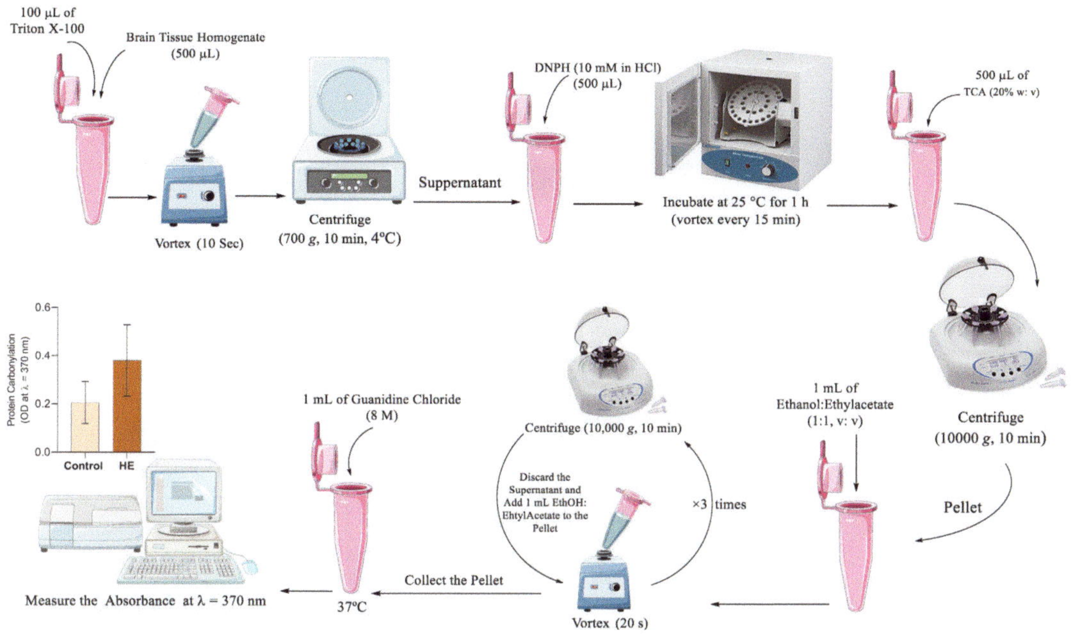

Fig. 5 Dinitrophenyl hydrazine (DNPH) reacts with oxidized protein carbonyl groups to form protein-DNP complexes (maximum absorbance at $\lambda = 370$ nm). It is important to remember that adequately washing the pellet with ethanol-ethyl acetate solution is critical to removing unreacted DNPH and preventing false-positive results. The final pellet also should be dissolved in guanidine chloride solution at 37 °C

7. Discard the supernatant and dissolve the pellet in 1 mL of ethanol: ethyl acetate (1:1 v: v) solution.
8. Centrifuge (10,000 *g*, 10 min, 4 °C) and repeat pellet washing thrice (Fig. 5).
9. Dissolve the final pellet in 1 mL of guanidine chloride solution (8 M) while heating at 37 °C.
10. Measure the absorbance at $\lambda = 370$ nm (Fig. 5).

3.3.3 Tissue Lipid Peroxidation

1. Add 1 mL of brain tissue homogenate (described in Subheading 3.3.1) to 1 mL of TBARs assay reagent and mix well (Fig. 6).
2. Heat samples at 100 °C for 45 min (in a water bath).
3. Add 1 mL of n-butanol and mix well (vortex).
4. Centrifuge samples (10,000 *g*, 10 min), add 100 μl of the n-butanol (upper) phase to a 96-well plate and measure the absorbance ($\lambda = 532$ nm) (Fig. 6).

3.3.4 The Total Antioxidant Capacity of the Brain Tissue

1. Add 100 μL of the well-mixed brain tissue homogenate (described in Subheading 3.3.1) to 900 μL of the freshly prepared FRAP working solution (Fig. 7).

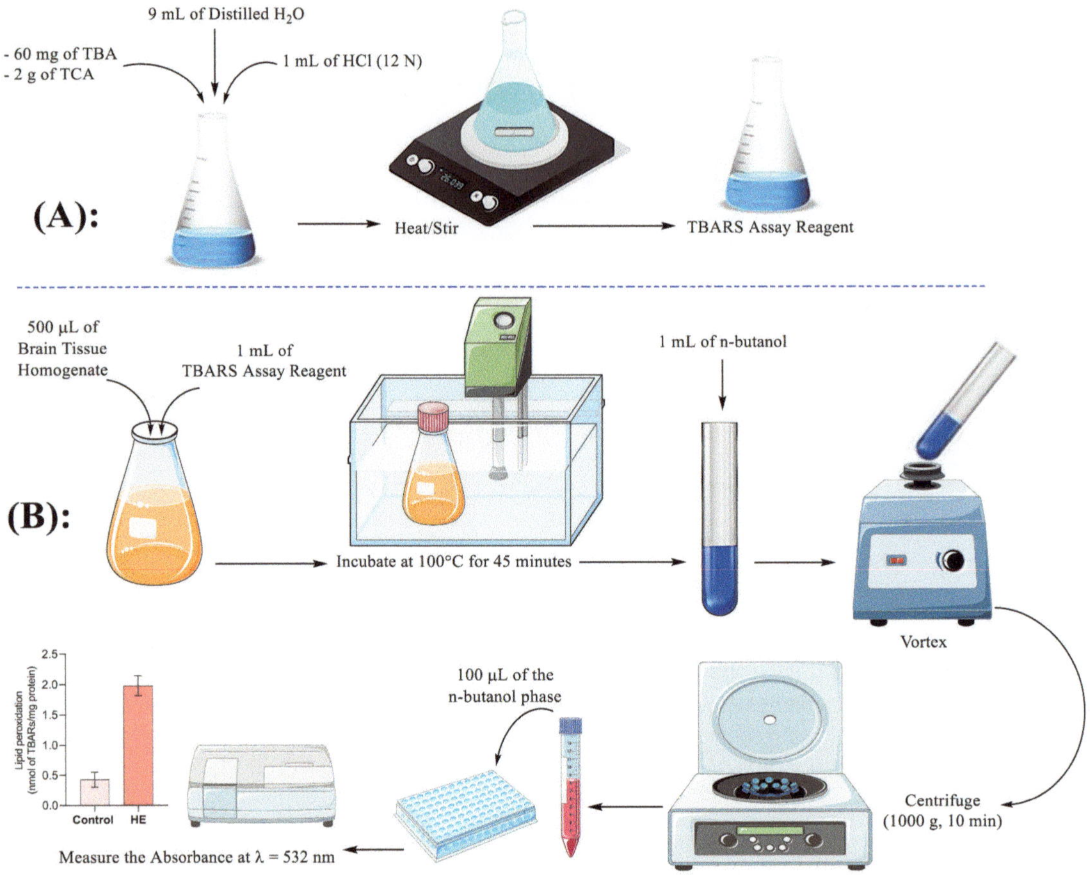

Fig. 6 Thiobarbituric acid (TBA) forms a stable complex with aldehyde product of biomembranes lipid peroxidation. The reaction occurs in an acidic environment (pH = 2) and high temperature (100 °C)

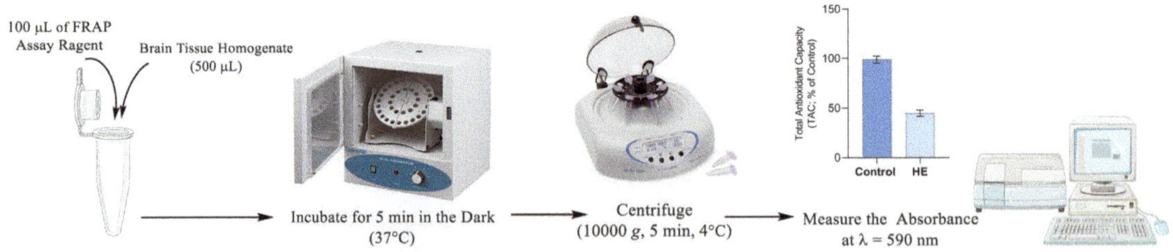

Fig. 7 The probe, 2,4,6-Tri(2-pyridyl)-s-triazine (TPTZ), reacts with ferrous iron (Fe^{2+}) to form a deep-blue complex. The capacity of tissue antioxidants to reduce ferric iron (Fe^{3+}) to Fe^{2+} is measured in the ferric reducing antioxidant capacity (FRAP) test

2. Incubate the reaction mixture for 5 min in the dark (37 °C shaker incubator).
3. Centrifuge (10,000 *g*, 5 min, 4 °C) and measure the absorbance at $\lambda = 590$ nm using a plate reader (Fig. 7).

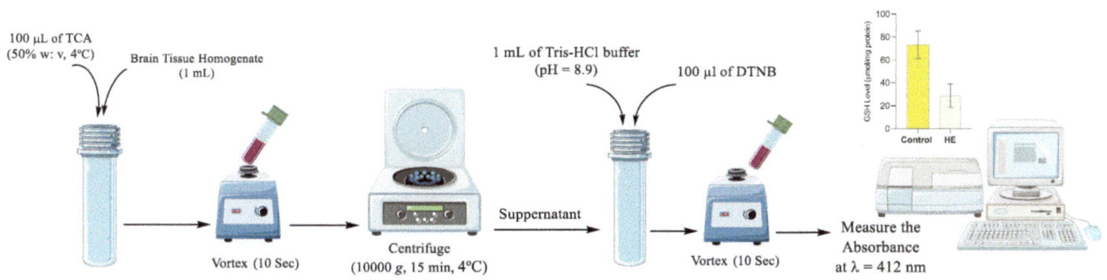

Fig. 8 Dithio-bis-nitro benzoic acid (DTNB) is a commonly used indicator for assessing reduced glutathione (GSH) levels in biological samples. DTNB is converted to GSTNB, and the yellow-colored $TNB^{\bullet-}$ (maximum absorbance at $\lambda = 412$ nm). A higher level of $TNB^{\bullet-}$ absorbance indicates a higher GSH content

3.3.5 Brain Reduced Glutathione (GSH) Content

1. Treat 1 mL of brain tissue homogenate (described in Subheading 3.3.1) with 100 μL of ice-cooled (4 °C) TCA (50% w: v) (Fig. 8).
2. Vortex (10 s) and centrifuge (10,000 *g*, 15 min, 4 °C).
3. Collect the supernatant and add 1 mL of cold 40 mM Tris–HCl buffer (pH = 8.9).
4. Add 100 μl of freshly prepared DTNB solution (10 mM) and mix.
5. Read the absorbance at $\lambda = 412$ nm (Fig. 8).

3.4 Isolation of Mitochondria from the Brain in HE Animals

Figure 9 illustrates the schematic depiction of the isolation process for brain tissue mitochondria using the differential centrifugation method. As a step-by-step procedure, follow the following instructions.

1. Anesthetize rats with intraperitoneal injection of thiopental (100 mg/kg) (Fig. 9).
2. Excise the whole brain (≈500 mg) and mince in a Petri dish containing ice-cooled (4 °C) mitochondria isolation buffer.
3. Homogenize the brain tissue in a Potter Elvehjem Teflon pestle homogenizer (1:10 w: v tissue: isolation buffer ratio).
4. Transfer the tissue homogenate to 50 mL tubes and balance with ice-cooled (4 °C) mitochondria isolation buffer.
5. Centrifuge (1000 *g*, 20 min, 4 °C) and collect the supernatant. Add fresh, ice-cold mitochondria isolation buffer medium.
6. Centrifuge (10,000 *g*, 20 min, 4 °C) and discard the supernatant. Re-suspend the pellet in a fresh mitochondria isolation buffer. Mix and centrifuge (10,000 *g*, 20 min, 4 °C) again. Repeat the recent step at least three times to increase the mitochondria yield (Brown pellet) (Fig. 9) (*see* **Note 14**).
7. Add incubation buffer and keep it on ice in the refrigerator (4 °C) (Fig. 3) (*see* **Note 15**).

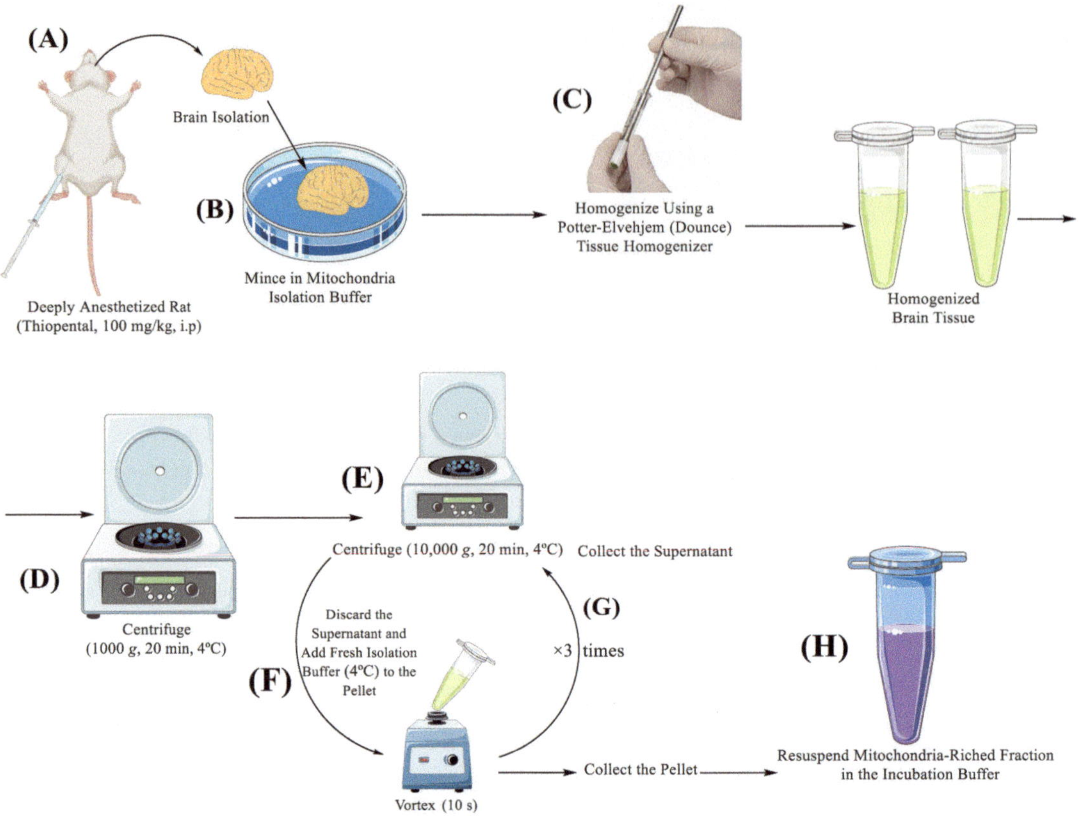

Fig. 9 Differential centrifugation is a convenient method to prepare mitochondrial fractions from biological samples. The process involves several steps of centrifugation at progressively higher speeds to separate cellular components based on size, density, and shape. For brain mitochondria isolation, the whole brain is first excised from deeply anesthetized animals (thiopental 100 mg/kg, i.p) (**a** and **b**). Then, cells are first broken open (lysed) to release their contents. This can be achieved using various methods, such as homogenization (**c**). The cell lysate is then centrifuged at a low speed (1000 *g*) to remove cell debris, large organelles, and nuclei (**d**). The resulting supernatant, containing smaller organelles and cytoplasmic components, is transferred to a new tube and centrifuged at a higher speed (10,000 *g*) to pellet down mitochondria and other organelles like lysosomes and peroxisomes. This step helps to further enrich mitochondria (**e**–**g**). The resulting pellet contains a mitochondria-rich fraction, which can be resuspended in a storage buffer (4 °C) and used directly for downstream experiments (**h**)

3.4.1 Measurement of Mitochondrial Dehydrogenases Activity

1. Suspend isolated brain mitochondria (0.5 mg protein/mL) in 1 mL of the incubation buffer (Fig. 10).
2. Add 50 μL of MTT solution (10 mg/mL) (*see* **Note 10**).
3. Incubate in the dark for 30 min (37 °C).
4. Centrifuge (10,000 *g*, 10 min). Discard the supernatant and dissolve the pellet in 1 mL of DMSO.
5. Mix well and centrifuge (10,000 *g*, 10 min).
6. Read the absorbance at $\lambda = 570$ nm.

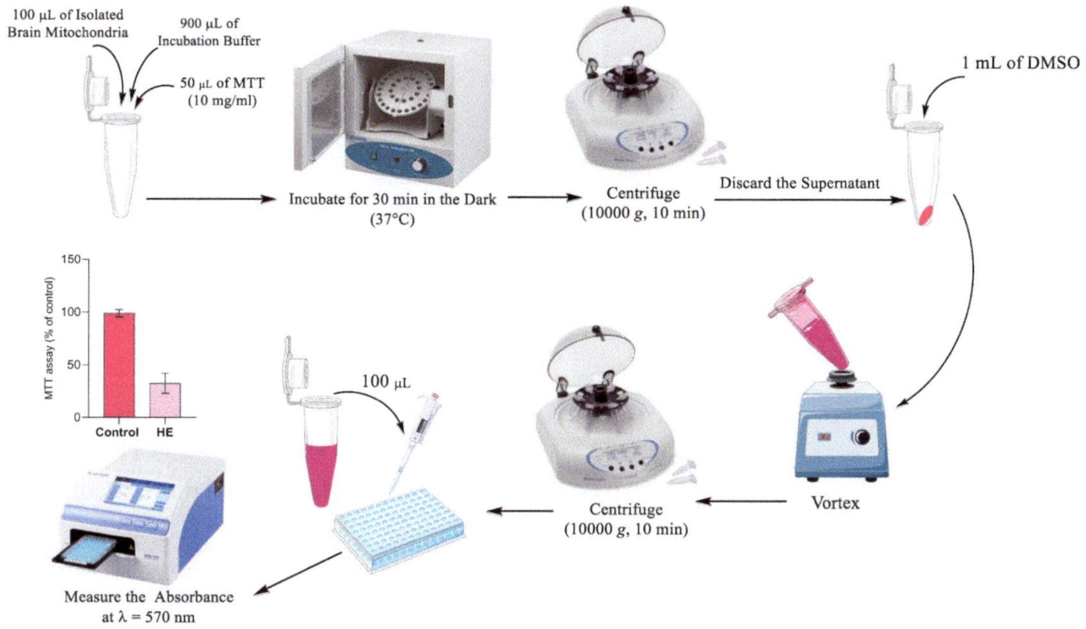

Fig. 10 Mitochondrial dehydrogenases activity is assessed by methyl-tetrazolium (MTT) test. MTT is reduced to purple-colored formazan dye (maximum absorbance at $\lambda = 570$ nm) by mitochondrial dehydrogenases enzymes (e.g., succinate dehydrogenase)

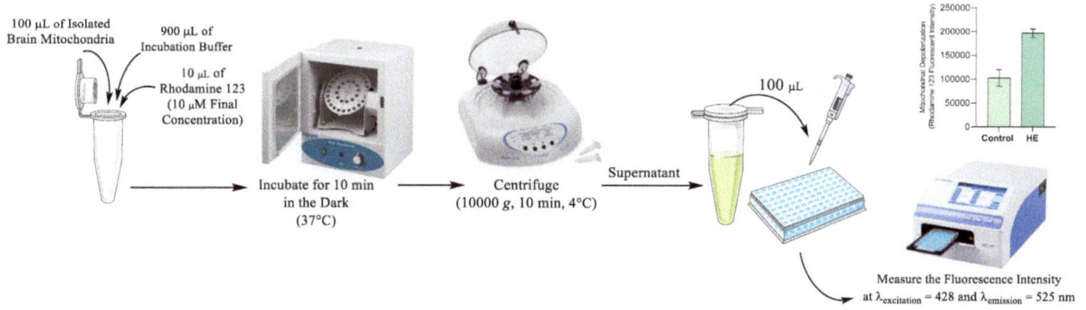

Fig. 11 Rhodamine 123 is a commonly used probe for estimating mitochondrial depolarization. Rhodamine 123 accumulates in intact mitochondria. Consequently, the intensity of this fluorescent agent is decreased in the supernatant. Rhodamine 123 should be freshly prepared on the experiment day and protected from light

3.4.2 Assessment of Mitochondrial Depolarization

1. Add 100 μL of isolated brain mitochondria (1 mg protein/mL) to 900 μL of mitochondria depolarization assay buffer (Fig. 11).
2. Incubate samples at 37 °C (20 min in the dark).
3. Centrifuge (15,000 *g*, 1 min, 4 °C).
4. Add 100 μl of the supernatant to a 96-well plate and measure the fluorescence intensity (FI) ($\lambda_{excitation} = 485$ and $\lambda_{emission} = 525$ nm (Fig. 11). An increase in the supernatant FI is directly connected with increased mitochondrial impairment [18].

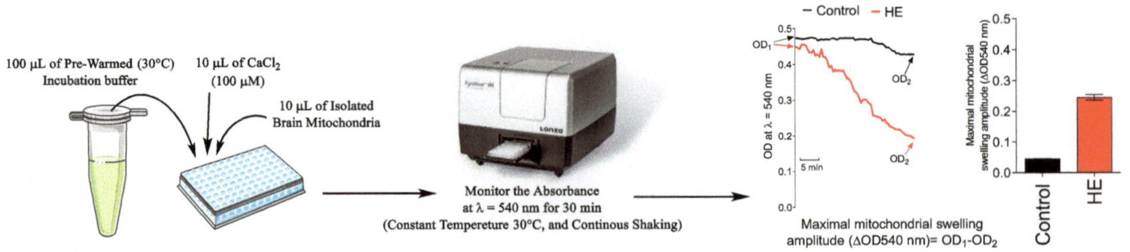

Fig. 12 Calcium (Ca^{2+})-induced mitochondrial permeabilization and swelling is a convenient method to assess mitochondrial functionality. The samples' optical density (OD) difference is associated with mitochondrial swelling amplitude

3.4.3 Mitochondrial Permeabilization and Swelling

1. Suspend isolated mitochondria (0.5 mg protein/mL) in pre-warmed (30 °C) mitochondria swelling buffer (Fig. 12).
2. Add 100 μL of isolated mitochondria and 400 μL of incubation buffer to a 96-well plate.
3. Add 10 μL of the calcium chloride stock solution (1 mM) to induce mitochondrial permeabilization and swelling.
4. Monitor the light absorbance (Optical density; OD) at $\lambda = 540$ nm (Fig. 12) using a plate reader (30 °C constant temperature, 30 min, Kinetic mode, 30-s intervals) (Fig. 12). It is widely acknowledged that the reduction in light absorbance correlates with an escalation in mitochondrial swelling and permeabilization [19].
5. Calculate the difference between two time points (maximal mitochondrial swelling amplitude) (*see* **Note 17**) (Fig. 12). The maximal mitochondrial swelling amplitude (ΔOD540 nm) = OD1–OD2 [20, 21].

3.4.4 Mitochondrial ATP Content

1. Add 100 μl of 0.5% w: v TCA solution to 500 μL of isolated brain mitochondria (1 mg protein/mL) (*see* **Note 18**) (Fig. 13).
2. Incubate on ice for 5 min.
3. Centrifuge samples (15,000 *g*, 10 min, 4 °C) and collect the supernatant.
4. Prepare ATP kit solutions based on kit instructions.
5. Add 100 μl of ATP detection kit content to 100 μl of the mitochondria supernatant and measure the luminescence intensity ($\lambda = 560$ nm) (Fig. 13).

3.4.5 Lipid Peroxidation in Isolated Brain Mitochondria

1. Mix 500 μL of stock isolated brain mitochondria with 1 mL of ice-cold MOPS-KCl (50 mM–100 mM; pH = 7.4) solution (Fig. 14).
2. Centrifuge (10,000 *g*, 10 min, 4 °C).

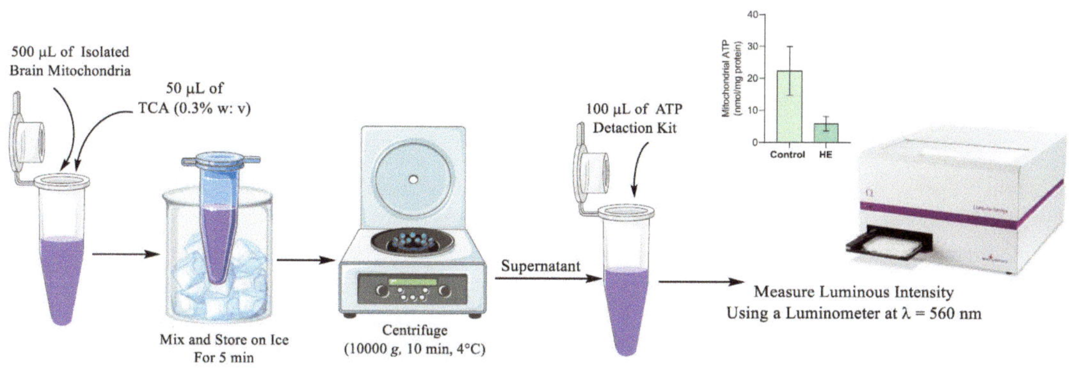

Fig. 13 The activity of ATP-dependent luciferase enzyme to produce luciferin and light is a convenient method for assessing ATP levels in biological samples (e.g., isolated brain mitochondria). Note: as trichloroacetic acid (TCA) could decrease the luciferase enzyme activity, minimum volume and concentration (50 μL of 0.3% w: v) should be applied in this test

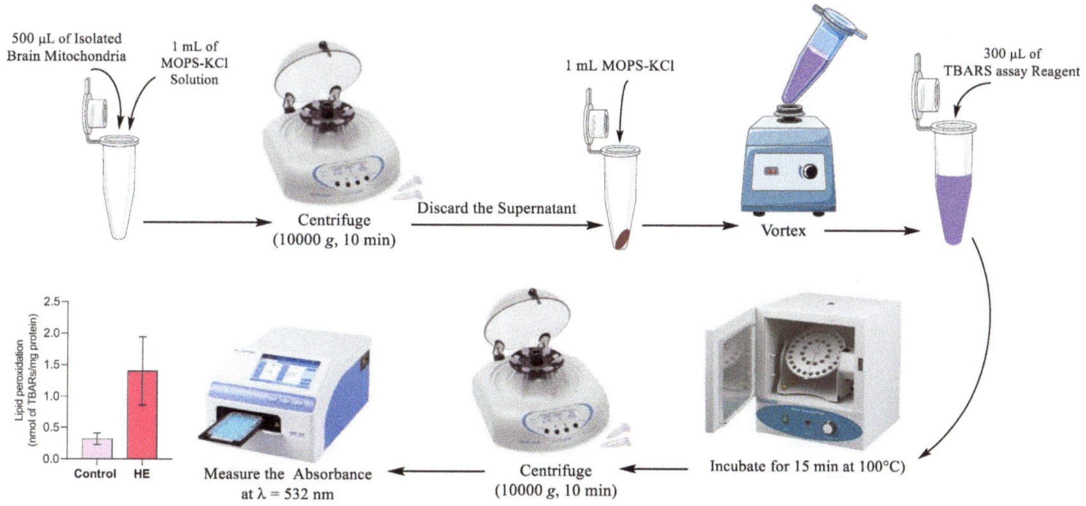

Fig. 14 Evaluating lipid peroxidation in isolated brain mitochondria. Lipid peroxidation could be determined by the thiobarbituric acid-reactive substances (TBARS) assay. However, as sucrose interferes with the TBARS assay, isolated mitochondria should be washed to remove sucrose from the medium. For this purpose, isolated brain mitochondria are suspended in a solution with MOPS-KCl (50 mM MOPS and 100 mM KCl, pH = 7.4) and centrifuged. Then, the pellet is resuspended in MOPS-KCl solution and undergoes TBARS assay

3. Resuspend the pellet in 1 mL of MOPS-KCl solution (4 °C).
4. Add 300 μL of TBARS assay reagent to each sample and mix.
5. Heat samples (100 °C) in a water bath for 15 min.
6. Centrifuge (10,000 *g*, 10 min, 4 °C) and measure the absorbance at λ = 532 nm.

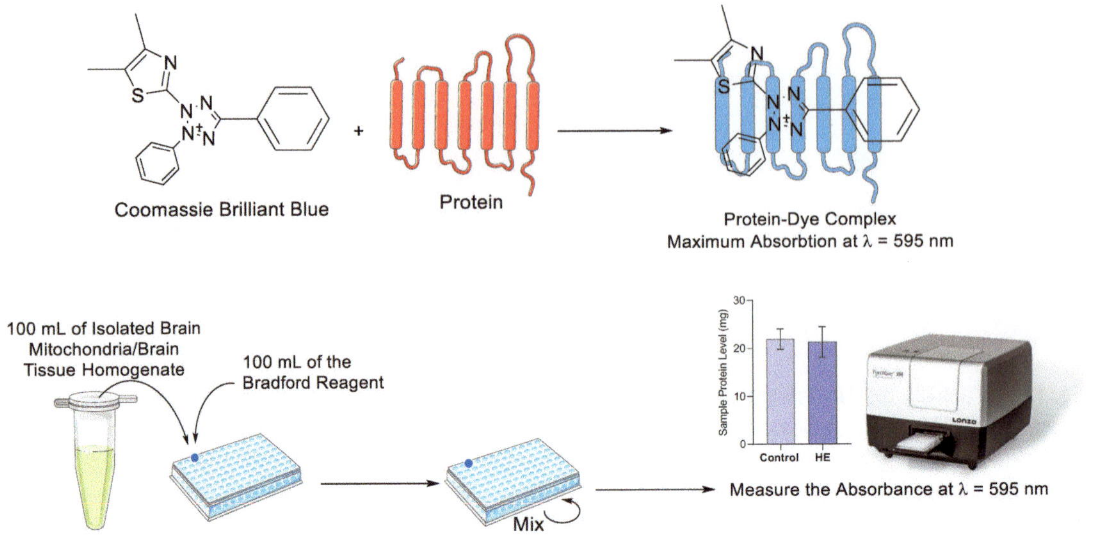

Fig. 15 The Bradford reagent (Coomassie Brilliant Blue) reacts with proteins to form a stable protein-dye complex (maximum absorbance at $\lambda = 595$ nm). The Bradford method is the most applied procedure for assessing samples' protein content, which is crucial for data standardization

3.4.6 Assessing Samples' Protein Content for Data Standardization

1. Mix 100 μL of brain tissue homogenate or isolated brain mitochondria samples with 100 μl of the Bradford reagent in a 96-well plate (Fig. 15).
2. Mix well and leave for 5 min (*see* **Note 18**).
3. Measure the absorbance at $\lambda = 595$ nm using a plate reader.
4. Use bovine serum albumin (BSA) to prepare the standard curve and convert the samples' absorbance to mg of protein (Fig. 15).

4 Notes

1. Work with phenol in a chemical fume hood.
2. For accurate results, the oxidizing agent should be freshly prepared on the day of the experiment.
3. The DCF-DA working solution (1 mM) should be freshly prepared on the day of the experiment by adding 10 μL of the stock solution to 990 μL of Tris–HCl buffer.
4. Dissolving DNPH requires vigorous stirring and slight heating.
5. The TBA solution requires a gentle application of heat to get fully dissolved (Fig. 6).
6. This solution is highly volatile. Be sure to store it in containers with tight lids, preferably in the refrigerator.

7. The FRAP working solution should be prepared on the day of the experiment and protected from light.
8. The precise estimation of GSH content relies heavily on the critical pH of the Tris–HCl buffer, which is set at 8.9. A higher pH level could result in a pseudo-yellow color, while a lower pH level may lead to white precipitations forming in the samples.
9. The Bradford reagent is gradually photodegraded. Hence, ideally, it is better to prepare it freshly. However, if stored in dark bottles and protected from light, it could be stored for up to 1 month.
10. Perform vigorous shaking for almost 10 min with the vortex before using the MTT solution. The solution should be protected from light.
11. The dextrose-KCl-NaCl (5%, 0.2%, and 0.45% w: v respectively) should be administered after each dose of thioacetamide to avoid weight loss, hypoglycemia, and renal failure and decrease the mortality rate in thioacetamide-induced liver failure and HE model.
12. Commercial kits could be used to assess biomarkers of liver injury (serum LDH, ALT, and AST) to ensure hepatic failure in thioacetamide-treated rats.
13. Samples (100 μL) such as urine, serum, and plasma could be directly treated with phenol, oxidizing agent, and sodium nitroprusside for assessing ammonia content.
14. The isolated mitochondria can be preserved on ice or in the refrigerator (4 °C) for at least 4 h without any notable alterations in the experimental outcomes. Additionally, the isolated brain mitochondria can be stored at −80 °C to analyze parameters like the activity of mitochondrial respiratory chain complexes.
15. The isolated brain mitochondria can be prepared at a final concentration of 20 mg protein/mL to serve as the stock mitochondria solution for evaluating mitochondrial functional parameters.
16. Based on our expertise, we recommend taking into account time points following a 10-min incubation period as OD1.
17. TCA is considered to be among the top agents for extracting ATP from biological samples. Additionally, this compound can deactivate enzymes that could break down ATP before analysis. Typically, a concentration of 0.5–2.5% TCA in a sample is sufficient for ATP extraction. It is important to note that an excessive amount of TCA may hinder the luciferase enzyme

reaction; therefore, it is recommended to use the minimum concentration required for effective ATP extraction.

18. It is recommended to measure the absorbance of the samples and standard between 5 and 10 min after the addition of the Bradford reagent, as an extended incubation period may reduce the precision of the assay.

Acknowledgments

This work was financially supported by a grant (#23195) from the Vice Chancellor of Research Affairs of Shiraz University of Medical Sciences, Shiraz, Iran.

References

1. Häussinger D, Dhiman RK, Felipo V, Görg B, Jalan R, Kircheis G, Merli M, Montagnese S, Romero-Gomez M, Schnitzler A (2022) Hepatic encephalopathy. Nat Rev Dis Primers 8(1):43
2. Heidari R, Ommati MM, Niknahad H (2023) 12 – Ammonia. In: de Oliveira MR (ed) Mitochondrial intoxication. Academic, New York, pp 249–270
3. Heidari R (2024) Cellular and molecular mechanisms of ammonia-induced neurotoxicity: a neurotherapeutic prospect. In: Oliveira MR (ed) Natural molecules in neuroprotection and neurotoxicity. Academic, pp 957–995
4. Heidari R (2019) Brain mitochondria as potential therapeutic targets for managing hepatic encephalopathy. Life Sci 218:65–80
5. Sokol RJ, Devereaux M, Dahl R, Gumpricht E (2006) "Let there be bile"—understanding hepatic injury in cholestasis. J Pediatr Gastroenterol Nutr 43(Suppl 1):S4–S9
6. Arduini A, Serviddio G, Tormos AM, Monsalve M, Sastre J (2012) Mitochondrial dysfunction in cholestatic liver diseases. Front Biosci (Elite Edition) 4:2233–2252
7. Mormone E, George J, Nieto N (2011) Molecular pathogenesis of hepatic fibrosis and current therapeutic approaches. Chem Biol Interact 193(3):225–231
8. Poli G (2000) Pathogenesis of liver fibrosis: role of oxidative stress. Mol Asp Med 21(3):49–98
9. Brookes PS, Yoon Y, Robotham JL, Anders MW, Sheu S-S (2004) Calcium, ATP, and ROS: a mitochondrial love-hate triangle. Am J Phys Cell Phys 287(4):C817–CC33
10. Kowaltowski AJ, Vercesi AE (1999) Mitochondrial damage induced by conditions of oxidative stress. Free Radic Biol Med 26(3):463–471
11. Bai G, Kv RR, Murthy CRK, Ks P, Ar J, Norenberg M (2001) Ammonia induces the mitochondrial permeability transition in primary cultures of rat astrocytes. J Neurosci Res 66(5):981–991
12. Heidari R, Jamshidzadeh A, Ghanbarinejad V, Ommati MM, Niknahad H (2018) Taurine supplementation abates cirrhosis-associated locomotor dysfunction. Clin Exp Hepatol 4(2):72–82
13. Heidari R, Jamshidzadeh A, Ommati MM, Rashidi E, Khodaei F, Sadeghi A, Hosseini A, Niknahad H (2019) Ammonia-induced mitochondrial impairment is intensified by manganese co-exposure: relevance to the management of subclinical hepatic encephalopathy and cirrhosis-associated brain injury. Clin Exp Hepatol 5(2):109–117
14. Ghanbarinejad V, Ahmadi A, Niknahad H, Ommati MM, Heidari R (2019) Carnosine mitigates manganese mitotoxicity in an in vitro model of isolated brain mitochondria. Adv Pharm Bull 9(2):294–301
15. DeMorrow S, Cudalbu C, Davies N, Jayakumar AR, Rose CF (2021) 2021 ISHEN guidelines on animal models of hepatic encephalopathy. Liver Int 41(7):1474–1488
16. Heidari R, Niknahad H, Sadeghi A, Mohammadi H, Ghanbarinejad V, Ommati MM, Hosseini A, Azarpira N, Khodaei F, Farshad O, Rashidi E, Siavashpour A, Najibi A, Ahmadi A, Jamshidzadeh A (2018) Betaine treatment protects liver through

regulating mitochondrial function and counteracting oxidative stress in acute and chronic animal models of hepatic injury. Biomed Pharmacother 103:75–86

17. Heidari R, Jamshidzadeh A, Niknahad H, Mardani E, Ommati MM, Azarpira N, Khodaei F, Zarei A, Ayarzadeh M, Mousavi S, Abdoli N, Yeganeh BS, Saeedi A, Najibi A (2016) Effect of taurine on chronic and acute liver injury: Focus on blood and brain ammonia. Toxicol Rep 3:870–879
18. Mohammadi H, Sayad A, Mohammadi M, Niknahad H, Heidari R (2020) N-acetyl cysteine treatment preserves mitochondrial indices of functionality in the brain of hyperammonemic mice. Clin Exp Hepatol 6(2):106–115
19. Heidari R, Arabnezhad MR, Ommati MM, Azarpira N, Ghodsimanesh E, Niknahad H (2019) Boldine supplementation regulates mitochondrial function and oxidative stress in a rat model of hepatotoxicity. Pharm Sci 25(1): 1–10
20. Caro AA, Adlong LW, Crocker SJ, Gardner MW, Luikart EF, Gron LU (2012) Effect of garlic-derived organosulfur compounds on mitochondrial function and integrity in isolated mouse liver mitochondria. Toxicol Lett 214(2):166–174
21. Niknahad H, Jamshidzadeh A, Heidari R, Zarei M, Ommati MM (2017) Ammonia-induced mitochondrial dysfunction and energy metabolism disturbances in isolated brain and liver mitochondria, and the effect of taurine administration: relevance to hepatic encephalopathy treatment. Clin Exp Hepatol 3(3): 141–151

Chapter 13

Nessler Histological Staining for Ammonia Detection in Formalin-Fixed Paraffin-Embedded (FFPE) Mammalian Tissue

Teresa C. Delgado, Javier de las Heras, and María Luz Martínez-Chantar

Abstract

Hyperammonemia, a life-threatening condition, is a common complication of a wide variety of disorders as well as a primary etiologic agent of hepatic encephalopathy (HE), a decline in brain functions secondary to advanced liver dysfunction and failure. Measurements of ammonia in blood were first described more than a century ago and since then, several new methods have been proposed for the determination of ammonia in biological fluids. The direct enzymatic determination of ammonia in plasma is nowadays the most commonly used ammonia assay in clinical chemistry. Here, we summarize the current recommendations for the quantification of ammonia in biological fluids, in particular in plasma, by listing sample pre-analysis handling requisites and principal potential contaminants.

Ammonia homeostasis is a whole-body dynamic process tightly regulated. Assessing the altered accumulation of ammonia and disrupted metabolism in the different organ/systems tissues is important for increasing our knowledge of the underlying disturbed molecular mechanisms in a series of pathologies. Therefore, the establishment of adequate methodologies for the detection of ammonia in mammalian tissue is of paramount importance. Herein, we provide a detailed description of a method for detecting ammonia in formalin-fixed paraffin-embedded mammalian tissue. This histological-based method takes advantage of the Nessler colorimetric reaction to detect ammonia analyzing this metabolite distribution in specific regions of interest of the mammalian brain, liver, and other organs of interest.

Key words Ammonia, Nessler, Formalin-fixed paraffin-embedded tissue, Hepatic encephalopathy, Liver, Brain

1 Introduction: Whole-Body Ammonia Homeostasis

In the human body, the ammonium ion is formed from ammonia gas and vice versa in an equilibrium reaction NH_3 (gas) + H^+ NH_4^+, with a pKa of about 9.2. At a physiological pH, 99% is in the form of the ammonium anion, NH_4^+, and only 1% is in the form of NH_3 [1]. Usually, the term ammonia refers to the total ammonia concentration (NH_3 + NH_4^+). Ammonia homeostasis is a whole-body dynamic process tightly regulated (Fig. 1),

Tiziano Balzano (ed.), *Experimental and Clinical Methods in Hepatic Encephalopathy Research*, Neuromethods, vol. 219, https://doi.org/10.1007/978-1-0716-4354-9_13,

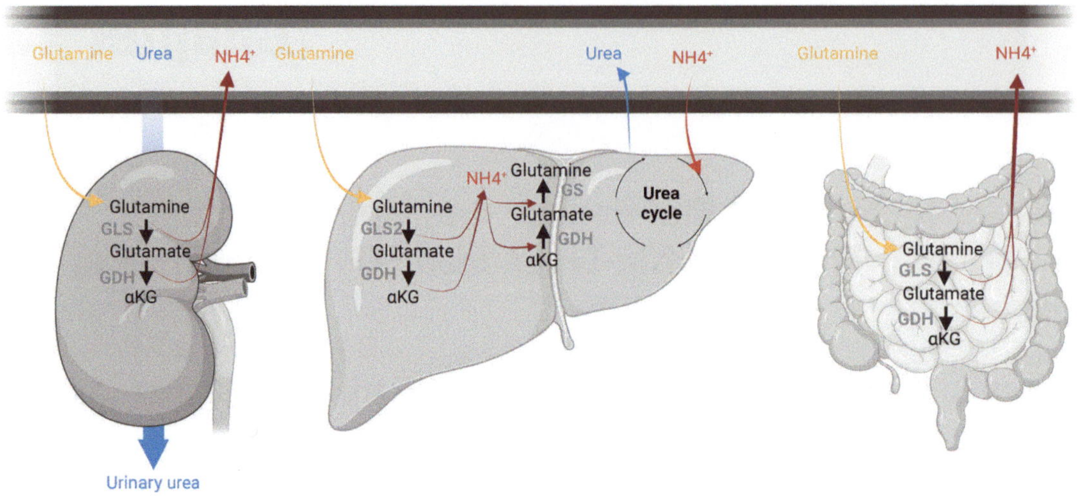

Fig. 1 Whole-body ammonia mammalian metabolism (created with BioRender.com). Whole-body ammonia metabolism is a tightly regulated process. Under healthy conditions, ammonia is produced mostly in the gut by three main mechanisms: hydrolysis of urea by bacterial urease, bacterial protein deamination, and intestinal mucosal glutamine metabolism. Although the colon is conventionally assumed to be the major site of gut ammonia production, recent evidence indicates that the stomach and small intestine are also involved. From the gut, ammonia enters the bloodstream where it is usually taken up by the liver through the peri-portal hepatic region. In the liver, ammonia is eliminated through the urea cycle, a biochemical cycle that produces urea (ureagenesis) from ammonia composed of five urea cycle enzymes: carbamoyl phosphate synthetase I (CPS-1), Ornithine transcarbamoylase (OTC), argininosuccinate synthetase (ASS), argininosuccinate lyase (ASL), and arginase 1 (ARG1) and 2 transporters. Released urea is secreted into the bloodstream and up taken by the kidney where it is expelled as urinary urea. This provides a high capacity for detoxification of the vast amount of surplus nitrogen. Whenever the hepatic urea cycle is saturated, α-ketoglutarate is converted into glutamate, by the enzyme glutamate dehydrogenase, and glutamate then fixes ammonia through its conversion to glutamine by the enzyme glutamine synthetase abundant in the perivenous hepatocytes. In this case, glutamine synthetase serves as a backup system with high affinity (but low capacity) to ammonia. The produced glutamine is released to the bloodstream and re-enters the liver where it is metabolized by the mitochondrial glutaminase (GLS2 isoform under healthy conditions and GLS in the injured liver) to glutamate and then to α-ketoglutarate to enter the tricarboxylic acids cycle, a process releasing ammonia. Blood glutamine can also be taken up by the intestine and the kidney, which breaks it to glutamate by glutaminase, usually the GLS isoform 1, and then to α-ketoglutarate releasing ammonia into the bloodstream. Neither the brain nor the skeletal muscle remove or produce ammonia under healthy conditions, but they can seize elevated blood ammonia during hyperammonemia releasing glutamine into the bloodstream. Also in skeletal muscle, deamination of adenosine monophosphate, particularly during physical exercise, results in ammonia production. *Abbreviations*: Ammonia (mostly composed of NH_4^+ at physiological pH), GLS Glutaminase isoform 1, GLS2 Glutaminase isoform 2, GDH Glutamate dehydrogenase, GS Glutamine synthetase, α-KG alpha-ketoglutarate

adapting the physiological response to nutrients, hormones, and metabolites [2, 3]. Whereas controlled levels of endogenous ammonia provide a gut-brain signaling basis for emotional behavior [4], hyperammonemia, elevated blood ammonia, is a common complication in a wide variety of rare inherited metabolic disorders (e.g., urea cycle disorders [5] and organic acid disorders [6], among others) and other conditions such as Reye's syndrome [7], urinary tract infection [8], leukemia patients undergoing chemotherapy [9] and sodium valproate therapy [10]. Hyperammonemia is also a primary etiologic agent of hepatic encephalopathy (HE), a reversible syndrome observed in patients with advanced liver dysfunction and failure [11–14].

Hyperammonemia is a life-threatening condition. Clinical manifestations of hyperammonemia are variable but common features, mostly related to adverse effects of high ammonia in the brain as a consequence of hyperammonemia, include vomiting, hyperactivity or lethargy, diarrhea, poor feeding, seizures, hypotonia, global development delay, psychiatric symptoms, respiratory distress, and in more severe cases hyperammonemia coma. As blood ammonia levels are predictors of the onset of clinical manifestations and eventually allow an earlier selection and regulation of therapeutic interventions, the development and advances in methodologies for the accurate and reproducible quantification of ammonia in biological fluids are highly valuable for clinical practice and improving the quality of life of patients.

1.1 Methods for Measuring Ammonia in Biological Fluids

The first technique used for measuring ammonia in blood was reported in 1895 [1]. Since then, several methods have been developed for the determination of ammonia in biological fluids including blood, plasma, erythrocytes, urine, saliva, sweat, cerebrospinal fluid, and bile. Common ammonia contaminants in biological fluids include hemolyzed plasma, smoking by technicians in the laboratory and patients before and during ammonia determination or during blood collection, heavy traffic contamination, the natural presence of bacteria or environmental bacteria that can hydrolyze urea in urine samples, and in addition patients should abstain from performing exercise before analysis as it increases ammonia in the blood by threefold [1]. The major limitations of conventional ammonia measurements of ammonia in biological fluids are the complexity involved in the proper drawing and handling of the sample, the time allowed between drawing and assaying, and finally, the assay itself. Most methods recommend collecting samples from patients who have fasted for at least 6 h but preferably patients should be fasted from 12 to 14 h before sample collection to avoid lipemia which interferes with the test [15]. Ideally, free-flowing (central in neonates and infants) venous or even arterial blood should be used. Donors' arms should be as relaxed as possible, because muscle exertion may increase ammonia levels. Capillary blood is not suitable for ammonia determination because tissue fluid and disrupted cells will lead to artificial elevations. However, some hospitals use capillary blood in their emergency room for the exclusion of hyperammonemia being the convenience of this approach it's use as a bedside test. In addition, the ammonia content in freshly drawn blood rises rapidly on standing because of the deamination of labile amides such as glutamine and also due to the release of ammonia from red blood cells. At room temperature, the ammonia content in blood can increase by a factor of 2 or 3 in several hours. As a general rule, blood ammonia from healthy subjects should be measured within 30–60 min of collection, whereas blood samples derived from people with liver disease should be analyzed within 15 min due to high content in blood gamma-glutamyl transferase, an enzyme that hydrolyzes glutamine [1]. Nowadays, the determination of ammonia in plasma is the most commonly used ammonia assay in clinical chemistry. EDTA is the anticoagulant of choice for the measurement of plasma ammonia, as heparin has been suggested to interfere with ammonia detection [16, 17]. Blood should be drawn into a chilled, EDTA vacuum tube, immediately placed on ice, and plasma preferentially separated within 15 min. When promptly separated from blood, plasma ammonia levels are stable at 4 °C for 4 h and for 24 h if stored frozen at −20 °C [18].

The choice of technique for ammonia determination in biological fluids depends predominantly on the available equipment. Historically, two steps were necessary for the measurement of the ammonia concentration in biological fluids. First, ammonia gas had to be liberated and secondly, the ammonia liberated quantified [1]. Liberation was achieved by distillation, aeration, ion exchange chromatography, microdiffusion, or deproteinization [1]. Otherwise, techniques for ammonia quantification include colorimetric/fluorometric assays, ion-selective electrodes (ISEs), and enzyme-based assays. Examples of colorimetric reactions include the blue color formed by the reaction of ammonia with phenol and hypochlorite—the indophenol reaction—and the Nesslerization where a brown/orange colored complex is formed by the reaction of ammonia with mercury (II) iodide in an alkaline solution, among others. In alternative, the *o*-phthalaldehyde method in which the reagent reacts with ammonia producing a fluorometric result proportional to the ammonia concentration in the sample. Nowadays, the direct enzymatic determination of ammonia in plasma has become the basis for the most commonly used blood ammonia assay in clinical chemistry. This assay is based on the reaction between α-ketoglutarate and ammonia in the presence of glutamic acid dehydrogenase and an analog of stabilized nicotinamide adenine dinucleotide phosphate (NADPH) to form glutamic acid. The stoichiometry of the reaction can be followed spectrophotometrically or fluorometrically by NADPH disappearance at 340/694 nm. This method offers advantages relative to the enzymatic determination of ammonia using NADH as a coenzyme used in earlier studies as other NADH-consuming systems are present in the blood resulting in an overestimation of the level of plasma ammonia.

Currently, although well-defined reference values for the considered healthy concentrations of plasma ammonia do not exist, the following limits, which depend on the age/gender of the patient and the assay method used, for use in clinical practice are used: (i) newborns (arterial cord blood): 50–150 micromoles per liter (micromol/L); (ii) infants and children: 24–48 micromol/L; (iii) adult (females): 11–48 micromol/L; (iv) adult (males): 15–-55 micromol/L [19]. As observed, except for newborns, plasma ammonia levels in healthy subjects are usually lower than 50 micromol/L. An increase of plasma ammonia levels to only 100 micromol/L can lead to changes in consciousness whereas plasma ammonia levels of 200 micromol/L are associated with coma and convulsions.

1.2 Methods for Measuring Ammonia in Mammalian Tissues

Ammonia is diffused and transported across all plasma membranes. Indeed, the concentration of ammonia within cells and in tissues is reported to be higher than systemic ammonia with ammonia being usually trapped as ammonium in cellular compartments with a low pH (such as in lysosomes and renal tubule cells). Aberrant accumulation of ammonia within tissues was detected in a series of pathologies. For example, our group and others have demonstrated the overt accumulation of ammonia in the liver during liver disease, including liver fibrosis and drug-induced liver injury [20, 21]. Even though some commercially available enzymatic methods-based kits, initially developed for the quantification of ammonia in biological fluids, have been adapted and tested for measuring ammonia in cell extracts and supernatants, their utilization to measure ammonia content in tissue biopsies is limited. A major concern resides in the fact that intra-tissue ammonia concentrations are expected to be highly modified by freezing delay time (for example the time between surgical resection and the sample being frozen) or repeated thaw and freeze cycles. To overcome the limitations of the use of fresh or frozen tissues for the detection of organ-specific ammonia accumulation, a histological-based method for ammonia detection in formalin-fixed paraffin-embedded (FFPE) mammalian tissue has been recently developed and validated by our group [22]. This method is based on the Nesslerization reaction and can be used to detect ammonia in FFPE mammalian tissues. Herein, we provide a thorough detailed description of the protocol for the histological detection of ammonia by the Nessler reaction in FFPE mammalian tissues.

2 Materials

2.1 Reactive

- 10% neutral buffered formalin (NBF) fixative.
- Ethanol.

- Xylene or in alternative histo-clear.
- Paraffin.
- Nessler's Reagent (K_2HgI_4).
- Distilled water.
- Mayer's hematoxylin solution.
- DPX mounting media for histology.
- Running water tap.

2.2 Small Material

- Tissue cassettes.
- Microscope slides with adhesion and 90° ground edges.
- Hydrophobic barrier PAP pen.
- Staining chamber for microscope slides with black lid.
- Pipette 20–200 μL.
- Pasteur pipettes.
- Plastic staining cuvette and staining racks or Coplin staining jars.
- Cover glasses.

2.3 Equipment

- Automatic or manual tissue processor.
- Embedding system station.
- Microtome.
- Histological water bath.
- Histology slide incubator drying oven.
- Light microscope.

3 Methods

3.1 The Nessler Method: Theoretical Background

Nessler's reagent, a commercially available reactive, consists of mercury (II) iodide or potassium iodide in an alkaline solution, either in sodium hydroxide or potassium hydroxide. Upon reaction with ammonia, Nessler's reagent forms a yellow-colored compound, which can be determined using a spectrophotometer at wavelengths of 410–425 nm. The color intensifies from yellow to yellow-orange/reddish-brown along with an increase in ammonia concentration. The overall reaction is shown in Eq. (1).

$$2K_2HgI_4 + NH_3 + 3KOH \rightarrow HgO \cdot Hg(NH_2)I + 7KI + 2H_2O \quad (1)$$

3.2 Fixation, Dehydration, Paraffin-Embedding, and Sectioning Protocols for Obtaining Formalin-Fixed Paraffin-Embedded (FFPE) Mammalian Tissue

Fresh tissues (human specimens or animal-related) are placed on adequate-size tissue cassettes and immediately immersed in a recipient containing 10% neutral buffered formalin (NBF) (*see* **Note 1**). After 24 h of fixation, NBF is replaced by a 50% ethanol solution until further processing which consists of dehydration of the tissue-fixated biopsies before paraffin embedding. The proposed consecutive protocol for tissue processing, by using an automated or manual tissue processor system, consists of the following:

1. 10 min 50% ethanol at room temperature (RT)
2. 90 min 70% ethanol at RT
3. 90 min 80% ethanol at RT
4. 90 min 96% ethanol at RT
5. 90 min 100% ethanol at RT
6. 90 min 100% ethanol at RT
7. 90 min 100% ethanol at RT
8. 45 min in Xylene or histo-clear at RT
9. 120 min in paraffin at 60 °C
10. 120 min in paraffin at 60 °C.

Afterward, tissue samples are embedded in fresh paraffin in an embedding station and if necessary, tissue oriented before it hardens. After initial trimming of the fixed tissues embedded in paraffin at a thickness of 10–30, the specimens are then sectioned at 5 μm in a standard microtome. Sectioned 5 μm samples are moved to a water bath kept at 38 °C by using tweezers that are also used to separate the sections. The hot water bath will allow the tissue to relax and smooth before being mounted on a slide as well as facilitating the adherence of the paraffin tissue to the slide (*see* **Note 2**). The separated tissue sections are picked from the hot water bath directly onto the microscope slide. Finally, the microscope slides with mounted tissues are left to dry for some minutes in the vertical position in a slide rack before continuing with the analysis.

3.3 Deparaffinization and Hydration of Formalin-Fixed Paraffin-Embedded (FFPE) Tissue

After 15 min treatment at 58 °C in a histology slide incubator drying oven, 5 μm tissue sections attached to microscope slides are deparaffinized according to the following protocol (*see* **Note 3**):

1. 10 min xylene or histo-clear
2. 10 min xylene or histo-clear (new solution)
3. 5 min 100% ethanol
4. 5 min 100% ethanol
5. 5 min 96% ethanol

6. 5 min 96% ethanol
7. 5 min 70% ethanol
8. At least 5 min in distilled water.

3.4 Nessler Staining of Formalin-Fixed Paraffin-Embedded (FFPE) Mammalian Tissue

Transfer the microscope slides to a staining chamber with a black lid by placing moistened kimwipes at the bottom to create a humid environment. Carefully but rapidly place the microscope slides on the histological tray on top of the rails that are topped with a polymer non-slip strip that holds the slides in place without marking them, even when the tray is at an angle. In each of the microscope slides with the mounted tissue specimen, round the tissue with the hydrophobic PAP pen and then add distilled water with a Pasteur pipette immediately to prevent artifacts due to the drying of the samples. To start the staining, remove the water from the microscope slides by angling almost to the vertical tray and wipe the slide (back and edges) with a kimwipe one slide at a time (move rapidly to avoid the slide having less chance to dry out). Afterward, add 100 μL of Nessler's reagent with a precision pipette to each sample slide to the tissue surrounded by the hydrophobic PAP pen and immediately close the chamber with its black lid to protect it from the light. Incubate with the Nessler reagent for 5 min (*see* **Note 4**). After 5 min, immerse each microscope slide by rapid back and forward movements for 15 s in a recipient containing distilled water while observing the development of color (*see* **Note 5**). Samples are now ready to be placed in a histology slide rack for counterstaining by immersing in a recipient with Mayer's hematoxylin for 1 min. Finally, transfer the rack to a recipient with water and wash it with running tap water for at least 30 s.

3.5 Dehydration of Formalin-Fixed Paraffin-Embedded (FFPE) Mammalian Tissue

To finish the process, slides are dehydrated in a two-step 5-s immersion in 100% ethanol followed by a two-step 30-s immersion in xylene or histo-clear. After the removal of xylene or histo-clear excess, by using filter paper, DPX mounting media is used to attach the coverslip to the microscope slide so the tissue is protected during microscopy and storage (*see* **Note 6**).

3.6 Microscope Image Acquisition

Upright light microscope coupled to a microscope digital camera can be used to acquire representative microphotographs of the Nessler staining in tissue. Usually, five to ten random images per sample are taken. The stained area percentage of each sample is calculated using FRIDA software (FRamework for Image Dataset Analysis) http://bui3.win.ad.jhu.edu/frida/. Some examples of acquired images are shown in Fig. 2.

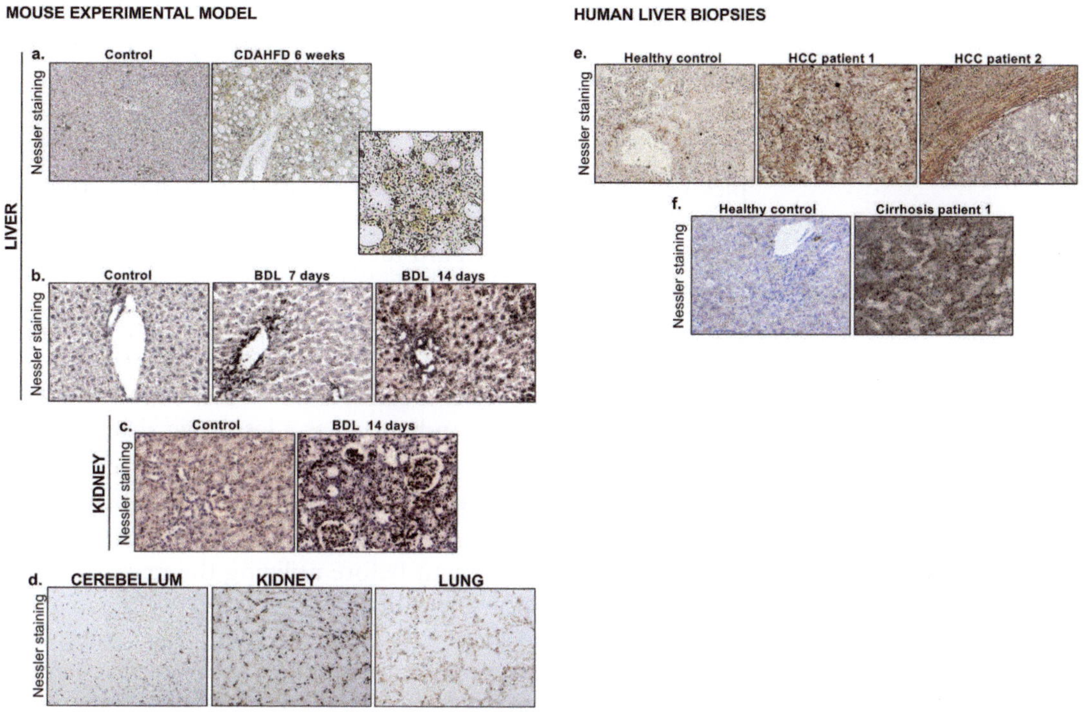

Fig. 2 Nessler histological staining of ammonia in formalin-fixed paraffin-embedded (FFPE) mammalian tissue. (**a**) Representative microphotographs of ammonia staining in liver biopsies of C57BL/6J mouse animal models of metabolic-associated steatotic liver disease (MASLD), fed for 6 weeks with an amino acid-defined high-fat diet (CDAHFD), relative to animals maintained on a standard chow diet (Control). (**b**, **c**) Representative microphotographs of ammonia staining in liver and kidney biopsies of C57BL/6J mouse animal models of liver cholestasis, animals submitted to bile duct ligation (BDL) surgery before and at different time points after surgical intervention. (**d**) Representative ammonia staining in tissue biopsies (cerebellum, kidney, and lung) of a C57BL/6 J healthy control adult mouse. Representative microphotographs of ammonia staining in liver biopsies of patients diagnosed either with (**e**) hepatocellular carcinoma (HCC) or (**f**) liver cirrhosis. Ammonia shows as yellow-orange/reddish-brown along with an increase in ammonia concentration during injury

4 Notes

1. 10% neutral buffered formalin (NBF) is the most commonly used fixative; other fixatives may be suitable depending on the tissue of interest to study. No fixative will penetrate a piece of tissue thicker than 1 cm. Inclusion in histological cassettes facilitates the formalin fixation procedure. The ideal fixation time will depend on the size of the tissue block and the type of tissue, but fixation between 18–24 h is suitable for most applications. Use a pencil to label the tissue cassette. Include in the label important information (e.g., ID, type of tissue, experiment, and others). The fixative should be 15–20 times more in

volume than the specimen. Work in toxic-fume hood. To date, several tissues have been validated by our group for the Nessler staining protocol using NBF: liver, brain, kidney, pancreas, and muscle.

2. Before sectioning in the microtome, paraffin-embedded samples may be placed in contact with ice as this allows thinner sections to be obtained by supporting harder elements to the tissue specimen. Water bath temperature may need to be lowered for fatty tissue, such as the adipose tissue, breast, or brain.
3. If a drying oven is not available microscope slides can be placed in a microwave oven at 600 W for 2 min. Deparaffinization and hydration may be performed either in an automized robot or manually. Histo-Clear can be used as a less toxic alternative to xylene without any change in protocol.
4. Before using the Nessler's reagent is important to avidly agitate the flask as it tends to precipitate. Then wait some minutes for the precipitates to sediment before utilizing the reactive which should be protected from light at all times. Do NOT move the chamber once the slides are loaded with the Nessler reactive or the movement can spill water onto the slides from the chamber wells. This protocol has to be carried out in a toxic fume hood as Nessler's reagent is harmful. Also, the reagent and the related samples should be properly handled and disposed of.
5. This water washing step is very important and the same washing time should be used for each sample. No more than five to six samples should be stained simultaneously to avoid longer expositions to the Nessler's reagent before the washing.
6. Let DPX-mounted microscope slides dry in a fume hood for at least 24 h before the microscope analysis.

5 Conclusions

Hyperammonemia is a life-threatening health problem associated with a wide spectrum of diseases such as rare conditions [5, 6], as well as other conditions and some drug-toxicity-related aspects [7–10]. Moreover, due to the important role played by the liver in the detoxification of ammonia, it is not surprising that hyperammonemia-related hepatic encephalopathy (HE), a major neuropsychiatric complication, is very common in advanced liver cirrhotic disease [11–14]. Due to the clinical gravity of hyperammonemia, it is important to be able to quantify ammonia levels in a very accurate and reproducible way as blood ammonia levels are potential predictors of the onset of clinical manifestations. Indeed, although blood ammonia levels have been shown not to have a role in the diagnosis or treatment of cirrhosis-associated HE, they have an important role in predicting liver-related events and in prognostication in cirrhotic patients. Under these circumstances, the disparity between blood ammonia levels and clinical severity probably resides in failure to take into account the influence of individual genetic variations in the expression of enzymes involved in glutamine metabolism, cell membrane proteins associated with the regulation of osmotic gradients in the brain, and the role of other unidentified neurotoxic substances [23]. Moreover, blood ammonia levels reflect systemic ammonia metabolism, and abnormal organ-specific ammonia accumulation is a factor to take into consideration at least under pathologic conditions. It is well known that ammonia is diffused and transported across all plasma membranes and is known that the concentration of ammonia within cells and in tissues is higher than systemic ammonia. In this regard, our group and others have shown that ammonia is increased in biopsies of liver fibrosis [21] and metabolic-associated steatotic liver disease (MASLD) characterized by steatohepatitis, the so-called metabolic-associated steatotic steatohepatitis (MASH) [24]. The accumulation of hepatic ammonia in early MASH is known to progress to systemic hyperammonemia as the disease advances [21], suggesting that ammonia is a pathogenic factor probably accounting for MASLD-related multiorgan dysfunction and disease progression [25]. Thus, on the one hand, there is a need for suitable techniques for the detection of ammonia levels in biological fluids, such as blood and plasma, to be used as potential predictors of the onset of clinical manifestations or eventually allow an earlier selection and regulation of therapeutic interventions, and on the other hand methodologies for the detection of ammonia in mammalian tissue, that can potentially be applied to experimental research as well as used in clinical assays whenever tissue biopsies are available.

The instability of ammonia in whole blood has been a major challenge for clinical laboratory analysis of blood ammonia concentration. Nowadays, the determination of ammonia in plasma by direct enzymatic determination is the most commonly used ammonia assay in clinical chemistry. Briefly, blood is drawn into a chilled, EDTA vacuum tube that is immediately placed on ice, and plasma is separated within 15 min of collection. Plasma ammonia is quantified based on the reaction between α-ketoglutarate and ammonia in the presence of glutamic acid dehydrogenase and an analog of stabilized NADPH to form glutamic acid, a reaction that can be followed spectrophotometrically or fluorometrically by NADPH disappearance at 340/694 nm. Moreover, is important to be aware that ammonia determination can be contaminated by hemolyzed plasma, smoking by technicians in the laboratory and patients before and during ammonia determination or during blood collection, heavy traffic contamination, and in addition, patients should abstain from performing exercise before analysis as it increases ammonia in the blood by threefold [1]. In the last years, a huge effort has been placed in the development of systems and methods that allow for the continuous, automated monitoring of changes in the quantity of total blood ammonia or in alternative a point-of-care (POC) device that may be used clinically and alert medical staff to an impending hyperammonemia condition. However, to date, there is still no approved ammonia POC or medical analyzer monitoring system for continuous ammonia monitoring.

Currently, available techniques for the detection of ammonia in mammalian tissue are limited. Indeed, as previously mentioned, the intra-tissue ammonia concentrations are expected to be highly modified by freezing delay time (for example the time between surgical resection and the sample being frozen) or repeated thaw and freeze cycles. To overcome this limitation, the histological staining for ammonia detection in formalin-fixed paraffin-embedded (FFPE) mammalian tissue, the methodology detailed here, is an attractive protocol that has been previously validated in different types of mammalian tissue, including the liver, brain, kidney, and other [22]. This method presents several advantages relative to the quantification of ammonia by commercially available kits in frozen tissue. Firstly, FFPE tissue biopsies are used and this is relevant for clinical research as these are the most widely available human tissue collections. Secondly, formalin fixation provides a way to stabilize ammonia in the tissue for longer periods than for example frozen samples where ammonia stability can be compromised by thaw-freezing. Thirdly, the Nessler staining protocol utilizes standard histological material and can be routinely performed in histological laboratories. Finally, the Nessler histological staining method can provide additional information on the specific localization of ammonia accumulation within the tissue biopsies relative to the commercially available kits. This can be observed for example in the representative staining of ammonia in Fig. 2b in liver biopsies, where during injury (BDL-induced liver injury), ammonia is mostly accumulated in the periportal region relative to the rest of the parenchyma. On the other side, one of the main disadvantages of the Nessler staining methodology is its limited quantitative potential. Whereas in the original protocol, a Nessler staining score is suggested by using a couple of negative and positive control samples, this quantification approach can only be used in one laboratory and not between laboratories. Alternatively, we can use specific software to measure the area of the specific signal given by the dark brownish Nessler staining, but only to compare samples where the staining was simultaneously performed and not between staining procedures performed at different points.

In summary, herein we provide a detailed description and specific notes to facilitate the implementation of the Nessler histological staining for ammonia detection in FFPE mammalian tissue.

This histological-based protocol can be applied to several types of mammalian tissue and on one hand can increase our current knowledge of whole-body ammonia homeostasis and its deregulation during human pathology and on the other hand, promote the study of novel drugs that act as ammonia chelators.

Acknowledgments

Teresa C. Delgado is funded by "Ayuda RYC2020-029316-I financiada por MCIN/AEI/10.13039/501100011033 y por El FSE invierte en tu futuro." This work was supported by the Gilead Sciences Research Scholars Program in Global Liver (to TCD), grant from Ministerio de Ciencia, Innovación y Universidades (MICINN: PID2022-139395OB-100 integrado en el Plan Estatal de Investigación Científica y Técnica e Innovación, con Fondos FEDER). This work was supported by grants from Ministerio de Ciencia, Innovación y Universidades MICINN: PID2020-117116RB-I00 integrado en el Plan Estatal de Investigación Científica y Técnica e Innovación, cofinanciado con Fondos FEDER for (MLM-C); Project funded by CIBEREHD. This work was partially funded by the Basque Department of Health (2023222005) (for JdH). JdH and TCD are members of the European Reference Network for Rare Hereditary Metabolic Disorders (MetabERN) Project ID No. 739543.

References

1. Huizenga JR, Tangerman A, Gips CH (1994) Determination of ammonia in biological fluids. Ann Clin Biochem 31(Pt 6):529–543
2. Gebhardt R, Matz-Soja M (2014) Liver zonation: novel aspects of its regulation and its impact on homeostasis. World J Gastroenterol 20(26):8491–8504
3. Jungermann K, Kietzmann T (1996) Zonation of parenchymal and nonparenchymal metabolism in liver. Annu Rev Nutr 16:179–203
4. Wang P et al (2023) Gut microbiome-derived ammonia modulates stress vulnerability in the host. Nat Metab 5:1986–2001
5. Mathias RS, Kostiner D, Packman S (2001) Hyperammonemia in urea cycle disorders: role of the nephrologist. Am J Kidney Dis 37(5):1069–1080
6. Haberle J et al (2018) Hyperammonaemia in classic organic acidaemias: a review of the literature and two case histories. Orphanet J Rare Dis 13(1):219
7. Glasgow AM, Cotton RB, Dhiensiri K (1972) Reye's syndrome. I. Blood ammonia and consideration of the nonhistologic diagnosis. Am J Dis Child 124(6):827–833
8. Kenzaka T et al (2015) Hyperammonemia in urinary tract infections. PLoS One 10(8): e0136220
9. Nott L et al (2007) Hyperammonemia encephalopathy: an important cause of neurological deterioration following chemotherapy. Leuk Lymphoma 48(9):1702–1711
10. Segura-Bruna N et al (2006) Valproate-induced hyperammonemic encephalopathy. Acta Neurol Scand 114(1):1–7
11. Kotsiliti E (2022) Hyperammonaemia in liver cirrhosis. Nat Rev Gastroenterol Hepatol 19(10):624
12. Jalan R, Lee WM (2009) Treatment of hyperammonemia in liver failure: a tale of two enzymes. Gastroenterology 136(7): 2048–2051

13. Jover-Cobos M, Khetan V, Jalan R (2014) Treatment of hyperammonemia in liver failure. Curr Opin Clin Nutr Metab Care 17(1): 105–110
14. Butterworth RF (2000) Complications of cirrhosis III. Hepatic encephalopathy. J Hepatol 32(1 Suppl):171–180
15. Kaplon J et al (2019) Improved assay protocol for measurement of ammonia on the Roche Cobas 8000 automated platform. Pract Lab Med 13:e00115
16. Wannasilp N et al (2006) Heparin is an unsuitable anticoagulant for the detection of plasma ammonia. Clin Chim Acta 371(1–2):196–197
17. Wannasilp N et al (2005) EDTA should be the anticoagulant of choice for the measurement of plasma ammonia: report of a problem sample. Clin Chim Acta 357(1):84–85
18. Barsotti RJ (2001) Measurement of ammonia in blood. J Pediatr 138(1 Suppl):S11–9;discussion S19–20
19. Haberle J (2011) Clinical practice: the management of hyperammonemia. Eur J Pediatr 170(1):21–34
20. Rodriguez-Agudo R et al (2022) Methionine cycle rewiring by targeting miR-873-5p modulates ammonia metabolism to protect the liver from acetaminophen. Antioxidants (Basel) 11(5):897
21. De Chiara F et al (2018) Urea cycle dysregulation in non-alcoholic fatty liver disease. J Hepatol 69(4):905–915
22. Gutierrez-de-Juan V et al (2017) A morphological method for ammonia detection in liver. PLoS One 12(3):e0173914
23. Ballester MP et al (2023) Development and validation of the AMMON-OHE model to predict risk of overt hepatic encephalopathy occurrence in outpatients with cirrhosis. J Hepatol 79(4):967–976
24. Mercado-Gomez M et al (2024) The lipopolysaccharide-TLR4 axis regulates hepatic glutaminase 1 expression promoting liver ammonia build-up as steatotic liver disease progresses to steatohepatitis. Metabolism 158: 155952
25. Thomsen KL et al (2023) Role of ammonia in NAFLD: an unusual suspect. JHEP Rep 5(7): 100780

INDEX

Tiziano Balzano (ed.), *Experimental and Clinical Methods in Hepatic Encephalopathy Research*, Neuromethods, vol. 219, https://doi.org/10.1007/978-1-0716-4354-9,

M

N

P

Q

S

W